Volume 1:
ALUMINA & BAUXITE

LIGHT METALS 2006 VOLUME 1: ALUMINA & BAUXITE

TMS Member Price: $67 TMS Student Member Price: $54 List Price: $94

Related Titles

- *Light Metals 2005*, edited by H. Kvande
- *Light Metals 2004*, edited by A. Tabereaux
- *Light Metals 2003*, edited by P. Crepeau

HOW TO ORDER PUBLICATIONS

For a complete listing of TMS publication offerings, contact TMS for a free copy of the Publications@TMS catalog or visit the on-line TMS Document Center (*http://doc.tms.org*). Through the TMS Document Center, customers will find:

- An easy and convenient way to purchase publications
- Access to TMS member resources, such as the on-line version of *JOM* and *TMS Letters*
- In-depth publications information, including complete descriptions, tables of contents, and sample pages
- Award-winning landmark papers and re-issued out-of-print titles
- A vast selection of resources from which to compile customized publications that meet the customer's unique needs

MEMBER DISCOUNTS

Members of TMS (The Minerals, Metals & Materials Society) receive a 30% discount off list prices for all TMS publications. In addition, TMS members also receive free monthly subscriptions to the journals *JOM* (both in print and on-line formats) and *TMS Letters* (published exclusively in on-line format), discounts on meeting registrations, and more. To begin immediate savings on TMS publications and to receive additional member benefits, become a TMS member today. Fill out a membership application when placing a publication order at the on-line TMS Document Center (*http://doc.tms.org*) or contact TMS for more information:

- Phone: 1 (800) 759-4867 (within the U.S.) or (724) 776-9000 (elsewhere)
- Fax: (724) 776-3770
- E-mail: membership@tms.org or publications@tms.org
- Web: *www.tms.org*

Light Metals 2006

Volume 1: ALUMINA & BAUXITE

Proceedings of the technical sessions presented
by the TMS Aluminum Committee
at the 135th TMS Annual Meeting,
San Antonio, Texas, USA
March 12-16, 2006

Edited by
Travis J. Galloway

A Publication of **TMS (The Minerals, Metals & Materials Society)**
184 Thorn Hill Road
Warrendale, Pennsylvania 15086-7528
(724) 776-9000

Visit the TMS web site at
http://www.tms.org

ISBN Number 978-0-87339-615-8

If you are interested in purchasing a copy of this book, or if you would like to receive the latest TMS publications catalog, please telephone (800) 759-4867 (U.S. only) or (724) 776-9000, EXT. 270.

LIGHT METALS 2006 VOLUME 1
TABLE OF CONTENTS

Alumina and Bauxite

Solids/Liquid Separation

Bauxite and Bauxite Characterization

Bayer Digestion Technology

Joint Session of Alumina and Bauxite & Aluminum Reduction Technology

Plant Design, Operation and Maintenance

Precipitation Fundamentals

LIGHT METALS 2006 VOLUME 2
TABLE OF CONTENTS

Aluminum Reduction Technology

Environmental Elements

Cell Development and Operations - Part I

Cell Development and Operations - Part II

Pot Control and Modeling

Inert Anodes - Part I

Cell Development Part III and Emerging Technologies

Fundamentals, Emerging Technologies and Inert Anodes - Part II

LIGHT METALS 2006 VOLUME 3
TABLE OF CONTENTS

Carbon Technology

Anode Raw Materials

Greenmill/Rodding

Anode Baking

Cathode Properties/Refractory Materials

Cathode Preheating/Wettable Cathodes

LIGHT METALS 2006 VOLUME 4
TABLE OF CONTENTS

Cast Shop Technology

Cast House Operations

Furnace Operation and Refractory Materials

Melt Treatment, Quality and Product Properties

Shape Casting and Foundry Alloys

Casting, Solidification and Cast Defects

Cast Processes and Chain Analysis

Recycling

Aluminum Recycling

PREFACE

The past year has presented significant challenges for the aluminum industry with dramatic increases in energy and raw materials prices compressing profit margins despite the positive impact of higher metal prices. This has occurred against the backdrop of continued brownfield expansions and greenfield feasibility studies and plans for both aluminum smelters and alumina refineries driven by world-wide demand led by China. In this environment, the strategic importance of innovation from R&D and implementation of technological improvements throughout the integrated production process from bauxite mining through refining, smelting and downstream applications can not be overemphasized. TMS continues to serve as the key forum for disseminating scientific and technological information to maintain the competiveness and sustainability of the aluminum industry.

For **Light Metals 2006**, the papers cover the full range from theoretical aspects to practical applications of technology for integrated aluminum manufacturing in keeping with the TMS Annual Meeting's theme of "Linking Science and Technology for Global Solutions". There is a special Joint Session of Alumina and Bauxite & Aluminum Reduction Technology with a focus on aspects of alumina quality of common interest to customer and supplier. There was also for the first time a Cast Shop Operations session for which authors made presentations but did not write formal papers. In total, 161 papers are included in this year's **Light Metals** proceedings, which compares very favorably with the number of previous years' contributions. Of this number, 80% percent were contributed by authors outside the United States, which continues the trend in globalization of aluminum technology.

In continuation of the tradition begun last year with the plenary session on the role of technology in the global primary aluminum industry, a plenary session was presented this year for Aluminum Fabrication featuring eight corporate leaders from around the world.

For the first time, the Light Metals proceedings are being published exclusively in CD format for attendees. Print volumes of selected symposia are also available. The on-line Conference Management System was expanded this year to a CMS-Plus version that greatly facilitated the preparation and editing process. New features included on-line editing and communication between authors and session chairs. While the CMS-Plus system continually evolves to meet the needs of our constituency, this facility greatly enhances the preparation of abstracts and manuscripts as well as the editing and session organization process.

Appreciation is expressed to the authors of the numerous papers and presentations and to their organizations, which provided support. Special acknowledgment and thanks are also due to the Symposium Organizers: Dr. Jean Doucet (Alumina and Bauxite), Stephen Lindsay (Reduction Technology), Morten Sorlie (Carbon Technology), and Dr. Rene Kieft (Cast Shop Technology and Operations) and to the Session Chairs for their tireless work in organizing and editing the papers. The TMS staff led by Alexander Scott, Executive Director, provided their traditional excellent support; particular acknowledgment is due to Stephen Kendall, Publications Manager, for his skills and dedication in producing this volume and to Christina Raabe, Manager- Technical Programming and Continuing Education, for her excellent facilitation in organizing the program. The continuing support of the Aluminum Committee in its oversight and strategic direction is also acknowledged and appreciated.

Travis J. Galloway
Vice-Chairman Aluminum Committee
Editor **Light Metals 2006**

EDITOR'S BIOGRAPHY

TRAVIS J. GALLOWAY
LIGHT METALS 2006 EDITOR

Travis J. Galloway is Technical & Engineering Director- Bauxite Mining/Alumina Refining for Century Aluminum. He began his career in 1964 with Reynolds Metals Company at the Hurricane Creek alumina refinery in Arkansas where he became Chief Process Engineer. He later served in various other technical management positions with Reynolds including Manager-Alumina Division Technology in Corpus Christi, Texas, Technical Manager- Worsley alumina refinery in Australia, and Technical Director- Raw Materials Division at the corporate headquarters in Richmond, Virginia. In 2001 he joined Century Aluminum at its Hawesville, Kentucky aluminum smelter. He obtained his B.S. from the University of Arkansas and M.S. from Rice University, both in chemical engineering. Additional graduate education includes an MBA from the University of Arkansas at Little Rock and a Master of Engineering Management from Old Dominion University. He is a Registered Professional Engineer and is a senior member of the American Institute of Chemical Engineers. He has served as a TMS Subject Chair and Session Chair for Alumina and Bauxite and became a member of the TMS Aluminum Committee in 2004.

PROGRAM ORGANIZERS

Alumina & Bauxite

Jean Doucet *obtained his PhD in Chemistry from the University of Montréal in 1974. He then began his career with Alcan in Jonquiére, Canada as a Research Chemist and then held various positions with technical development groups working in Alcan's Jonquiére Alumina Refinery, their specialty chemical group and at an Aluminium Fluoride plant. Jean then became responsible for the Research and Development program for the Bauxite and Alumina Division; in 1995, Jean transferred to Montréal to become the Technology Licensing and Intellctual Property Manager for the same division. From 2002 to 2005, Jean operated out of Brisbane, Australia and in January 2006 returned to Montréal as Director Knowledge Managment and Intellctual Property.*

Dag Olsen *holds a Bachelors degree in Petrochemistry from the Telemark Technical College (1976) and a Masters degree in Chemical Engineering from the Norwegian University of Science and Technology (1982). He joined Norsk Hydro in 1984 as a process engineer at the petrochemical complex at Rafnes, Telemark. In 1990 he joined the newly established Alumina & Bauxite group at Hydro's Corporate Research Centre in Porsgrunn where he worked on various projects and studies in addition to R&D work on Alumina Quality. Dag is currently working in the Alumina & Bauxite department of Hydro Aluminium Primary Metals where his main responsibilities are technology follow-up, R&D cooperation and business development projects.*

Aluminum Reduction Technology

Stephen J. Lindsay *has served in numerous technical and managerial capacities at Alcoa's locations in Massena, NY, Alcoa, TN and Knoxville, TN over the past 26 years. He currently has technical support responsibilities that include; technical support for European operations, plus alumina and metal purity responsibilities that span Alcoa's smelting operations worldwide. He is currently based in Knoxville, TN.*

His articles on alumina and metal purity have been published in ***Light Metals 2005****, the proceedings of the* ***8th Australasian Smelting Technology Conference*** *in 2004 and the* ***7th International Alumina Quality Workshop*** *in 2005. Steve has also contributed to the TMS Industrial Electrolysis Course and Short Courses as well as acting as a guest speaker for the UNSW-UA Graduate Level programs in 2003 and 2004.*

Carbon Technology

Morten Sorlie *graduated MSc in Extractive Metallurgy, Norwegian University of Science and Technology, Trondheim, Norway in 1974 and received a PhD in Inorganic Chemistry, same place, 1978. Worked as Post Doctoral Fellow at Oak Ridge National Laboratory, TN, 1978-1980. Spent the following year at Institute of Inorganic Chemistry, Norwegian University of Science and Technology, Trondheim, Norway, starting up the cathode materials research there. Employed by Elkem since 1982. Present position is Corporate Specialist in Elkem Aluminium ANS, stationed at the Elkem research facilities in Kristiansand, Norway, and Adjunct Professor at Institute of Materials Technology, Norwegian University of Science and Technology, Trondheim, Norway. Responsibilities in Elkem Aluminium include cathodes and cathode materials, anodes and anode raw materials. Morten Sorlie has authored/co-authored more than 100 papers in international journals and conference proceedings, including several in TMS Light Metals. He is also a co-author of the textbook* ***Cathodes in Aluminium Electrolysis****.*

Todd W. Dixon *received his B.S. in Chemical Engineering from the Colorado School of Mines in 1990 and is licensed Professional Engineer in Louisiana. He has worked with Conoco (now ConocoPhillips) since graduation. In his career at Conoco, he has been involved with the design, operation, and troubleshooting of delayed cokers and rotary kilns processing petroleum coke. Currently, he is a Process Team Leader at the ConocoPhillips refinery in Borger, Texas. Previously, he was the Operations Manager at the Venco - Lake Charles calcining plant. He has also served as a TMS Session Chair.*

Cast Shop Technology

Rene Kieft *received his Ph.D in Mechanical Engineering from the Eindhoven University of Technology, The Netherlands in 2000 where he worked heat and fluid flow. He joined Corus Research & Technology to work on modelling of aluminium casting, focussing first on mnetal distribution systems and mechanical deformation during DC casting. After that he was involved in the European projects on building a virtual aluminium production chain (VIR[*] project). Besides casting he also worked on purification of aluminium. Within the European Molten Aluminium Purification project he worked together with different partners on different purification concepts. In 2003 he became knowledge group leader of the Corus Research group on Molten aluminium Processing.*

Gerd-Ulrich Gruen *received his diploma in geophysics from Technical University of Clausthal (Germany) in 1982. After some project work regarding flow in porous media related to deep drilling research he then joined VAW aluminium AG in 1990, where he was responsible for various research activities in the DC casting area mainly focusing on process modeling. This is documented in several papers he presented at previous TMS conferences. One recent activity was the co-ordination of the European wide model development project VIR[CAST], which brought together major European Aluminum producers and leading institutes and universities in the field of microstructure research. He is now with Hydro Aluminium, R&D and actually Program Manager Sheet Ingot in the Competence Center Casting, Alloys & Recycling. Since 2001 Gerd-Ulrich Gruen is a member of TMS.*

Recycling

Gregory Krumdick *obtained his M.S. degree in Bioelectrical engineering from the University of Illinois at Chicago, focusing on complex control systems. In 1990, he joined Argonne National Laboratory and has worked as an Engineer in designing and managing the construction of numerous pilot plant systems for the Process Evaluation section. Starting with industrial control systems, his field of interest moved to metallurgical processes where Greg has working on numerous aluminum projects including electrodialysis of aluminum saltcake waste brines, molten aluminum oxidation and the development of inert anodes for the production of aluminum.*

Cynthia K. Belt *is Energy Engineer for Aleris International working with the plants from the former IMCO Recycling, Commonwealth Aluminum, ALSCO Metals, Alumitech, and Ormet in identifying Best Practices in energy, throughput and recovery and implementing them throughout the corporation. Cindy is trained as a Black Belt in Six Sigma. She has worked for Aleris and the former Barmet and Commonwealth Aluminum since 1996. Prior to this, Cindy worked in engineering project management at Ekco Housewares, Crown Cork & Seal, and The Timken Company. She has over 24 years of engineering experience.*

Cindy earned her Bachelor of Science, Mechanical Engineering degree from Ohio Northern University with additional graduate work in materials at Case Western Reserve University and Akron University. She has written two published papers on energy. Cindy is a member of the Recycling Committee for TMS.

AULUMINUM COMMITTEE 2005-2006

Chairperson
Halvor Kvande
Hydro Aluminium AS
Oslo, Norway

Vice-Chairperson
Travis J. Galloway
Century Aluminum Co.
Hawesville, KY, USA

Past Chairperson
Alton T. Tabereaux
Alcoa Inc.
Muscle Shoals, AL, USA

LMD Chairperson
Ray D. Peterson
Aleris International
Rockwood, TN, USA

JOM Advisor
Pierre P. Homsi
Pechiney Group
Voreppe, France

TMS Staff Liaisons
Warren Hunt and Gail Miller
TMS
Warrendale, PA, USA

MEMBERS THROUGH 2006

Lars Arnberg
Norwegian Univ. of Sci. & Tech.
Trondheim, Norway

Andre R. Bolduc
Alcan Inc.
Monreal, QC, Canada

Jay N. Bruggeman
Alcoa Primary Metals
North Charleston, SC, USA

John J.J. Chen
University of Auckland
Auckland, New Zealand

David D. DeYoung
General Motors Corp.
Pontiac, MI, USA

Les C. Edwards
CII Carbon LLC
New Orleans, LA, USA

Seymour G. Epstein
Aluminum Association
Silver Spring, MD, USA

Ann Marie Fellom
Light Metal Age
S. San Francisco, CA, USA

Wayne R. Hale
SUAL Holding Management Co.
Davis, CA, USA

Peter V. Polyakov
Non-Ferrous Metals & Gold Academy
Krasnoyarsk, Russia

Wolfgang A. Schneider
Hydro Aluminium AS
Bonn, Germany

Martin Segatz
Hydro Aluminium GmbH
Neuss, Germany

Geoffrey K. Sigworth
Alcoa Inc.
Rockdale, TN, USA

Fiona J. Stevens McFadden
University of Auckland
Maungaraki, New Zealand

Murat Tiryakioglu
Robert Morris University
Moon Township, PA, USA

Barry J. Welch
Welbank Consulting
Orakei Aukland, New Zealand

MEMBERS THROUGH 2007

Johannes Aalbu
Hydro Aluminium AS
Ovre Ardal, Norway

Yousuf Ali Mohammed Alfarsi
Dubal Aluminium Co. Ltd.
Dubai, United Arab Emirates

Milind V. Chaubal
Sherwin Alumina Co.
Corpus Christi, TX, USA

Paul N. Crepeau
General Motors Corp.
Pontiac, MI, USA

David B. Kirkpatrick
Kaiser Aluminum & Chem. Corp.
Gramercy, LA, USA

Amir A. Mirchi
Alcan Inc.
Montreal, QC, Canada

Gary B. Parker
Wise Alloys LLC
Muscle Shoals, AL, USA

Tor Bjarne Pedersen
Elkem Aluminium ANS
Farsund, Norway

Ramana G. Reddy
University of Alabama
Tuscaloosa, AL, USA

Michael Hal Skillingberg
Aluminum Association
Arlington, VA, USA

Mark Taylor
University of Aukland
Auckland, New Zealand

MEMBERS THROUGH 2008

Thomas R. Alcorn
Noranda Aluminum Inc.
Florence, AL, USA

Chris M. Bickert
Pechiney Group
Neuilly Sue Seine, France

Corleen Chesonis
Aloca Inc.
Alcoa Center, PA, USA

James W. Evans
University of California
Berkeley, CA, USA

Markus W. Meier
R&D Carbon Ltd.
Sierre, Switzerland

David V. Neff
Metaullics Systems Co. LP
Solon, OH, USA

Ray D. Peterson
Aleris International
Rockwood, TN, USA

Alton T. Tabereaux
Alcoa Inc.
Muscle Shoals, AL, USA

MEMBERS THROUGH 2009

Todd W. Dixon
ConocoPhillips Inc.
Bouger, TX, USA

Gerd Ulrich Gruen
Hydro Aluminium AS
Bonn, Germany

Dag Olsen
Hydro Aluminium Primary Metals
Porsgrunn, Norway

ALUMINA & BAUXITE

PROGRAM ORGANIZERS
Jean Doucet
Alcan Inc.
Brisbane, QLD, Australia

Dag Olsen
Hydro Aluminium Primary Metals
Porsgrunn, Norway

Solids/Liquid Separation

SESSION CHAIR
Monique Authier
Alcan Inc.
Jonquiere, QC, Canada

DEVELOPMENT OF NEW POLYACRYLATE FLOCCULANTS FOR RED MUD CLARIFICATION

Everett C. Phillips and Kevin L. O'Brien

Nalco Company; 1601 W. Diehl Rd., Naperville, IL 60563-1198, USA

Keywords: Flocculant, Red Mud Clarification, Polyacrylate

Abstract

Conventional polyacrylate flocculants have been widely used to settle red mud in the Bayer process over the past two decades. A new manufacturing process has led to the development of higher molecular weight polyacrylate flocculants that outperform existing polyacrylate flocculants by 20% or more. Implementation of the new polyacrylate flocculant technology has allowed plants to reduce overall consumption of flocculant. Laboratory test results and plant applications of these new flocculants at several refineries are summarized in this paper. Additional benefits of these new flocculants are currently under investigation.

Introduction

Conventional polyacrylate flocculants are used in the Bayer process to settle red mud in the clarification of digested bauxite slurries. Acrylate/acrylamide copolymers are also used very effectively in the mud washing circuit. Industry trends include increasing production targets, processing more problematic bauxite deposits, and rising raw material and energy costs. These trends increase the importance of improving the performance and cost-performance of the additives used in the Bayer process.

The use of polyacrylate flocculants for primary clarification of red mud slurries was shown to be effective in the mid 1960's.[1] Actual use of polyacrylate flocculants in the alumina industry became common place in the mid to late 1970's.[1]

Acrylic acid is a major raw material in the production of all red mud flocculants. The global economic recovery and increased consumption of super-absorbent polymers (SAP) in disposable diapers (especially in Europe and Asia) has caused an acrylic acid and propylene global supply and demand imbalance. This has led to a rapid doubling in the price of acrylic acid. These changes in the acrylic acid market have emphasized the need for further improvements in the cost performance of the polyacrylate flocculants used in the Bayer process.

In the Bayer process, red mud slurries from digestion are first clarified in a primary decantation step using high molecular weight polyacrylate based flocculants. The decanter overflow is filtered to reduce mud solids to less than 10 mg/L and the resulting Liquor is sent to the alumina precipitation circuit. Underflow from the primary decanter (or Settler) is washed multiple times using a counter current decantation (CCD) process to recover alumina and caustic soda from the slurry. Polyacrylate and poly(acrylate/acrylamide) flocculants are used to settle the mud in each step of the CCD circuit. As the caustic soda of the wash circuit decreases the optimum anionic charge of the flocculant decreases (the ratio of the acrylate to acrylamide in the copolymer is reduced). Therefore, typically more than one flocculant is used in a mud settling and washing circuit. The decision to use more than two flocculants, to minimize overall flocculant costs must be weighed against the capital costs for additional flocculant make-up and storage equipment.

Typical polyacrylate flocculants used in the Bayer process have a molecular weight exceeding 1,000,000 Daltons. The settling performance of a flocculant has been directly correlated to the molecular weight of the polymer.[1] Increasing the polymer molecular weight reduces the flocculant dose required to achieve the desired mud-settling rate in a particular vessel. The added benefit to using more efficient polymers, especially for primary decantation, is there is a reduced likelihood of residual flocculant in the overflow. Residual flocculant in the overflow has been shown to impede liquor filterability at levels as low as 50 ug/L.[2]

Controlling the process under which polyacrylate polymers are made can greatly influence the characteristics of the polymer in terms of both its average molecular weight and overall polymer conformation in solution (e.g., linear, branched and or cross-linked). Recent changes in the manufacturing process of these types of polymers has led to a more "Rigid" flocculant conformation. This "Rigid Rod Architecture" (RRA) technology, has been extended to a complete platform of polyacrylate flocculants with varying anionic charges, assuring optimum flocculant chemistry and activity across the entire red mud washing circuit.

Polyacrylate Polymer Chemistry

The typical commercial manufacturing processes for polyacrylate polymers utilize what is called chain-growth polymerization.[1] This polymerization process involves three distinct steps; these are shown in Figure 1-3. The first step or initiation step involves forming the free radical on the vinyl group of the monomer. The second step is the propagation step, which requires a long sequence of identical reactions repeated several times with a rate constant of Kp. The termination step stops the propagation step from repeating and terminates the polymer chain. Controlling the rates of the three steps is critical to making high molecular weight polymers.

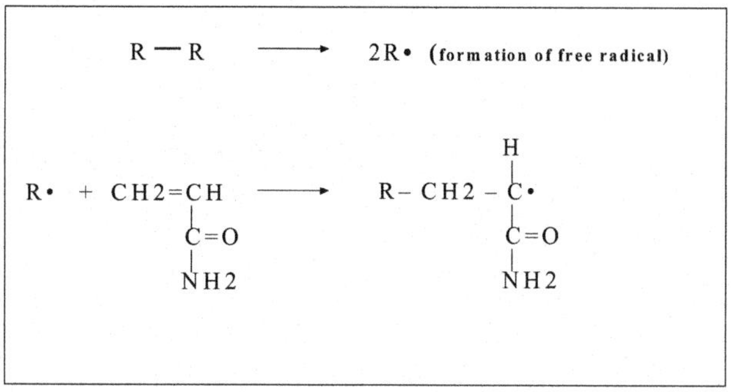

R — R ⟶ 2R• (formation of free radical)

R• + CH2=CH(C=O)(NH2) ⟶ R- CH2 - C•(H)(C=O)(NH2)

Figure 1 Initiation Step

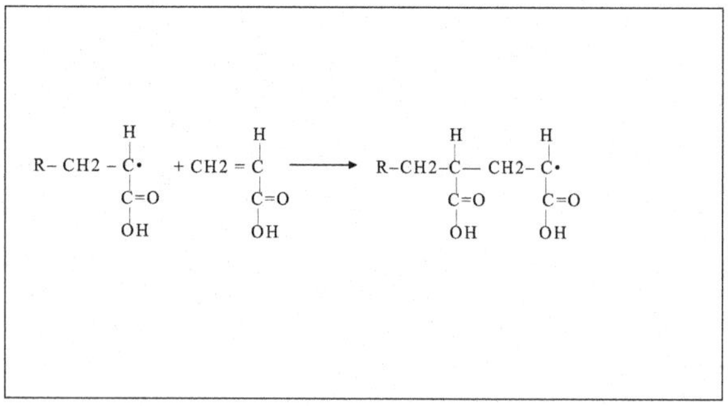

R- CH2 - C•(H)(C=O)(OH) + CH2 = C(H)(C=O)(OH) ⟶ R-CH2-C(H)(C=O)(OH)— CH2- C•(H)(C=O)(OH)

Figure 2. Propagation Step

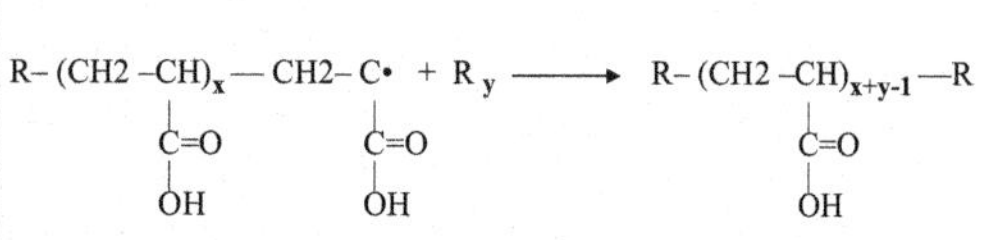

R- (CH2 -CH(C=O)(OH))$_x$— CH2- C•(C=O)(OH) + R$_y$ ⟶ R- (CH2 -CH(C=O)(OH))$_{x+y-1}$—R

R$_y$ – May be an iniator radical, polymer radical or other molecule capable of chain transfer

Figure 3. Termination Step

Laboratory Testing of New Rigid Rod Polyacrylate Flocculants

Preparation of Polymer Solutions

All flocculants tested were water-in-oil latex emulsions which were inverted to 1% in 20 g/L NaOH using a cage mixer @ ~700 rpm for 45-60 minutes (the primary dilution step). These primary solutions were diluted to 0.1% solutions in Deionized water prior to use in cylinder testing.

Standard Cylinder Test Procedure

Cylinder testing was completed on a number of Bayer process slurries from different bauxites, spanning a large alkalinity range. Graduated cylinders (1000 mL) were filled with the appropriate red mud slurry. The 0.1% flocculant solution was typically split dosed (50/50%) using a syringe and after both the first and second dose of flocculant, the slurry was mixed with an equal number of plunges. The settling time was recorded for the mud interface to settle from 900-700 mL; and the settling times were converted to settling rates in ft/hr. Settling rates were measured at several dosages of flocculant and the results plotted to generate dose response curves. Product replacement ratios were then calculated for dosages required of the new flocculant relative to the standard flocculant program, typically at target settling rates of 25ft/hr and 50 ft/hr (unless otherwise specified).

Test Results from Plant A

Cylinder tests were conducted on Blow-off slurry at Plant A to compare the settling performance of the new Rigid Rod flocculant 85252 RRA to the current Nalco 9779 program. Results showed that 85252 RRA had superior settling performance as evidenced by product replacement ratios of 0.71 and 0.64, respectively (at 25 and 50 ft/hr settling rates) vs. Nalco N9779 (Figure 4). Thus, 85252 RRA reduces the flocculant dose by 30 % or more.

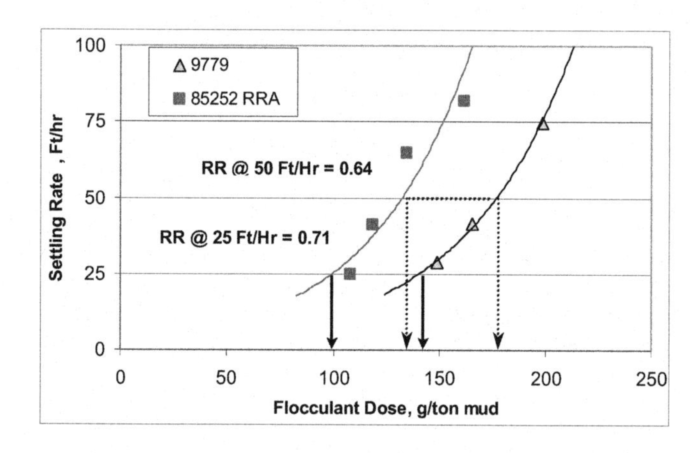

Figure 4. Results from Blow off Slurry at Plant A

Cylinder tests were also completed using slurries collected from the Back-end washers in Plant A (Figure 5). The current plant program in this case involved Nalco 85144 and thus it was used as the benchmark. For direct comparison, the Rigid Rod flocculant 85292 RRA was tested. The results showed superior settling performance for 85292 RRA with replacement ratios ranging from 0.76 to 0.85 (at 25 and 50 ft/hr mud settling rates, respectively).

Additional tests were conducted on slurries having different soda concentrations to further examine the potential to reduce 85144 flocculant usage across the CCD circuit. Indeed, 85292 RRA product replacement ratios ranged from 0.62 to 0.89 vs 85144. The results plotted against the slurry soda concentrations are depicted in Figure 6. These results suggest overall flocculant usage in the CCD circuit could be reduced by as much as 20-25 % by replacing 85144 with 85292 RRA.

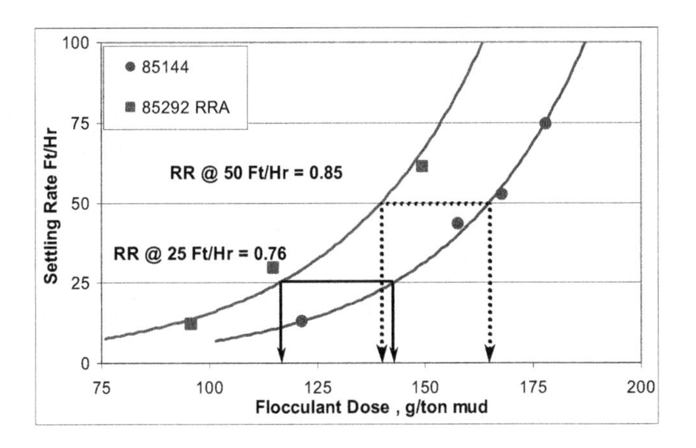

Figure 5. Results from Back End Washer Slurry at Plant A

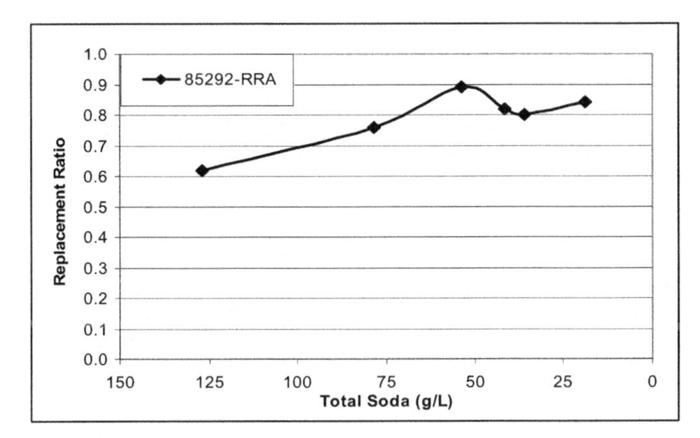

Figure 6. 85292 RRA vs. 85144 Product Replacement Ratio at Total Soda levels between 145 to 10 g/L as Na_2CO_3.

It is interesting that the 85292 RRA replacement ratios trend lower as the soda concentration of the slurry increases. It is postulated that the exceptional performance of the 85292 RRA at higher soda levels is a result of a more rigid polymer conformation in solution and a higher average molecular weight compared to current 85144.

Test Results from Plant B

The relative performance of the new 85252 RRA and 85282 RRA flocculants were compared to the current N9779 and 85130 products on a series of plant slurries with Total Soda between 245 and 12 g/L as Na_2CO_3 (Figure 7). Table I. summarizes the product replacement ratios obtained for settling rates between 60-100 ft/hr for typical blow-off slurries and between 25-60 ft/hr for slurries collected from the washer circuit.

Table I. RRA Replacement Ratios versus Total Soda for Slurries Collected at Plant B.

Total Soda, g/L as Na_2CO_3	245	145	96	59	41	12
Slurry Solids, g/L	30	35	35	35	48	36
* RRa Product Replacement Ratios vs. Plant Flocculants at Target Settling Rates						
	60 - 100 ft/hr	25 - 60 ft/hr				
85252 RRA vs. N9779	0.70 - 0.75	0.73 - 0.80	0.66 - 0.71	0.72 - 0.85	0.77 - 0.95	0.97 - 1.00
85282 RRA vs. N9779	-	-	0.50 - 0.54	0.44 - 0.46	0.77 - 0.84	0.76 - 0.78
85282 RRA vs. 85130	-	-	0.64 - 0.66	0.58 - 0.70	1.00	0.90 - 0.95

** results from multiple samples and tests*

Again the new Rigid Rod Flocculants display superior settling performance and their use throughout the circuit could lead to overall dosage reductions between 20 to 30 %.

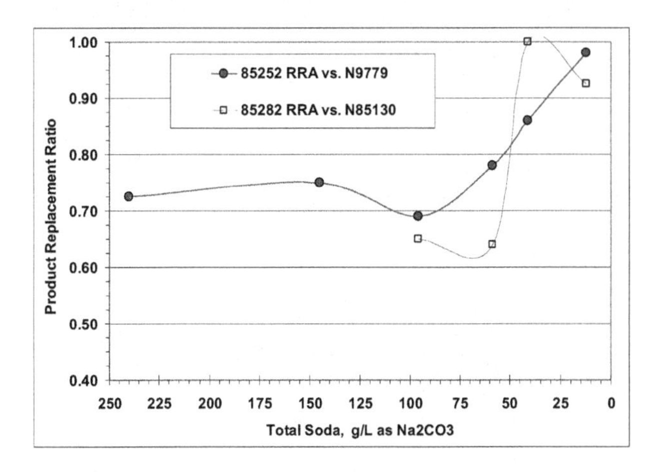

Figure 7. 85252 RRA vs N9779 and 85282 RRA vs. 85130 Product Replacement Ratios at Total Soda levels Between 245 to 10 g/L as Na_2CO_3.

The effect of mud solids concentration on the efficacy of 85252 RRA was studied at TS ~ 245 g/L. Test slurries containing between 55 and 100 g/L solids were prepared and the mud-settling dose response curves were obtained using 85252 RRA and N9779. At each solids level the flocculant dose required to achieve a 65 ft/hr target settling rate was determined and the results are displayed in Figure 8.

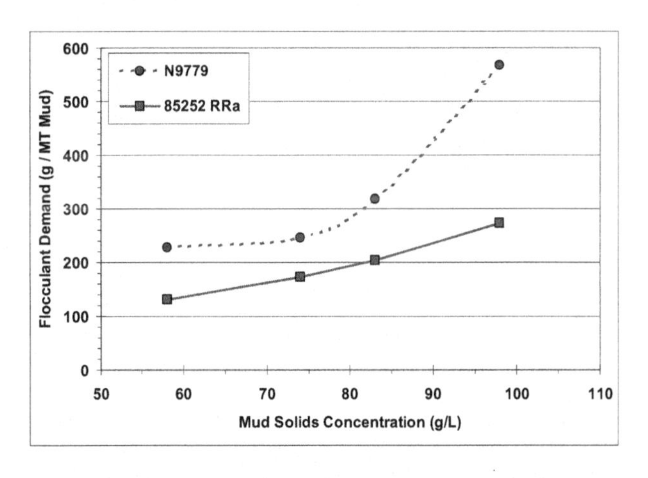

Figure 8. N9779 and 85252 RRA Flocculant Dosages Needed to Achieve a 65 ft/hr Settling Rates on Blow-off Slurries Containing 50 to 100 g/L Mud Solids.

The differences in these two curves defines the replacement ratio at each feed solids level given in Table II.

Table II. 85252 RRA Replacement Ratios vs. N9779 at Various Mud Solids Concentrations.

Mud Solids (g/L)	58	74	83	98
RR @ 80 Ft/hr	0.58	0.74	0.64	0.48

In general, the replacement ratio remains relatively constant between 0.6-0.7 at lower mud solids. Interestingly, the superior settling behavior of 85252 RRA is more pronounced in the higher mud solids environment (98 g/L). This suggests that the new flocculant should be more tolerant of changes in mud solids concentrations. Sudden changes in solids concentration can happen in blow-off slurries as a result of upsets in the flash train.

Test Results from Plant C

The relative performance of the 85252, 85282 and 85292 RRA flocculants were compared to the current N9779 and 85144 flocculants on typical front, middle and last stage washer slurries. Product replacement ratios were determined from the settling response curves at both 25 and 50 ft/hr mud settling rates. The results are summarized in Table III.

Table III. RRA Product Replacement Ratios for Washer Slurries Collected at Plant C.

RRA Flocculant		Product Replacement Ratios at Settling Rate Target 25 ft/hr			50 ft/hr		
		2nd Stage Washer	4th Stage Washer	8th Stage Washer	2nd Stage Washer	4th Stage Washer	8th Stage Washer
85252	vs. N9779	0.68	0.58	0.89	0.77	0.61	0.87
85282		>1.0	0.39	0.67	>1.0	0.45	0.68
85292		-	0.74	0.55	-	0.88	0.54
85282	vs. 85144	-	0.50	0.91	-	0.49	0.88
85292		-	0.94	0.75	-	0.96	0.69

Once again the new 85252 RRA samples gave superior settling performance versus N9779 on the front and middle washer slurries, with RR = 0.6 to 0.7. The best flocculant for the middle stages is 85282 RRA and 85292 RRA is the best flocculant for the last stage. These results are consistent with matching the optimum polymer charge with the slurry alkalinity.

Plant Trials at Plant B

Two plant trials with 85252 RRA were conducted at Plant B. In early 2005 a short one week test was completed replacing N9779 on a single washer (Figure 9). For the two week period prior to the test the average flocculant consumption on the washer was 0.18 lbs floc per metric ton of alumina produced. This decreased by ~18 % to 0.145 during the week-long test with the new 85252 RRA flocculant.

Following this the N9779 flocculant was replaced with 85252 RRA on four front-end washers (Stages 2 to 5). This second trial was performed over three months and the monthly averages for flocculant consumption were compared for each vessel.

Operational performance of each vessel was maintained (e.g., mud bed levels, underflow density, and rake loads) with no upsets during the trial period. The average monthly flocculant usage per vessel for the 5 months prior to the trial and the three months of the trial are plotted in Figure 10.

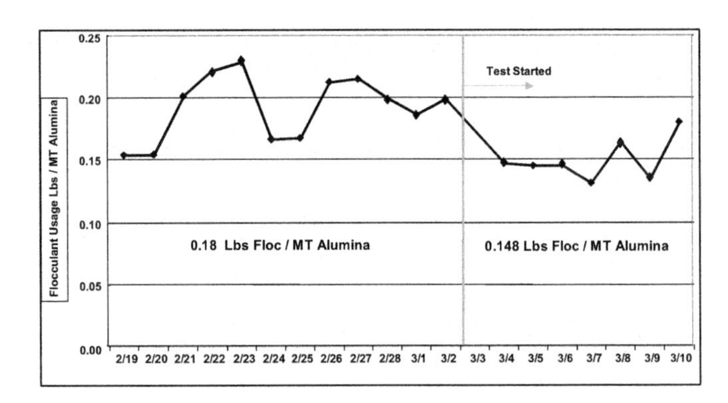

Figure 9. Flocculant Consumption Before and During a Trial of 85252 RRA on a Single Front-end Washer at Plant B.

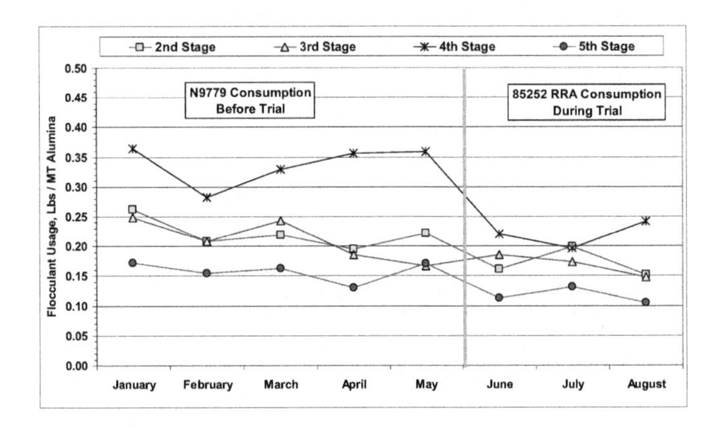

Figure 10. Flocculant Usage Before and During the 85252 RRA Trial on Washer Stages 2 to 5 at Plant B.

The average flocculant consumption on the second stage washer dropped ~23 % from the Jan-May average, 0.221 Lbs/MT to 0.171 Lbs/MT over the three month 85252 RRA Trial. On the 3rd stage vessel the flocculant consumption decreases 19.7 % from a Jan-May average of 0.211 to 0.169 Lbs/MT. On the 4th stage vessel the flocculant consumption decreases 35 % from a Jan-May average of 0.338 to 0.219 Lbs/MT. Finally, on the 5th stage vessel the flocculant consumption decreases 26.6 % from a Jan-May average of 0.158 to 0.116 Lbs/MT. The overall flocculant consumption for the 4 test vessels is summarized in Figure 11.

Further analysis of the results in Figure 11 indicates the overall flocculant use in Jan. appears to be higher than the four months prior to the test. This could be due to changes in bauxite quality and or digestion conditions, however, including this data in the analysis tends to inflate the overall flocculant consumption prior to beginning the test. Thus, using data beginning in February lowered the total N9779 usage on the four test vessels to 0.90 Lb flocculant per metric ton of alumina produced. During the three month trial with 85252 RRA this consumption was reduced by ~ 25 % to an average value of 0.675 Lbs/MT.

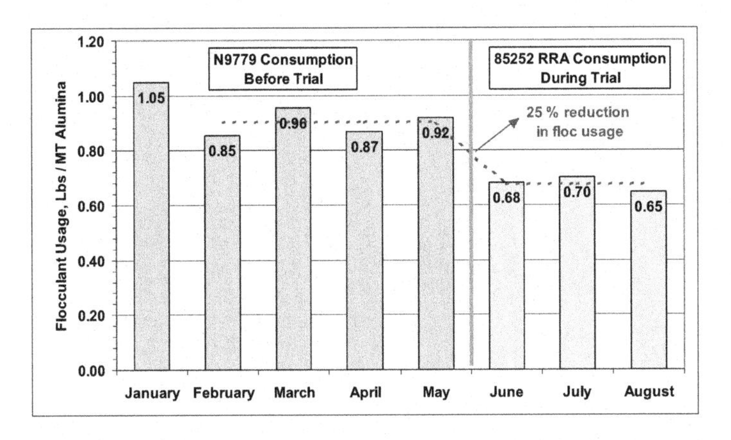

Figure 11. Total Flocculant Usage Before and During the 85252 RRA Trial on Washer Stages 2 to 5 at Plant B.

The 85252 RRA was also applied intermittently on the first Washer in the plant at various times during the trial period improving control of mud bed levels and underflow density. The N9779 flocculant consumption on this vessel just prior to start-up of the 85252 RRA trial was 0.147 Lbs/MT. This consumption was reduced by 49 % to 0.075 Lbs/MT for the period 85252 RRA was applied in the month of July.

Conclusion

A new manufacturing process has led to the development of a new platform of Bayer Process flocculants with unique structure and higher molecular weight. The new Rigid Rod Architecture technology can be used to prepare a variety of acrylate and acrylate/acrylamide copolymers for optimum performance and cost-performance benefits across the entire mud settling and washing circuit. Laboratory and full-scale plant usage show that this technology reduces flocculant consumptions across the red mud circuit by 20 % or more.

References

1. L.J. Connelly, D.O. Owens and P.F. Richardson, "Synthetic Flocculant Technology in the Bayer Process", Light Metals (1986), 61-68.

2. J. T. Malito, “Improving the Operation of Red Mud Pressure Filters”, Light Metals (1996), 81-85.

EFFECT OF FLOCCULANT MOLECULAR WEIGHT ON RHEOLOGY

Donald Spitzer and Qi Dai
Cytec Industries, 1937 West Main Street, Stamford, Connecticut 06904 U.S.A.

Keywords: red mud, rheology, settling

Abstract

Any polymeric flocculant used to settle suspended mud solids in reasonable times increases underflow mud yield stress, making the mud more difficult to flow. Yield stress (at a given solids concentration) always increases with polymer dosage, but depends somewhat on the type of polymer used. Primary emphasis of this paper is on the effects of molecular weight and the finding that, over quite a wide range, rheology does not depend on molecular weight. Thus, for lowest possible yield stress, molecular weight should be as high as possible, since this will give the lowest dosage for the required settling rate.

Introduction

In high solids slurries of any kind, rheology is determined not by the liquid phase, but by the movement of the solid particles against one another. As solids level in the mud goes up, it naturally becomes more difficult to move particles past one another and yield stress goes up.

Effects of high molecular weight polymers, i.e., flocculants, on mud rheology have been previously published [1, 2]. The most notable effect is that polymers have a large effect on mud yield stress, which is a measure of how readily the mud will flow into the underflow pump. For example, Fig. 1 shows a typical stress vs. shear rate curve for a mud settled with no flocculant. The yield stress is indicated by the shoulder at which the mud begins to flow readily and is only ~ 30 N/m^2 for this mud. In contrast, Fig. 2 shows a stress vs. shear rate curve for a sample of actual plant underflow red mud with somewhat lower solids, but yield stress is indicated by the high peak at ~200 N/m^2.

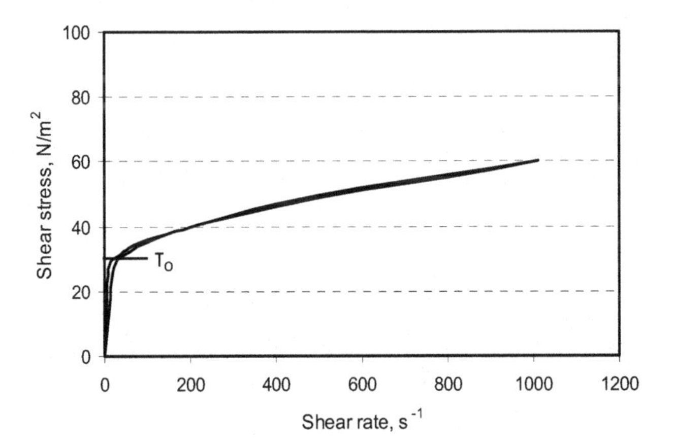

Fig. 1. Stress vs. shear rate for a red mud at 42.0 % solids with no flocculant added

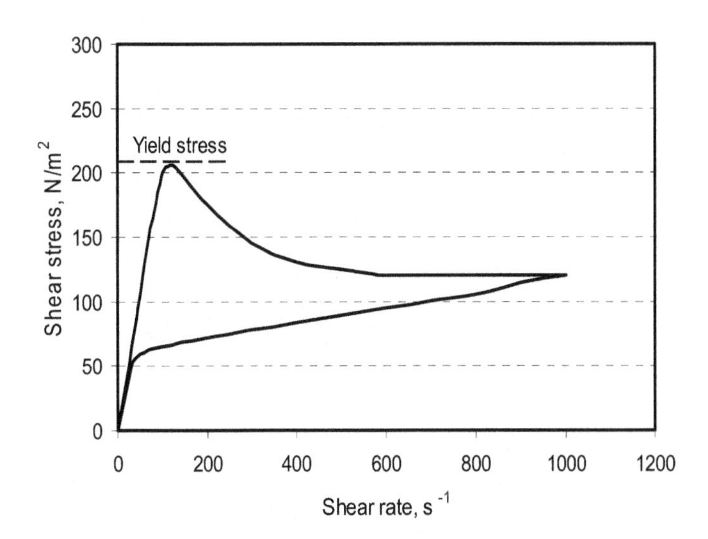

Fig. 2. Stress vs. shear rate for a flocculated red mud at 39.5 % solids.

It is clear that the easiest mud to handle would be generated using no flocculant at all. It is equally clear, however, that mud will not settle at any reasonable rate without adding a flocculant. Settling and subsequently raking and moving the mud thus become a balance between adding enough of the right type flocculant to get rapid settling and throughput but also mud that will still be fluid enough to flow into an underflow pump so that it can be moved to the next stage of processing. It was demonstrated in the earlier publications that the limiting step is the flow of the mud into the underflow pump, which is related to the yield stress; once the mud is in the pump, flocc structure is extensively broken down and there is no problem in moving the mud to the next stage.

This effect of flocculant on yield stress is emphasized in Fig. 3, where it is seen that the difference with and without flocculant increases rapidly as percent solids increases (the yield stress numbers with and without flocculant happen to be quite close to those in Figs. 1 and 2 although the Fig. 3 data was obtained on quite a different mud). Looking at a constant yield stress, the differences without and with flocculant can be equivalent to differences in solids of as much as 10 % (e.g., in Fig. 3, yield stress is ~50 N/m^2 at ~44 % solids without flocculant compared to ~34 % solids with flocculant).

As expected from the above discussion, it was confirmed earlier that yield stress increases as flocculant dosage increases. This dependence of yield stress on dosage applies to any high solids slurries and has been also reported for quite different systems [e.g., 3]. In addition, it is known that, at equivalent dosages, yield stress will depend on the type of polymer used as flocculant (Fig. 4) and this has been experienced in actual plant operations [4].

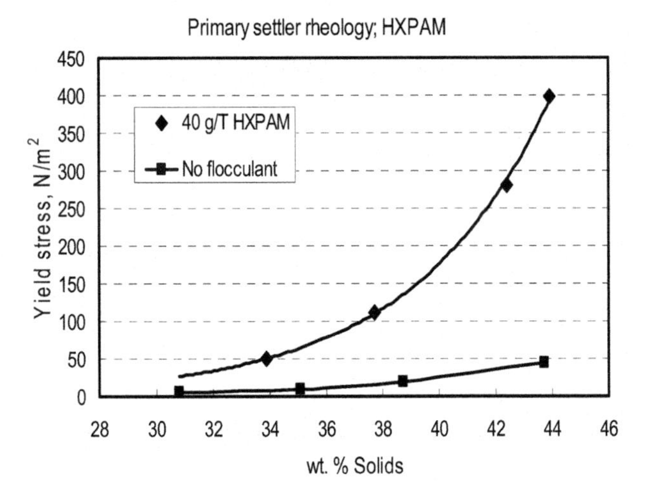

Fig. 3. Effect of flocculant on yield stress as a function of % solids.

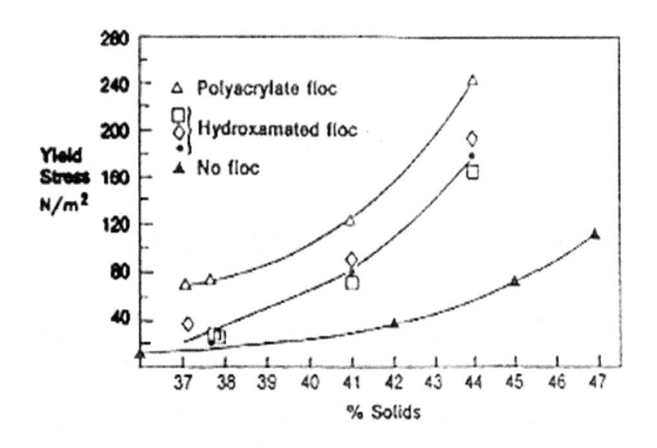

Fig. 4. Effect of polymer type on rheology (data from Ref. 2)

Experimental Procedures

Initially, measurements of yield stress were determined with a Fann model 39B rheometer. This gives a complete shear stress/shear rate curve, from which the yield stress is indicated either by the sharp change in slope, as in Fig. 1, or a definite peak as in Fig. 2. Later, a procedure was developed in which a Brookfield viscometer was modified by replacing the usual cylindrical spindle with a flat paddle. When operated at low rpm, this device gave excellent correlation between yield stress (as had been determined using the Fann rheometer) and the percent of scale reading on the Brookfield.

The muds were actual plant muds from discharge of a last washer. These were re-dispersed in appropriate liquor and resettled with the various flocculants. Increasing percent solids were obtained by simply allowing more time for compaction. At the beginning of the work, however, yield stress measurements showed excessive scatter. It was determined that this scatter was due to the fact that these last washer muds had already been repeatedly flocculated as they went through the settler and washer train in the plant. To get rid of the pre-existing flocculant, the muds were dried and heated to 500°C for two hours – this is sufficient to decompose all organics, but this temperature is too low to cause sintering and changes in particle size. Such heat-treated muds were readily dispersed in liquor using a Waring blender; the very small amount of coarse particles that remained was removed by pouring the slurry though a 60 mesh screen. Muds from two different plants were used in the work reported here.

Results

Last Washer Tests

Tests were done under conditions that simulated those of a last washer, frequently the sixth washer. As noted above, however, the mud had first been "cleaned" of flocculant that had been previously added, so we are seeing only the effects of the flocculants added in these tests. In this first set of tests on mud from plant #1, three 70 % anionic flocculants were used with molecular weights (as estimated from viscosity measurements) of 8.8, 5.2, and 1.9 million. The highest molecular weight (m.w.) sample was an emulsion product, the sample with 5.2 M m.w. was a dry product, and the lowest molecular weight sample was made by shearing a solution of the emulsion product.

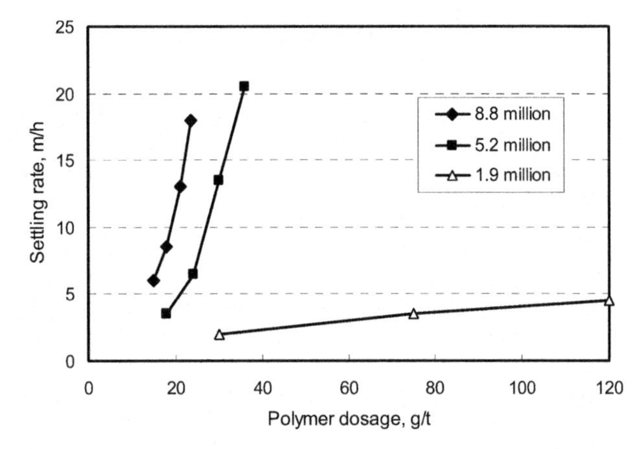

Figure 5. Settling rate vs. dosage for 70% anionic washer flocculants of various molecular weights; 6th washer tests.

Settling rate tests show the expected decrease in effectiveness as molecular weight decreases. The intermediate molecular weight still gives reasonably good settling, but the lowest molecular weight gives settling that is very poor compared to the higher molecular weights (Fig. 5).

In contrast to the large range of settling behavior, rheology of the muds produced by these three different flocculants at the same dosage showed no significant dependence on molecular weight (Fig. 6).

For the above set of samples where measurements were done with the Fann rheometer, it was noted that although yield stresses were essentially independent of molecular weight over the entire range of solids, there was some dependence of the shear rate required to reach maximum shear stress (yield stress) on molecular weight. Comparing the three polymers at approximately equal yield stresses (or equal % solids) gives Table I. As molecular weight increases, slightly more shear is required to break up flocc structure, but the effect is fairly small compared to the changes in molecular weight

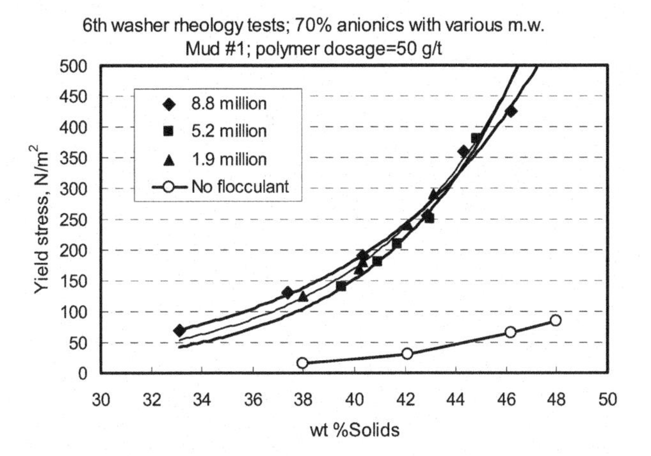

Figure 6. Yield stress vs. % solids for washer flocculants of various molecular weights.

Table I

yield stress	shear rate corresponding to yield stress peak m.w.=1.9 M	m.w.=5.2 M	m.w.=8.8 M
~100	82	96	110
~200	92	99	126

Additional experiments were done with another mud, from plant # 2. Last washer tests were again done as above, but also primary settler tests. For the last washer tests, the flocculants were different from those used with the first mud. This set of three flocculants started with a high molecular weight emulsion product and the two lower molecular weights were both obtained by shearing a solution of the high molecular weight polymer for various times.

Settling tests (Fig. 7) again show a wide range of performance, although the differences are not quite as great as with the first mud.

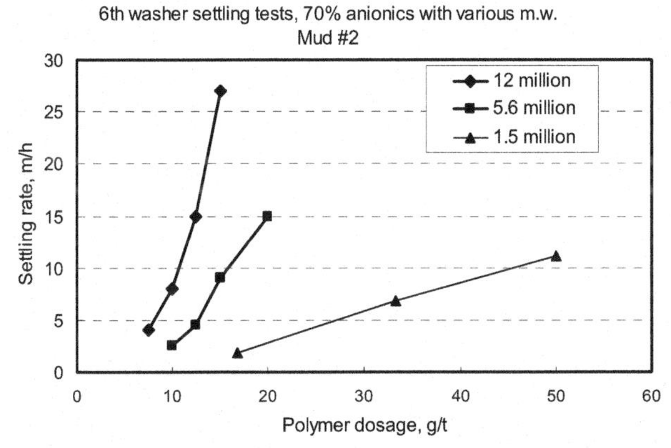

Figure 7. Settling rate vs. dosage for washer flocculants of various molecular weights.

The rheology results (Fig. 8), which are the averages of two or three series of tests, are somewhat different than those from the first mud (note that this is a different mud than used in the first set of tests and also that the polymer dosage used is lower, to be closer to what might actually be used in a plant). The two highest molecular weight polymers, which gave fair to good settling rate performance, gave muds with the same rheology, within experimental error. The lowest molecular weight polymer, which gave quite poor settling rates compared to the other two polymers, gave a yield stress vs. % solids curve that was definitely slightly below that from the other polymers. Nevertheless, the molecular weight effect on rheology is far less than on settling behavior.

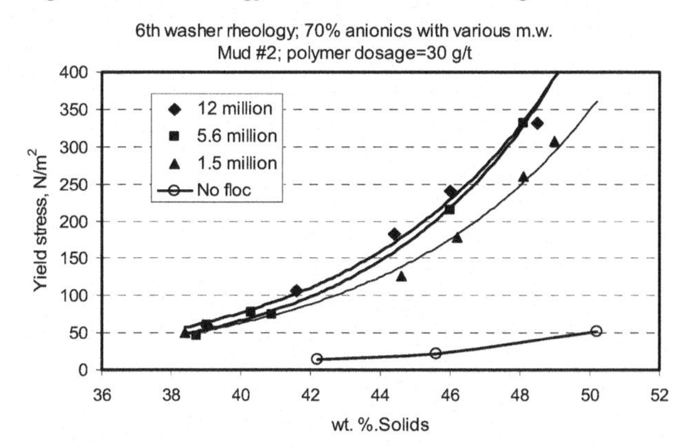

Figure 8. Yield stress vs. % solids for washer flocculants of various molecular weights.

Note again in both Figures 6 and 8 the large difference in yield stresses between no flocculant and any kind of flocculant.

Primary Settler Tests

The dependence, or lack of dependence, of rheology on flocculant molecular weights is not peculiar to the last washer, but was observed on all stages, from the primary settlers through all of the washers. As an additional example, primary settler tests using hydroxamated polyacrylamide (HXPAM) flocculants are shown below. The HXPAM with highest molecular weight was a standard emulsion product and the lower molecular weight samples were made by shearing a dilute solution of this product for various times. These tests used the mud from plant #2, as for the second set of last washer tests discussed above.

Settling rates vs. polymer dosage (Fig. 9) show the expected dependence on molecular weight, with good performance for both of the two highest molecular weights, but considerably poorer performance for the lower molecular weights.

Rheology results (Fig. 10) show the data points for the three highest molecular weight polymers scattering around a common line, while the lowest molecular weight sample gave measurably lower yield stresses. Tests with the highest molecular weight sample at half the dosage gave results comparable to those from the lowest molecular weight sample. Again, even the low dosage gives yield stresses that are much higher than those obtained with no flocculant.

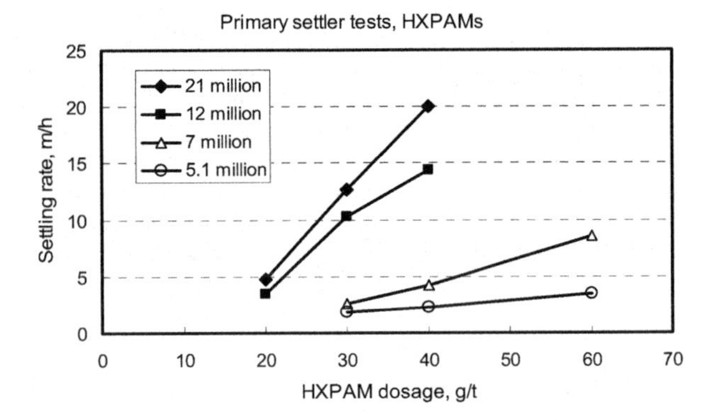

Figure 9. Settling rate vs. dosage for HXPAMs of various molecular weights.

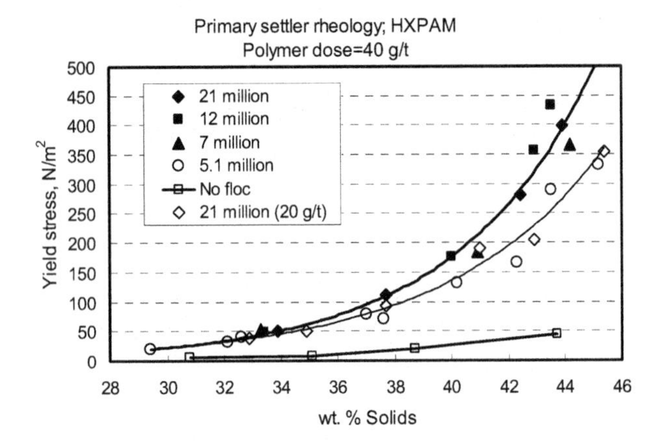

Figure 10. Yield stress vs. % solids for primary settler tests with HXPAMs of various molecular weights.

The effects of flocculant molecular weight for the various measurements are summarized below (Fig. 11) as yield stress at one reasonable solids level (~40%). This clearly emphasizes the lack of dependence of yield stress on molecular weight for all of the polymer-mud combinations examined.

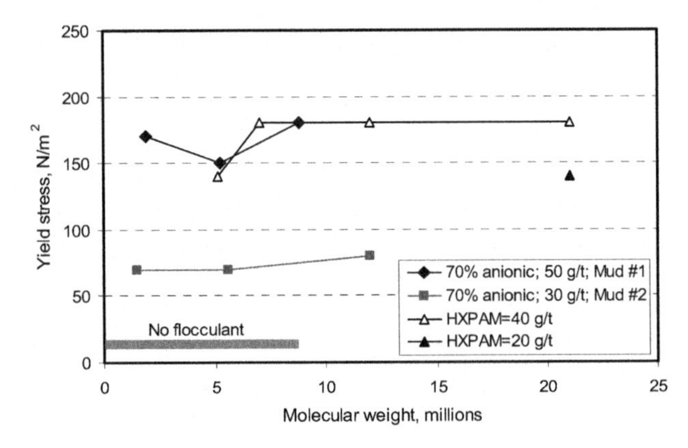

Fig. 11. Yield stress vs. molecular weight for muds with 40 wt. % solids.

Discussion

The most striking effect of polymeric flocculant is the large effect on the ability of the mud to flow, i.e., on the yield stress (Fig. 3). This is consistent with the fact that in the absence of flocculant, only the small primary particles need to move past one another, whereas in the flocculated mud, much larger, extended and irregular aggregates have to be moved past one another. Since the use of a flocculant is necessary for rapid settling, however, the problem becomes one of minimizing the damage done by the flocculant to the flow properties of the mud while maintaining rapid settling and compaction.

The amount of flocculant required for a large effect on rheology is surprisingly small. As little as 20 g/T reduces percent solids at a given yield stress as much as ~10 % (Fig. 10). Since the surface area of most muds is in the range of 2-20 m^2/gm, a dosage of 20 g/T corresponds to ~0.01 mg/m^2 – 0.001 mg/m^2. It is calculated from molecular dimensions that close packing of most simple organic molecules will be closely equivalent to ~ 1 mg/m^2, so that the coverage indicated above corresponds to only 0.1 – 1 % of the surface area.

The observed non-dependence of rheology on molecular weight at high molecular weights was initially unexpected. Intuitively, since higher molecular weights are associated with more viscous solutions, it might be expected that muds made with very high molecular weight polymers would be "stickier" and have higher yield stresses. It is now clear, however, that the observations of both rheology and settling rate are quite consistent with floc structures.

Settling rates show the expected dependence on molecular weight. That is, settling rates increase as molecular weight increases because the larger polymer molecules make larger floccs possible. It is the very large floccs that are directly responsible for high settling rates. On the other hand, flocc density measurements [e.g., 5] have shown these very large floccs to consist of quite loose aggregates that have densities barely more than that of the liquid phase, i.e., they consist mostly of liquid. Consequently, the large floccs that were responsible for rapid settling do not survive as the mud compacts.

The only floccs that survive through the compaction and raking process are the small, relatively dense floccs, which are known from flocc density measurements to be less than ~200-400μ in diameter for red muds [6]. It is these small, dense floccs that determine rheology. At low shear rates, they can be moved against one another only with difficulty, especially as underflow densities get high, and hence the yield stresses are high. At high shear rates, these floccs are broken into smaller pieces that can be moved more easily, as happens, e.g., upon going through the underflow pump.

The present results indicate that the structure of these small, dense floccs is not sensitive to molecular weight, at least at the highest molecular weights. The picture of a dense flocc is like an aggregate of primary particles that are loosely stuck together with the polymeric flocculant as a kind of glue, but with few points where the particles are "glued" together. As molecular weight increases, each molecule contains more of the active functional groups that are responsible for bonding, resulting in stronger "glue spots" holding particles together. It appears, however, that as long as molecular weight is high enough to give reasonably strong flocc structures further increase in molecular weight might serve to make these structures slightly stronger (as indicated, e.g., by the shear rate data in Table I), but does not significantly change their form. On the other hand, as molecular weight decreases below some minimum, the "glue spots" holding the floccs together will get weaker and some actual breakdown in flocc structure is expected. These smaller aggregates will naturally lead to lower yield stresses.

In tests not reported here, we have also noted that as molecular weight gets lower and lower – far below that required for a flocculant – yield stresses (at constant solids and polymer dosage) continue to decrease, but are always above those for the no flocculant case.

The above picture is consistent with all of our observations. Namely, above some minimum molecular weight, mud rheology is essentially independent of molecular weight. As molecular weight decreases, some decrease in flocc structure may occur and lead to lower yield stresses; this decrease is expected to depend on the particular combination of flocculant and mud characteristics.

Based on these results, optimization of a settler or washer should a) use the highest molecular weight flocculant available in order to minimize the required dosage and b) compare the various types of available flocculants, always at the highest possible molecular weight. Based on these principles, we believe that many plants can significantly increase underflow densities without increasing yield stresses, resulting in better separations throughout, with benefits such as higher overflows and improved mud washing, soda and alumina recovery.

The most obvious candidate for optimization would be the primary settlers. As shown in Table II, a reasonable increase in underflow solids of 2-3 % can increase the overflow rate by ~1 %, which would lead directly to an alumina recovery improvement at very little cost.

Table II

Overflow/underflow relationships for a primary settler, assuming:
Feed flow=1000 m^3/h and 50 g/L solids
Liquor density=1.3, solid density=3.0

Underflow U/F rate m^3/h	Solids g/L	Solids wt%	Density g/L	Overfow O/F rate m^3/h	O/F rate % increase
100	500	31.6	1583	900	0.0
95	526	32.9	1598	905	0.6
90	556	34.4	1615	910	1.1
85	588	36.0	1633	915	1.7
80	625	37.8	1654	920	2.2
75	667	39.7	1678	925	2.8
70	714	41.9	1705	930	3.3

Conclusions

The flocc structure that determines rheology is very different from that which determines settling rate. High settling rates are related to large, but weak, aggregates that break up as the mud compacts. Rheology is subsequently determined by the small, relatively dense floccs that are strong enough to withstand low shear.

Rheology does not depend noticeably on molecular weight in the range typical of good flocculants.

Rheology does depend on flocculant dosage (increases with increasing dosage) and also depends on the type of flocculant used.

By comparing rheology of flocculated muds with muds settled without flocculant, it is clear that there is still a very great deal of room for improvement in flocculated mud rheology. Using observations reported here, we believe that significant benefits can be obtained, either in actual increased alumina recoveries or improved mud washing efficiencies.

Settling tests are traditionally used to optimize settlers or washers. The present results, however, show that settling and rheology are not necessarily related. In optimizing a settler, therefore, underflow rheology should be considered as well as settling rate and clarity. It is quite possible that the best overall flocculant may not simply be the one that gives fastest settling and good clarity.

References

1. D. P. Spitzer and P. V Avotins,"The Effect of Flocculants on Rheological Properties of Thickener Underflow," *SME Annual Meeting,* Phoenix, AZ, 1992, pp.245-250.

2. D. P .Spitzer and P. V. Avotins, "Effects of Polymers on Flow Properties of Bayer Process Underflow Muds," *Light Metals,* 1994, pp. 1225-1230.

3. M. Pawlik and J. S. Laskowski, "Evaluation of Flocculants and Dispersants through Rheological Tests," *Polymers in Mineral Processing,* ed. J. S. Laskowski, pp. 541-555.

4. R. G. Ryles and P. V. Avotins, "Superfloc HX, A New Technology for the Alumina Industry," *Fourth International Alumina Quality Workshop*, Darwin, Australia, June 1996, pp. 206-215.

5. J. B. Farrow and L. J. Warren, "The Measurement of Floc Density – Floc Size Distributions," in *Flocculation and Dewatering*, ed. B. M. Moudgil and B. J. Scheiner, Engineering Foundation, New York, 1989, pp. 153-156.

6. M. J. Gagnon, A. Leclerc, G. Simard, and G. Péloquin,, "A Fractal Model for the Aggregate Size Distributions Generated during Red Mud Flocculation," *Light Metals*, 2003, pp. 105- 111.

A FRACTAL MODEL DESCRIBING THE AGGREGATION OF VARIOUS MINERAL MATERIALS

M.J. Gagnon[1], A. Leclerc[1], G. Simard[1], R. Verreault[1] and G. Peloquin[2]

[1]Université du Québec à Chicoutimi, 555, de l'Université, Chicoutimi, Québec, G7H 2B1, Canada
[2]Alcan International Limited, Arvida R&D Centre, 1955, Mellon Blvd., P.O. Box 1250, Jonquière, Québec, G7S 4K8, Canada

Keywords:

Abstract

The agglomeration of the particulate materials is necessary in many mineral processes to enhance the settling of the solid fine particles. Depending on physical and chemical parameters, the aggregates generated demonstrate specific density-size distributions that reflect the dominant agglomeration process. A fractal model of the agglomeration process has been previously presented. This empirical model was initially constructed to describe the red mud flocculation process using simple variables such as density, particle sizes and fractal order. The application of the model to other data sets found in the scientific literature demonstrates a wider utilization of the model and was helpful in the development of a criterion that should facilitate the identification of the dominant mechanism by which the aggregation occurs. In this paper, data from different sources, various mineral materials and different agglomeration processes is analyzed and compared in the light of the proposed model and the proposed criterion. All the above data has been fitted with a mathematical method of linearization. The values obtained with the fit are discussed in relation with the coagulation or flocculation processes, the fractal dimension and resulting fractal structures.

Introduction

For the mineral industry, the solid-liquid separation processes are fundamental steps in the overall treatment of the minerals. In bauxite treatment plants, gravity sedimentation in thickeners and clarifiers is widely spread. Among all solid-liquid separation processes occurring in mineral industries, the agglomeration of the particulate materials is necessary to enhance the settling of the solid fine particles. Some interesting physical properties of aggregates generated during coagulation and flocculation are the density and size. The relationship between the density and size of the particles has been previously described and the agglomeration process can be proposed in material such as glass spheres, fluorite or hematite [1-3]. Red mud aggregates have been characterized in terms of their densities, shapes, and sizes [4].

It has been suggested [1,3] that the floc growth of large aggregates occurs by cluster addition, and that small flocs and coagula generally possessing a more compact structure may have been partly formed by single particle addition. The concept of floc growth through random addition of clusters rather than individual solid particles was introduced by Klimpel and Hogg [5]. The density decreases with increasing floc size. Also the floc density-size relationships are probably determined more by the distribution of microfloc sizes and densities than by the primary particle size. The floc formation under various system conditions all yield similar density relationships [6]. The aggregates generated demonstrate specific density-size distributions that reflect the dominant agglomeration process. A fractal model of the agglomeration process has been previously presented [7].

In this paper, data from different sources, various mineral materials and different agglomeration processes are analyzed and compared in the light of the proposed model and the proposed criterion. The data has been fitted with a mathematical method of linearization and the results are discussed in relation to the coagulation or flocculation processes, the fractal dimension and resulting fractal structures. At first sight, the model seems applicable to the flocculated particles. However, from the data presented here, it is obvious that the model can be extended to describe aggregation by different mechanisms.

The Fractal Model

The original model was built to describe the density-size distribution of the flocs for the red mud. The first mathematical steps in the development of the model were previously presented and the fundamental assumptions pertaining to the mechanism supported by the model can be found elsewhere [7].

The density of the macrofloc of order n=1 is given by:

$$\rho_n = X^n \rho_0 + (1-X^n)\rho_l = X^n(\rho_0-\rho_l) + \rho_l \quad (1)$$

Let X represents, at each stage of agglomeration, the volume fraction of agglomerates from the previous stage. The fractal growth implies that the ratio (Y) of the diameter of a floc of order n to that of a floc of order n-1 is constant:

$$d_n = Y\, d_{n-1} = Y^n d_0 \quad (2)$$

Solving for n as a function of the diameter of the floc d_n, one obtains:

$$n = \log(d_n/d_0) / \log(Y) \quad (3)$$

Fractional values less than 1.0 corresponds to the agglomeration of activated particles. Fractal values greater than 1.0 corresponds to the agglomeration of microflocs. Equations (1) and (3) may be used to calculate the density of a floc of diameter d_n provided that the constants X, Y, d_0, ρ_0 and ρ_l are known. Alternatively, these constants may be obtained from the fit of an experimental curve relating floc density and diameter. The linear equation is:

$$\log((\rho_n - \rho_l)/(\rho_o - \rho_l)) = (\log X/\log Y) \log(d_n) - (\log X/\log Y) \log(d_0) \quad (4)$$

$y = m\,x + b$ *(linearity)* where
$y = \log((\rho_n - \rho_l)/\rho_o - \rho_l)$; $x = \log d_n$
$m = (\log X / \log Y)$; $b = -(\log X / \log Y) \log d_0$

$$\log((\rho_n - \rho_l)/(\rho_o - \rho_l)) = \log(d_n/d_0)(\log X / \log Y) \quad (5)$$

to which conventional least square techniques may be applied. From data fit, we can obtain :

1) $\log(d_o) = -b/m$ particle diameter (from least square fit)
2) the best ratio of $\log X/\log Y = m$.

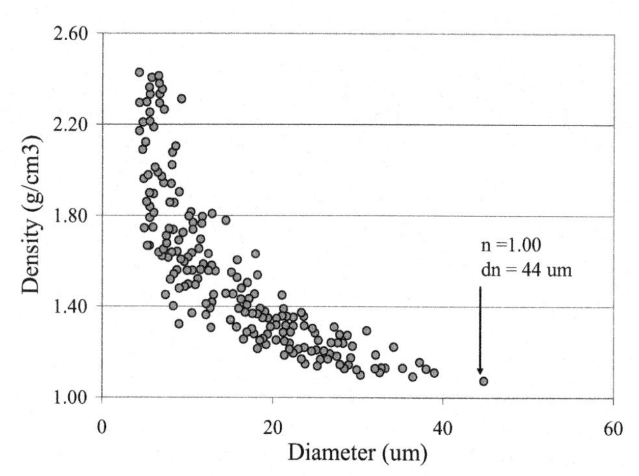

Figure 1. Density size distribution of aggregates for glass spheres [1].

Experimental Conditions

In spite of all the studies that have been previously presented in the scientific literature, there are only few references in which it is possible to find the original data on density size relationship. In many cases, the authors present the best-fit curve without the data. The data used in this paper originated from three sources in the literature [1,3]. The data analyzed here arise from published graphs and all the data points had to be determined from the adequate positioning on the y and x axis. Some data from our previous work [4] are also included. The characteristics of the materials used in this study are described in Table I. Different materials were used for the model: glass sphere, fluorite ore, hematite and red mud. The pH, the ionic strength, the agitation level and other experimental conditions vary considerably for each material. For more details on the material, refer to the original work cited in Table I. The density-size distribution for each material is presented in Figures 1 to 4. The particle diameter range is between 0.1 to 20.0 μm.

Table I. Characteristics of Materials Used for the Model. (Copolyacrylamide-Polyacrylate, PAM)

Material	Density (g/cm^3)	Size (μm)	Flocculating agent name	Reference number
Glass sphere	2.45	1.0-4.0	cationic	[1]
Fluorite ore	3.18	0.1-3.0	PAM	[2]
Hematite	5.20	>0.4-18.8< 80%	wheat starch or no flocculant	[3]
Red mud	3.00	>1.0-20.0< 90%	PAM	[4]

For each distribution, the primary particle diameter (d_0) was calculated from the original data and is presented in Table II. For the glass spheres, in spite of the fact that the particle size distribution lays between 1.0 and 4.0 μm, the initial particles diameter was fixed to 5.5 μm from the data in Figure 1.

In fact, the original authors [1] mentioned the possibility of doublets as the smallest particles found. For the fluorite ore, the average size by weight was determined to be 1.0 μm [1]. However in this study, a particle diameter of 2.0 μm is selected to obtain a better fit [2]. For hematite, the 50 % cut-off was 2.7 μm [3], which includes both coagulation and flocculation processes, and for the red mud, the mean particle size was calculated as being 13.7 μm using data from an alumina plant [4]. The particle sizes used in this study and other pertinent data are presented in Table II.

Table II. Values Found for the Model Variables Such as the Volume Fraction of Agglomerates from the Previous Stage (X), the Diameter of a Floc of Order n to that of a Floc of Order n-1 (Y), and the Diameter of the Original Material (d_0).

Material	X	Y	d_0 (μm)
Glass sphere	0.08	8	5.5
Fluorite ore	0.10	8	2.0
Hematite	0.15	8	2.7
Red mud	0.20	8	13.7

The aggregated particle size distributions vary considerably from one material to the other. Narrow aggregate size distributions are found with the glass spheres and the fluorite ore (Figures 1 and Figure 2) since the larger aggregate diameters reach only 50 μm for the glass spheres and 80 μm for the fluorite ore.

The larger aggregate size for the hematite reaches 180 μm (Figure 3). The largest aggregates found for the red mud can reach up to 1500 μm as shown in Figure 4. The flocculating agents are different in each case study (Table I). A cationic flocculent is used with the glass spheres, an anionic co-polymer is used for the fluorite ore and the red mud. In the case of hematite, the original material was coagulated or flocculated using starch [3]. In this last case, all the data points from both experiments have been placed on the same graph as seen in Figure 3.

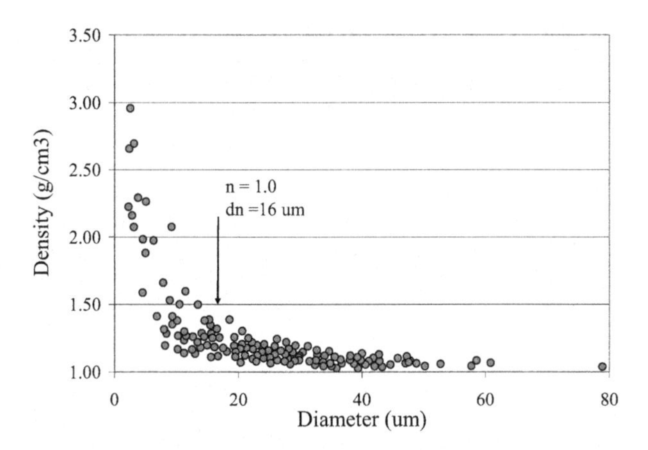

Figure 2. Density size distribution of aggregates for fluorite ore [2].

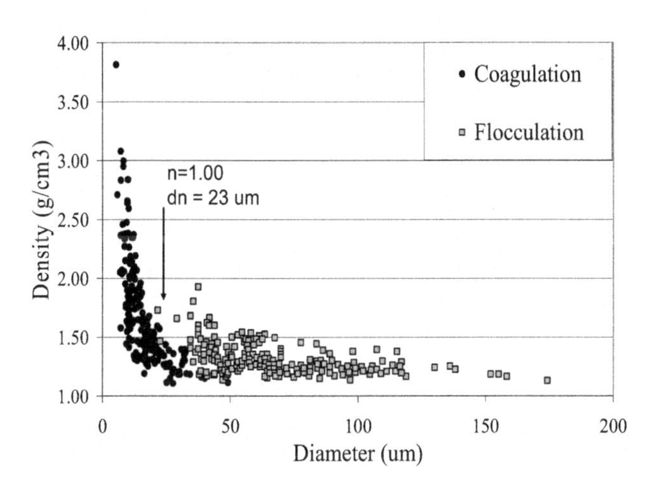

Figure 3. Density size distribution of aggregates for hematite [3].

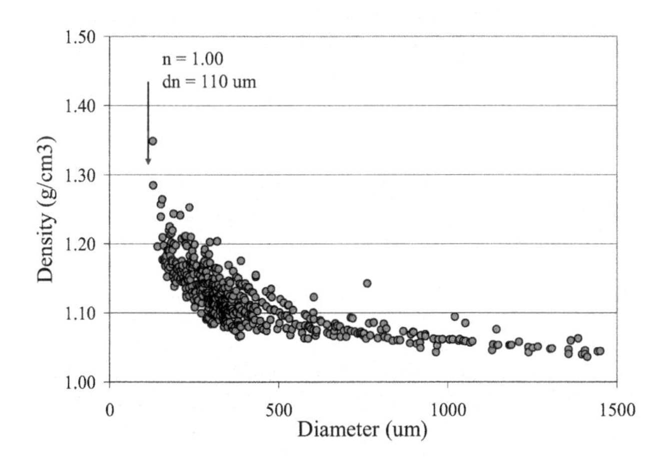

Figure 4. Density size distribution of aggregates for red mud [4].

Results and Discussion

For each material, equations (1) and (3) were fitted by a graphical method [7] to the original data and the results are listed in Table III. In this way, a value of 8 is found for the variable Y independently of the material used. This constant value of 8 found for the parameter Y may indicate a certain level of similarity in the overall agglomeration processes. In fact, this may indicate that the agglomeration is limited by the effective volume in which a group of particles (initial, activated, microfloc or floc) is capable of efficient agglomeration. The term volume can easily be replaced by kinetic and probability terms as many authors have indeed been using. On the other hand, the value of variable X seems to vary slightly with the nature of a material. The volume fraction of solid varies from 0.08 to 0.20. Independently of the material used, a relatively low solid content is found. No correlation seems to exist between the initial size of the particles and the volume solid ratio of the aggregates. The cause to effect relationship may be more dependent on the type of aggregation, the presence and properties of the polymer.

Table III: Values of the Linear Regression Parameters y = mx + b.

Materiel	Slope (m)	Intercept (b)	R^2
Glass sphere	-1.0341	0.6488	0.82
Fluorite ore	-1.0866	0.2999	0.79
Hematite	-0.5624	0.1704	0.56
Red mud	-0.9446	1.0659	0.85

Application of the Model

The values of the two model variables were found merely by manually fitting the best curve through the data points and evaluating parameters such as average particle size. A better way to proceed is to linearize the data and then use least square techniques to automatically fit the data. The linear regression equations have already been described above. As one can observe, the initial averaged particle size (d_0) can be obtained from the ratio of –b/m and can therefore be evaluated by calculating the slope of the linear regression and the intercept on the y axis. On the other hand, the individual values of the X and Y parameters cannot be obtained directly from the data, since they appear as a ratio (log X/ logY = m), this ratio is equivalent to the slope m and can be evaluated. Figure 5 presents an example of the linear curves obtained after treatment of the data. The results obtained from the linear regression analysis for all the different materials are presented in Table III. Interestingly, good fits ($R^2 > 0.79$) are obtained for three of the materials studied: red mud, fluorite and glass sphere. A low level of correlation of 0.56 was obtained for hematite. In the case of hematite, as previously stated, the original data was collected under two different experimental conditions [3].

From previous work, it has been shown that a certain consistency in the value of the parameter Y can be noticed after manually fitting the data. In fact, a constant value of Y = 8 is obtained, as presented in Table II. All the results presented in Table II were obtained either from the original data of the material or from the best curve fits obtained by graphical method. By using the value of the slope obtained by the linear regression and by fixing Y at a value of 8, the parameter X can be evaluated. For some materials, there is good agreement between the actual particle sizes and the value calculated from the model (glass sphere, fluorite and red mud), but for material such as hematite, the agreement is not as good. In fact, one notices that the agreement is even better for material with a very narrow range of particle size distribution, as for glass sphere and fluorite.

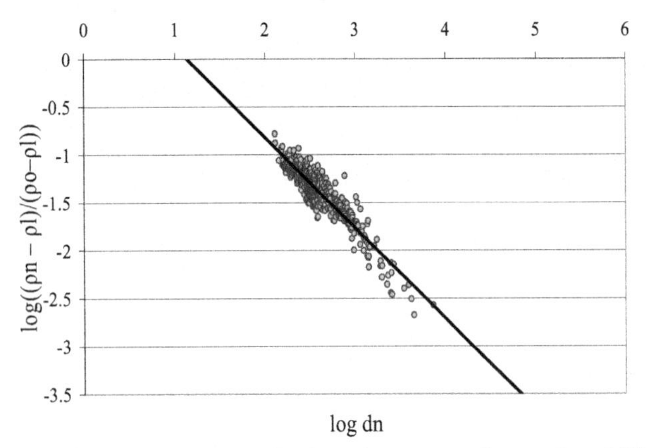

Figure 5. Density size distribution of aggregates from red mud [4].

In the case of red mud, the initial particle size distribution varies from 0.5 μm to 60 μm with 80 % found between 1.0 and 20 μm. In this case, a linear regression value of 13.4 μm for the particle compares well to a measured average particle size of 13.7 μm. A value of 5.6 μm was used in previous works [7] which corresponds to the median of the particle size distribution and yields a lower limit of 44 μm for the size of the microfloc. The limit of detection of the optical apparatus was evaluated at 60 μm but floc sizes below approximately 100 μm were rarely detected. In all cases, very good agreement exists between the average particle size of the untreated particles and the model predicted average size as presented in Table IV.

Table IV. Values for the Parameters Calculated from the Linear Regressions.

Material	Average particle (μm)	Calculated average particle size (μm)	logX/logY	Solid content
Glass sphere	1.0-4.0	4.2	-1.0341	0.12
Fluorite	0.1-3.0	1.9	-1.0866	0.10
Hematite	0.4-18.0	2.0	-0.5624	0.31
Red mud	1.0-20.0	13.4	-0.9596	0.14

Fractal Dimension:

To help evaluate the applicability of the model, analysis of the fractal dimension for the density-size distributions of the different materials was done. The data collected on the fractal level was obtained from the fit of the data points for each material. A power function was used to obtain the fractal level of each system as previously described [4]. The fractal dimension can be used to characterize changes in aggregate mass density [5].

From the density-size distribution of the power function (a, b, c), the fractal dimension can be calculated. Table V presents the fractal dimension for each material. A value of 1.00 represents a line of material grains and value of 3.00 represents a complete tri-dimensional packing (such as a perfect sphere). As can be seen from Table V, natural materials rarely reach a value of 3.00. However, coagulated material might possess a high fractal order since particles must be closely packed considering the physical mechanism involved in coagulation. In fact, it has been suggested that coagula grow by single particle addition and that flocs grow by cluster addition. It is well known that coagula are smaller than flocs [3].

Table V. Fractal Dimension for Each Material as Evaluated Using Power Fit Functions of the Density-size Distribution of the Particulate Materials.

Material	Power function fit	Fractal dimension	Mechanism	Size cut-off (μm)
Glass sphere	0.7912	2.56	coagulation	44
Fluorite	0.8451	2.13	flocculation	16
Red mud	0.7650	2.13	flocculation	110
Hematite	0.6816	1.27	both	27
Hematite coagulat	0.6316	1.70	coagulation	
Hematite floc	0.3000	1.86	flocculation	

Small flocs and coagula have more compact structures and may have partly, or initially at least, been formed by a single particle addition mechanism, before being involved in collision with other small aggregates to form larger more open aggregates [3]. From Table V, one can see that, based on the model presented, the aggregation of the glass sphere occurs via a coagulation mechanism since the agglomerate size needed to flocculate is 44 μm, a size never reached as shown in Figure 1. Furthermore, the fractal dimension for this material has a value of 2.56, which represents a close packed material, an unusual situation for flocculation. Fluorite and red mud both have a well defined distribution typical of a flocculated material as the majority of the floc sizes are well above the size cut-off necessary to start having flocculation mechanism (16 μm for fluorite and 110 μm for red mud) as presented in Figures 2 and 4. In both cases, the fractal dimension value is 2.13, showing a more open aggregate, as one should expect for a typical flocculation mechanism. In the case of fluorite, the few particles smaller than 16 μm, found in the distribution, could be non-flocculated larger particles of fluorite but it is not possible to find that out with certainty from the original paper [2]. In regards to red mud, it is interesting to note that the distribution of particle sizes stops at a lower limit of approximately 120 μm and that not many particles were detected in the range below 120 μm, in spite of a limit of detection of the optical apparatus of 60 μm [8]. In fact, most of the solid materiel

(> 99%) had already been detected when the particles of this size reached the optical system.
Hematite is a special case since the aggregates were generated by using both coagulation and flocculation methods [3]. The floc size distribution shows that the average floc size is much smaller using starch. Nevertheless, it was still possible to predict the size of the microfloc from the initial average particle size. A value of 27 μm was obtained. It fits admirably well with the results obtained by the author [3] who observed a limit of particle size around 20 μm for the separation between the coagulated and flocculated particulates. On the other hand, the level of correlation obtained for the linear regression was quite low compared to the other materials ($R^2 = 0.56$) since the regression was made on the entire data set that was obtained by combining both flocculated and coagulated hematite. Separating the data set increases the level of correlation for the coagulated hematite to a value of ($R^2 = 0.67$) but drastically reduces the level for the flocculated hematite ($R^2 = 0.27$). The large spread of the data might partially explain such low correlation values. However, the information obtained is still useful to calculate some parameters. The initial hematite particle size distribution is quite large (from 0.4 to 19 μm) with an average size of 2.7 μm. The model predicted a value of 2.0 μm in good agreement with the data reported by the authors [3]. The solid ratio of the microfloc is higher than the ratio for other types of materials with a value of 0.31 compared to a range of 0.10 to 0.14 for the others. The high density does not affect the calculation since the ratio is based on the volume. Since it is not possible at this time to obtain the original data, no definitive conclusion on this high solid content can be reached. The two mechanisms are present for hematite. Coagulation has a tendency to form smaller particles than flocculation. It has been previously demonstrated that the cut-off between the two phenomena can be evaluated by considering the level $n = 1.00$, since flocculation necessitates an organization of the particles with a flocculating agent and the formation of microflocs [6]. Based on the calculation made with this model, the size of the microfloc evaluated at $n = 1.00$ was estimated to be 27 μm as presented in Figure 3. The actual model seems to adequately predict the actual cut-off between coagulated and flocculated mechanism. This has been observed and discussed in the literature [3] but no model previous to this one was able to predict the size limit of the two mechanisms.

Conclusion

The density-size distributions of the aggregates formed are well described by the model, and this independently of the mechanism and conditions of formation. In the case of a fractal order of $n = 1.00$, the model permits the calculation of the size of the microfloc necessary to generate the flocculation process. A value of $n > 1.00$ describes the higher level of organization of the flocs. For values of $n < 1.00$, the model is actually able to describe the aggregation occurring under conditions typically found for the coagulation of particles. The proposed empirical model seems to adequately describe the aggregation mechanism from the density-size distribution and permits to distinguish between flocculation and coagulation mechanisms. The linearization of the model gives the capacity to calculate an initial averaged particle size and the solid content expected to be found in the first level of flocculation (microfloc).

Acknowledgements

The authors would like to thank Alcan for supporting this work and granting permission for its publication. This project was also supported by a research grant from la Fondation de l'Université du Québec à Chicoutimi (FUQAC).

References

[1] J.B. Farrow and. L.J. Warren, "The measurement of floc density-floc size distributions". In B.M. Moudgil and B.J. Scheiner Eds, Flocculation and Dewatering, Engineering Foundation, New York, 1989, pp. 153-166.

[2] J.B. Farrow and L.T. Warren, "A new technique for characterizing flocculated suspensions". In Proceedings Dewatering and Technology Conference, Brisbane, Australia. Inst Min Metal, Parkville. 1989, pp. 61-64.

[3] P.K. Weissenborn, L.J. Warren and J.G. Dunn, "Selective flocculation of ultrafine ore. 2. Mechanism of selective floculation". Colloids and Surfaces A: Physicochemical and Engineering Aspects., 99, 1995, pp. 29-43.

[4] M.J, Gagnon, G. Simard, A. Charrette and G. Peloquin, "Characterization of red mud aggregate populations generated under shear conditions". CIM Bulletin, 95, No. 1062, 2002, pp. 87-91.

[5] R.C. Klimpel and R. Hogg, "Effects of flocculation conditions on agglomerate structure". J. Colloid Interface Sci. , 113 (1), 1986, pp. 121-131.

[6] M.J. Gagnon, G. Simard, A. Leclerc and G. Peloquin, "Behavior of red mud agglomerated with various flocculants under shear conditions". Light Metals 2002, pp. 107-114.

[7] M.J. Gagnon, A. Leclerc, G. Simard and G. Peloquin, "A fractal model for the aggregate size distribution generated during red mud floculation". Light Metals, 2003, pp. 105-111.

[8] M.J. Gagnon, A. Leclerc, G. Simard, and G. Peloquin "Effect of flocculating agent dosages on the performance of red mud flocculation under shear conditions". Light Metals 2002 Métaux Légers., Ed. T. Lewis. 2002. pp. 13-24.

STUDY ON THE CLARIFICATION OF A RED MUD SLURRY DURING FLOCCULATION

Mélanie Normandin[1], Michel J. Gagnon[1], André Leclerc[1], Guy Simard[1], René Verreault[1] and Guy Péloquin[2]

[1]Université du Québec à Chicoutimi, 555, de l'Université, Chicoutimi, Québec, G7H 2B1, Canada
[2]Alcan International Limited, Arvida R&D Centre, 1955, Mellon Blvd., P.O. Box 1250, Jonquière, Québec, G7S 4K8, Canada

Keywords: Flocculation, Clarification, Bayer process

Abstract

One of the important steps in the Bayer process is the solid-liquid separation by which the particulate materials are removed from the feed slurry. One of the ultimate goals of the alumina industry is to obtain a perfect overflow clarity. In this study, the parameters that can influence the clarity of the overflow liquor resulting from the flocculation of the red mud slurry are systematically evaluated. Several laboratory tests are done by combining commercial synthetic flocculants and new polymers. By varying the concentration of polymer solutions, the quantity and the sequence of introduction of polymeric materials, extremely low levels of turbidity were found in the overflow liquor. This represents a definite step towards the objective of reaching a perfectly clear overflow liquor right at the decanter stage.

Introduction

In the Bayer process, which extracts alumina from a bauxite ore, flocculants are used to separate the red mud from the liquor in the primary settler or thickener, as well as to maximize the recovery of the aqueous phase containing alumina and soda from the red mud material in the washer circuit. Unfortunately, this separation process does not remove all solid particles in the overflow. A filtration step is necessary to remove the remaining particles. In an industrial process, this step is often problematic and also limiting when the concentration of suspended solids in overflow is high.

The main objective of this study is to reduce the suspended solids in overflow, where the concentration of suspended solids varies from 50 to 500 mg/L depending on the operating conditions. After the filtration step, this concentration is around 10 mg/L. Naturally, one of the ultimate goals of the industry is to obtain an almost perfect overflow clarity.

More and more, polymers are used in solid-liquid separation processes, water purification, mineral processing, oil recovery and paper making. For many years, conventional flocculation processes have depended on the use of polymeric flocculants. Polymers can be used alone or combined with other flocculant aids such as inorganic salts, surfactants, coagulants or even a second polymer. Researchers [1] found that a combination of two polymers would produce synergism in the flocculation. In some applications [2], a two-stage coagulation-flocculation system may be more effective to remove the suspended solids. For fine and ultra fine solid suspensions, use of double flocculant systems seems to offer a promising route for enhanced solid-liquid separation.

In this study, all the parameters that can influence the clarity (expressed as mg/L) of the overflow liquor resulting from the flocculation of the red mud slurry are systematically evaluated. Several laboratory tests are done by combining commercial, natural and synthetic flocculants and new polymers.

The first step of the global program is to carry out the laboratory tests with mud from the last washer. The experimental conditions are thus easier (lower temperature and caustic concentration) and allow testing to take place in university laboratories. Future work will take place at primary decanter conditions where the most promising findings are sought.

Experimental

In all experiments, the physical and chemical parameters used are those typically found in the last mud washer of the Bayer process: 10 g/L of caustic (expressed as NaOH), 16 g/L of sodium carbonate and 50 g/L of red mud slurry. The temperature of red mud slurry is ~ 21°C.

Jar Test Methodology

The various tests are carried out in one-Liter standard glass cylinders according to Alcan standard method for settling tests [3]. For each test, ten cylinders are used sequentially under similar operation conditions. The flocculating agent solutions and the red mud slurry need to be initially prepared. The slurry is used to fill each of the ten cylinders that have been previously placed in a temperature-controlled bath. When the adequate temperature of the slurry is reached, the slurry is mixed using a plunger made of a perforated plastic disk of proper diameter. Three strokes are sufficient for good mixing. Subsequently, half of the polymer solution volume required for the test is injected in the slurry using a plastic syringe and the slurry is mixed again using three plunger strokes. The rest of the polymer solution is added and the same mixing procedure is performed. Following this manipulation, an interface appears between the mud layer and the relatively clear supernatant. This interface progressively moves down the test cylinder with time. The resulting clear solution is kept without agitation for one hour. The supernatant is usually sampled by removing the first 300 mL of clarified solution at the top of the cylinder.

Reproducibility

The reproducibility of the methodology was evaluated from the standard deviation of a reference test. The standard deviation is calculated from settling velocities for ten repetitive tests. The standard method for settling tests was modified [4] to reduce variability of settling velocity measurements. A standard deviation lower than 5% can now be easily obtained for settling velocities.

Method for Preparation of Polymer Solutions

A solution containing 10 g/L NaOH is prepared by dissolving 20 g of NaOH pellets in a 2.0-Liter volumetric flask. Precisely 99.50 g of this solution is weighed and poured into a 250-mL glass beaker. Initially, a quantity of 0.50 g of flocculating agent is added progressively and stirred using a small impeller at 300 rpm during 5 minutes. Subsequently, the solution is transferred into a 500-mL plastic container and vigorously stirred for 8 hours using a magnetic stirrer. The polymer solution is then stocked in a refrigerator for a period of 72 hours. This solution can be used for flocculation tests for a period of up to 6 days and it is then discarded.

Method for Preparation of Red Mud Slurry

A plastic container (28 cm diameter x 36 cm height) is filled with 18 Liters of distilled water and 45 mL of concentrated Bayer liquor. The solution is mixed using an axial stirrer (9 cm diameter). The tip of the stirrer is placed 10 cm above the bottom of the container. Then, 930 mL of red mud is added and the resulting 50 g/L slurry is agitated for 30 minutes.

Turbidity Measurements

Turbidity was measured using a turbidimeter (H.F. Instruments Limited, Model DRT-1000). It is a nephelometric instrument that measures reflected light from scattered particles in suspension and direct light passing through a liquid. The turbidity is expressed in NTU units.

Clarity Measurements

The clarity was evaluated by filtration and is expressed in milligram per Liter (mg/L). A sample of overflow (300 mL) is removed and filtered over a 0.45-μm membrane filter (Millipore device). The filter is washed (with an organic solvent) and dried into a drying oven (at 40°C) to remove water. The filter is then weighed.

Results and Discussion

Several tests were done to study the parameters that influence the red mud slurry clarification by flocculation. More than 3500 jar tests have been carried out in the laboratory at Université du Québec à Chicoutimi (UQAC). The variables studied are flocculant dilution, flocculating agent, dosage, sequence of injection of polymers and red mud slurry concentration.

Each point showed in all the graphs presented hereafter represents the mean of 10 cylinder tests carried out under the same experimental conditions.

Several laboratory tests are done by combining commercial, natural and synthetic flocculants and new polymers. The selection criterions for the choice of flocculants are based on the level of anionicity and also on their capacity to interact with minerals surface. For confidentiality reasons, these flocculants will be denoted A, B, C, etc. All flocculants were studied individually in preliminary tests to evaluate the individual efficiency. It was the first stage of the selection process prior to the mixed-polymer addition. The results are presented in Table 1.

Table 1 Turbidity obtained with individual flocculants.

Flocculant	Dosage (g/t)	Turbidity (NTU)
A	32	210
B	24	300
C	16	385
D	951	160
E	1004	56
F	1004	185
G	2062	690
H	2008	760
I	1000	53
J	1965	940

Flocculant Dilution

The effect of polymer concentration was studied first. The results of the cylinder tests are shown in Figure 1, in terms of the overflow turbidity as a function of polymer concentration. As already indicated in Table 1, the dosage of polymer was fixed for all the tests at 32 g/t for polymer A, 24 g/t for polymer B and 16 g/t for polymer C. These dosages are obtained from the preliminary tests and correspond to the optimal dose observed when minimum turbidity is obtained. The variation of polymer concentration allowed to observe a gradual increase of the turbidity as a function of increased polymer concentration. This observation seems linked to the viscosity of the polymer solution [5, 6]. At lower viscosity, the dispersion of polymer through the suspension is easier and allows a reduction of the overflow turbidity. At this viscosity, the volume added to suspension is higher, for the same dosage. Thus, the increase of surface contact between polymers and solid particles allows a higher agglomeration of particles in suspension.

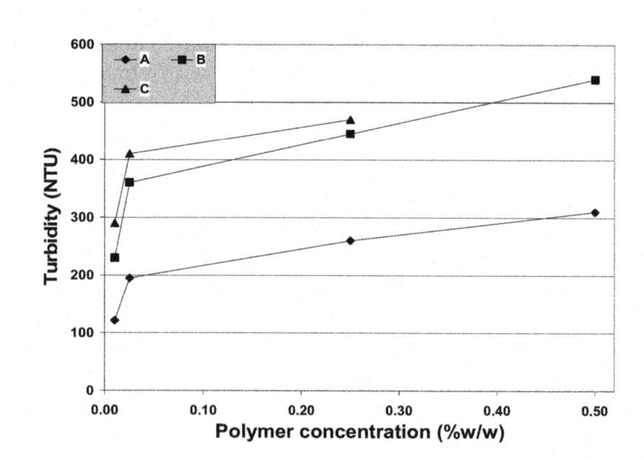

Figure 1. Turbidity of overflow
as a function of polymer concentration for 3 polymers.

Based on these results, a polymer concentration of 0.01% w/w was chosen for all experiments. At this concentration, the turbidity obtained is lower for all three polymers.

Furthermore, Figure 1 shows a difference between the efficiency of different polymers. The efficiency of the polymer is evaluated in relation to the turbidity results. Then, polymer A is more efficient to reduce the turbidity of overflow. Polymers B and C are less efficient but comparable.

Double-Polymer Addition

In the double-polymer addition tests, polymers are added in two sequences (AB/BA, BC/CB, AC/CA) using the same three different polymers (A, B and C). The relative ratios tested (expressed in %) of the different polymers in the mixture are respectively 50-50, 25-75 and 75-25. All possibilities were achieved in accordance with a two-way frequency table. To evaluate the quantity of polymer necessary for the mixture, the volume of polymers required to reach a hindered settling velocity within 10 and 20 m/h was determined. The corresponding dosage is subsequently calculated.

To evaluate the efficiency of double-polymer addition, the results of turbidity are compared with the results obtained from a single-polymer addition. The results obtained are listed in the right corner of Figure 2. As noted in the previous section, polymer A is more efficient to reduce the turbidity of overflow and 210 NTU is our reference value to evaluate the efficiency of flocculation for the double-polymer addition.

In Figure 2, the effect of two polymers, on the turbidity, is compared with that of the single polymer. The mean turbidity of the overflow solution as a function of the ratio of polymers in the mixture is presented. The mixture AB (75-25) led to the lowest turbidity. The dosage of polymers is 26 g/t for the polymer A and 6 g/t for the polymer B. The mean turbidity measured for this mixture is 123 ± 2 NTU, corresponding to a solid concentration in the overflow solution of 47 mg/L. The double-polymer addition allowed a reduction in the overflow turbidity from 210 (polymer A more effective) to 123 NTU. This is a ~40 % reduction, for an equivalent dosage (32 g/t), compared with the turbidity achieved with polymer A alone.

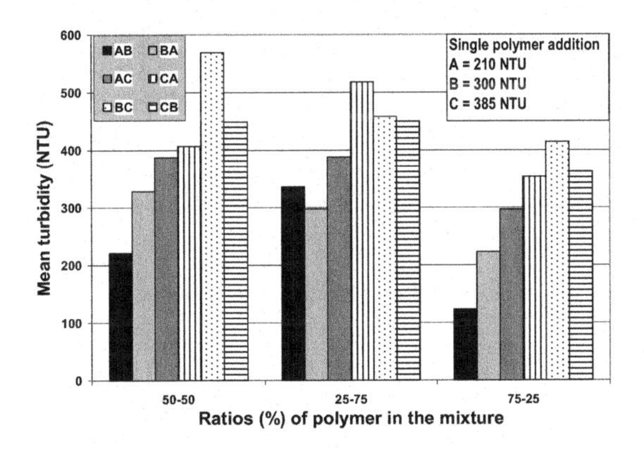

Figure 2. Mean overflow turbidity as a function of the ratios of polymers in the mixture for a double-polymer addition.

The mixture AB (75-25) is highly effective to reduce the turbidity of red mud slurry. In this sequence of introduction of polymers, the polymers (A and B) seem to have interactions (polymer-polymer) that increase the efficiency of individual polymers [1]. All other sequences of introduction of polymers are not very effective since the overflow turbidity is not reduced when compared to a single-polymer addition of polymer A.

Triple-Polymer Addition

After the preliminary tests were performed to evaluate the individual efficiency of each flocculant, another series of tests were carried out to evaluate the interactions between flocculants. Figure 3 presents the five best tests obtained for a triple-polymer addition. The combination IAB is the most effective to reduce the turbidity of red mud slurry. The dosage of different flocculants is 302 g/t for flocculant I, 10 g/t for polymer A and 6 g/t for polymer B. The mean turbidity obtained, for ten repetitive tests, is 8.0 ± 0.8 NTU, corresponding to a solids overflow concentration of 2.2 mg/L. This is ~94 % reduction in overflow turbidity, when compared to the best turbidity obtained for a double-polymer addition.

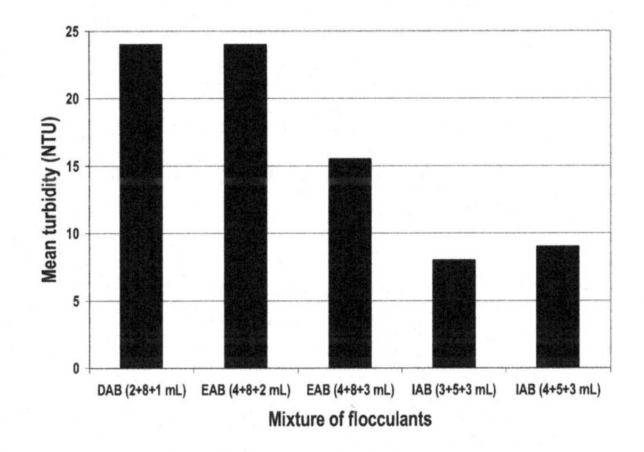

Figure 3. Mean overflow turbidity as a function of mixture of flocculants for a triple-polymer addition.

The sequence of introduction of polymers AB is always more effective (like in a double-polymer addition). This strongly suggests that polymers A and B may have polymer-polymer interactions that increase the efficiency over a single polymer addition.

The addition of flocculant I has led to a reduction in turbidity from 123 to 8.0 NTU. The dosage of polymer B was at the same level as the one used in a double-polymer addition but the dosage of polymer A was reduced to 16 g/t. As shown in Table 1, the simple-addition of flocculant I has a good efficiency only at high dosage (1000 g/t). The addition of polymers A and B has allowed a considerable reduction in the dosage of flocculant I (1000 to 302 g/t), this compounded by an end-effect leading to a very good turbidity. This suggests that flocculant I may have interactions with polymers A and B that increase the efficiency over a double-polymer addition.

Red Mud Slurry Concentration

The effect of red mud slurry concentration was studied using the best mixture found for a triple-polymer addition (IAB). The dosage of flocculants was kept constant (dosages listed in the previous section) for all the tests. The results of the cylinder tests are shown in Figure 4 in terms of overflow solids concentration as a function of red mud slurry concentration. An increase in red mud slurry concentration led to a decrease in overflow solids concentration, for example by as much as half from 50 g/L to 75 g/L.

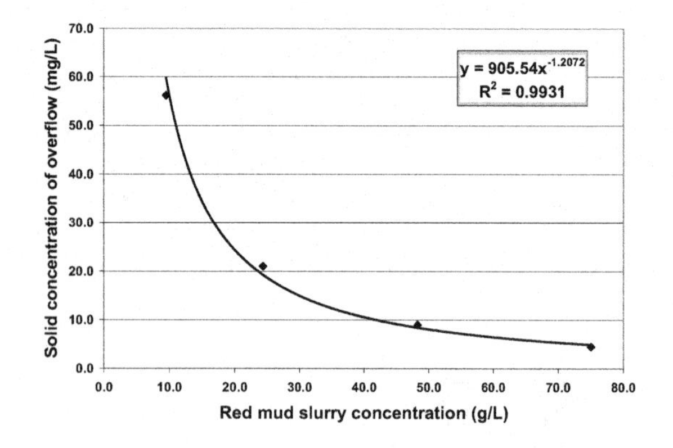

Figure 4. Solids concentration in the overflow as a function of red mud slurry concentration for a triple-polymer addition (IAB).

This observation may be related to the increase in particle surface area which allows to agglomerate more particles in suspension [7] as the slurry concentration increases.

Preliminary Results Obtained under Primary Decantation Conditions

Additional laboratory jar tests have been carried out in the laboratory at Alcan Arvida Research and Development Center (ARDC) to confirm the efficiency of the method that was developed. In all the experiments, the physical and chemical parameters used are those typically found in the primary decanter of the Bayer process: 220 g/L of caustic, 35 g/L of sodium carbonate and 35 g/L of blow-off solids. The temperature of the blow-off is ~105°C.

Based on the results obtained under the last washer conditions, the positive effect of a triple-polymer addition was also observed under these new operating conditions.

The experimentation allowed a reduction in the overflow turbidity to 31 ± 1 NTU, corresponding to an overflow solids concentration of 12 mg/L. The triple-polymer addition is so effective, under these operating conditions, that it allows to reach the target originally set within this work and approach a level of clarity close to that found in the plant after a filtration stage.

Additional tests are underway to fully characterize the observed behaviors and to translate these laboratory results into plant applications.

Conclusion

Sequential use of two or three polymers has been shown to significantly enhance the flocculation of red mud slurry. The enhanced flocculation obtained with double-polymer addition is probably due to polymer-polymer interaction, which results in excellent interparticle bridging. The mode of polymer addition also has a measurable effect on the flocculation. The triple-polymer addition allowed to reach a level of clarity approaching the one found after a filtration stage in a plant.

Acknowledgements

The authors wish to thank François Girard, UQAC, and Rock Lemyre, Alcan, ARDC, for their valuable contribution during the laboratory tests.

References

1. Y. Xiang, P. Somasundaran, "Enhanced flocculation with double flocculant," Colloids and surfaces. A. Physicochemical and engineering aspects, 81 (1993), 17-23.

2. A. Rushton, A.S. Ward, and R.G. Holdich, *Solid-Liquid Filtration and Separation Technology* (Federal Republic of Germany, Wiley-VCH, 2nd edition, 2000), 587.

3. G. Fulford, P. Ferland, "Settling tests on flocculant/red mud slurries" (Alcan Method 1234-90, Arvida Research and Development center, 1990).

4. M. Normandin et al., "Efficiency of the cylinder test for the evaluation of red mud slurry flocculation and residual turbidity" (Proceedings of the international symposium on Light Metals, Calgary, Canada, 2005) 495-505.

5. L.J. Connelly, D.O. Owen, P.F. Richardson, "Synthetic flocculant technology in the Bayer process", *Light Metals*, 1986, 61-68.

6. A.T. Owen et al., "The impact of polyacrylamide flocculant solution age on flocculation performance," *International Journal of Mineral Processing*, 2002, 124-143.

7. B. Yarar, "Evaluation of flocculation and filtration procedures applied to WSRC sludge", Colorado School of Mines, 2001, 1-34.

Light Metals 2006 *Edited by Travis J. Galloway* ***TMS (The Minerals, Metals & Materials Society), 2006***

WASTEWATER TREATMENT METHODS

Dana Smith[1], Fred Williams[2], Scott Moffatt[3]

[1]AWA Atlantic; PO Box 101; Point Comfort, TX 77978, USA
[2]CMIS Corp; 103 Woodglen Drive; Victoria, TX 77904, USA
[3]Cytec Industries, Inc.; 1937 West Main Street; Stamford, CT 06904-0060, USA

Water Treatment, Bayer Lakewater, Ferric Sulfate, Total Organic Carbon

Abstract

Alcoa's Point Comfort, Texas industrial facility is a combination of a bauxite refining plant utilizing the Bayer process and an aluminum fluoride production plant. Due to the location's use of dry stack technology for bauxite residue disposal, the pond surface areas for evaporation are minimal compared to the rainfall catchment areas. This results in the periodic need to reduce accumulated volumes of storm water at the Residue Disposal Area (RDA).

This paper describes the options for treating a combination of wastewaters from the RDA. The current water treatment method utilizes ferric sulfate for total organic carbon (TOC) and metallic ion adsorption. The precipitated solids are separated and the treated water neutralized prior to discharge. Experimental work will also be presented for the treatment of Bayer process water alone.

Introduction

Point Comfort Facility

The Alcoa World Alumina Atlantic (AWA Atlantic) facility at Point Comfort, Texas is located in the Gulf Coast region of the United States. The plant property is bordered by the brackish water bodies of Lavaca Bay to the west and Cox Bay to the south. Along the eastern perimeter lies the Cox Creek watershed. The water in this watershed is impounded for plant fresh water use prior to excess water overflowing into Cox Bay. The industrial facility includes a Bayer Processing plant for production of 2.3 million metric tonnes of alumina per year and a plant for converting alumina trihydrate into 64 thousand metric tonnes per year of aluminum fluoride.

Bayer Process

The Point Comfort Bayer plant processes primarily Boke' bauxite and utilizes a combination of low and high temperature digestion units. After digestion the bauxite residue is washed in a series of counter-current decantation washers. It is then pumped as thickened slurry from the last washer to waste disposal sites located within the facility's Residue Disposal Area (RDA).

At the RDA the bauxite residue is dry stacked at elevations above the exterior containment dikes. The water runoff from the dry stack areas is stored in a separate water reservoir for use in the Bayer Process or transferred to the aluminum fluoride wet disposal area as described below. Other inputs may include rainwater collected off the plant site, water released from the thickened bauxite residue slurry and leachate collected from the dry stack underdrains. This water will be referred to as Bayer Lakewater.

Aluminum Fluoride Process

The first step in the aluminum fluoride process is to react fluorspar (CaF_2) with sulfuric acid (H_2SO_4) to produce hydrogen fluoride (HF) gas and byproduct gypsum (anhydrous $CaSO_4$). The next and final processing step is to react the HF gas with alumina trihydrate to form aluminum fluoride (AlF_3).

The gypsum waste material is slurried and transported to a separate area of the RDA using a recycling flow of water. The slurry returning to the RDA is pH 2 but the body of water in the disposal site is maintained above a pH of 7 by the addition of Bayer Lakewater. This transport water will be referred to as Chemicals Lakewater.

Background

Water Management Issues

With the conversion of older wet disposal lakes to dry stacking of bauxite residue, water management in the disposal area becomes an important issue because the rainfall catchment area is now much greater than the evaporative area. This imbalance results from the location objective to maximize the amount of residue stored in the former wet lake areas, creating more open area than is necessary solely for dry stack management. As a result it will be a number years before the site water balance becomes manageable without frequent periods of treatment and discharge of water.

Annual rainfall totals measured at the plant site have ranged from 13.8 to 75.9 inches, with an average of 43.7 inches. Annual pan evaporation has ranged from a minimum of 57.0 to 90.5 inches, with an average of 73.5 inches. Due to dry stacking, the rainfall catchment area in the RDA is 80% greater than the evaporative area. The evaporative area will be further diminished as more acreage is utilized for bauxite residue storage. Ultimately, 90% of the area will be catchment area. There are no typical annual wet or dry seasons for this area, but at any time significant amounts of rain may fall during brief, intense storm events. The Texas Gulf Coast is subject to occasional tropical storms and hurricanes.

Wastewater Discharge Permits

The Federal National Pollutant Discharge Elimination System (NPDES) and the Texas Pollutant Discharge Elimination System (TPDES) permits authorize the discharge of process wastewater and storm water from Point Comfort Operations to Lavaca Bay. The specific effluent characteristics, discharge limitations, and monitoring requirements are shown in Table I. In Table I, the term "flow MGD" refers to the allowable daily discharge flow in millions of gallons. TOC and TSS are acronyms for total organic carbon and total suspended solids, respectively. Concentrations are stated in units of milligram per Liter (mg/L). The permit limit criteria given in mass loadings have been converted to mg/L based on an outfall flow of 2.88 MGD.

Table I. Water Permit Limitations

	Daily Average	Daily Maximum
Flow MGD	<4.32	<6.48
pH	6.0-9.0	6.0-9.0
TOC mg/L	55	75
TSS mg/L	101.8	209.8
As mg/L	0.1	0.3
Cr mg/L	0.10	0.36
F mg/L	37.2	78.2
Ni mg/L	0.14	0.47
Se mg/L	0.1	0.3
Zn mg/L	0.37	1.20

The concentration ranges of the lake waters are shown below in Table II for comparison. From Table II it can be seen that the Bayer Lakewater is significantly higher in pH, alumina, TOC, As, Cr and Se. As the Bayer Lakewater is transferred to the Chemicals Lakewater for pH control, metallic ions and TOC are reduced through the mechanism of pH swings, alumina precipitation and gypsum precipitation occurring in that lake system.

Table II. Plant Solution Concentrations

Parameters	Chemicals Lake	Bayer Lake Water
pH	7.2-9.6	12.0-12.3
Al2O3 g/L	0.004-0.10	1.5-2.5
TC g/L		2-7
TA g/L	0.2-1.0	12-22
SO4 g/L	20-60	0.6-3.1
TDS g/L	30-100	12-28
TOC mg/L	80-135	190-370
As mg/L	0-0.2	1.6-3.5
Ca mg/L	20-600	0.8-2.8
Cr mg/L	<0.01	0.05-0.09
F mg/L	30-70	30-90
Ni mg/L	0.05-0.15	<0.01
Se mg/L	0.09-0.10	0.40-0.70
Zn mg/L	<0.05-0.09	<0.05

Wastewater Treatment Methods

Historically, Point Comfort Operations has had to treat and discharge water to maintain water volumes in the RDA areas. In the early 1990's Chemicals Lakewater was treated and discharged in order to convert a wet lake to a dry stack area. The parameters of concern were TOC, pH, and TSS. TOC levels in the water ranged from 100 to 120 mg/L. References [1,2,3] are related to treatment of wastewater for TOC removal. The treatment system consisted of injecting ferric sulfate solution into a stream of lake water in a pipeline. The amount of ferric sulfate was controlled to achieve a pH of 5.0. At that pH condition, a high percentage of the TOC absorbed onto the precipitated iron. The pipeline discharged into a series of three shallow settling ponds. Water from the last pond was blended with caustic prior to being pumped to a settler. The settler overflow water passed through media bed filters before being discharged through the designated outfall to Lavaca Bay.

In the late 1990's, Chemicals Lakewater was treated and discharged due to an accumulation of water in the RDA system because of heavier than normal periods of rain. The element of concern was again TOC but this time the concentration was closer to the allowable discharge levels of 55 mg/L. The treatment system consisted of pumping the water through a series of two activated carbon beds and then into the settler. The settler overflow went directly to the designated outfall without any further filtration.

With the recognition of the water balance issues in the RDA, a permanent water treatment facility has been installed. Chemicals Lakewater is pumped into a stirred reactor where a combination of ferric sulfate and 50% caustic is added to maintain proper pH control for TOC with a target pH of 4.5. The iron precipitate generated is flocculated and allowed to separate in lamella plate settlers. The clear overflow water flows into a second reactor where the final pH control of 6.5 to 7.5 is achieved by addition of 50% caustic. The second reactor overflows into a holding tank and the water is pumped to the previously used settler before discharge to the Bay. The sludge from the lamella settler underflow is pumped back into the RDA.

Additional wastewater treatment methods that have proved unsuccessful with Chemicals Lakewater include pilot plant evaluations using organoclay absorptive media and ultra (micro) membranes for TOC removal and on a Dissolved Air Flotation (DAF) lamella settler for iron solids separation. Filter cartridges upstream of a reverse osmosis treatment process using nanofiltration membranes successfully removed the organic carbon but also removed the sulfate ion. Recycling this reject water back into the Chemicals Lakewater system would result in an unacceptable increase in sulfate concentration.

The purpose of this paper is to present laboratory data that supports the present method of Chemicals Lakewater treatment as well as data on treating Bayer Lakewater separately to meet the permitted outfall criteria presented in Table I. The treatment processes are primarily directed at lowering TOC using an iron precipitate as an adsorption media. Additional steps discussed for treatment of Bayer Lakewater include neutralization with either sulfuric acid or carbon dioxide, and both physical and chemical methods for enhancing alumina settling characteristics. Reductions of the arsenic and selenium ion concentrations are also discussed and references [4-8] are related to removal of these ions from wastewater.

Experimental

Commercial ferric sulfate solutions were used as a treatment chemical. These solutions nominally contained 12% Fe by weight with essentially no free acid. Additional reagent grade chemicals used included dry solid forms of CaO, $CaCl_2$, $FeCl_3$, NaOH, and $FeSO_4 \cdot H_2O$. Reagent grade sulfuric acid was used as well as CO_2 from a compressed gas cylinder.

A variety of anionic and cationic flocculants from Cytec were also tested for settling the precipitated solids.

In general, the experiments were carried out in stirred beakers containing an electronic pH probe to monitor solution pH. The volume of water to be treated in individual experiments varied from 250 mL to 2 Liters. Procedures specific to the individual experiments will be discussed along with the data for those experiments.

Point Comfort laboratory's automated titration equipment measured the concentrations of alumina (Al_2O_3), total caustic (TC), and total alkalinity (TA). TC and TA are the North American convention of expressing soda concentrations in terms of sodium carbonate. Also, the location's Dohrmann DC-80 was used to report TOC concentrations. Analyses of other ions were made by local accredited laboratories.

Data and Results

Chemicals Lakewater Treatment

1.0 Neutralization Experiments Using Ferric Sulfate - A series of beaker experiments were performed by adding incremental amounts of the acidic ferric sulfate solution to the initially basic Chemicals Lakewater to determine the TOC removal response through changing pH. Figure 1 plots the response of TOC as a function of pH. The g/L Fe needed for treatment is also shown on the secondary axis curve.

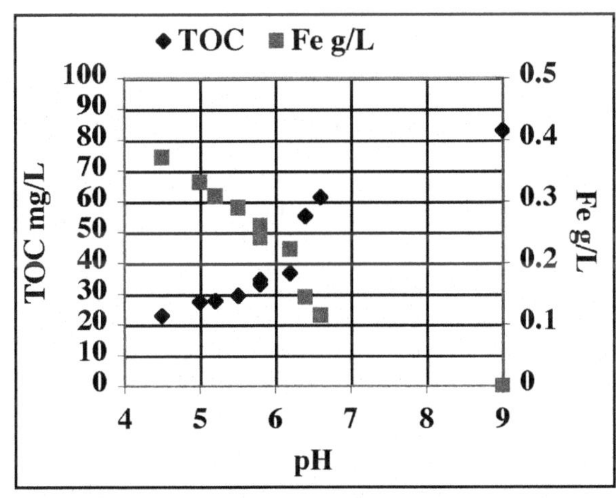

Figure 1. Chemicals Lakewater Iron Treatment Response Curve

TOC removal is the result of adsorption on the iron precipitate formed at the pHs shown. The point at which the maximum amount of TOC is removed is around a pH of 4.5 but with an initial TOC of 83 mg/L, the iron is effective at reducing the TOC to below 55 mg/L in the 4.5 to 6 pH range. Above pH 6.5 there is a sharp upward break in the curve. The plotted curve for TOC implies, and further experiments confirmed, that the adsorption of TOC on the iron precipitate is reversible which necessitates removal of the iron precipitate before neutralization of the treated water back to pH 7 for discharge. Not shown on the graph is that below pH 4 the iron floc re-dissolves and releases all of the TOC back into solution.

2.0 Ferric Treatment For TOC Removal - The Chemicals Lakewater was treated in laboratory experiments with both ferric sulfate and ferric chloride. Over the timeframe of this experimental work, the TOC concentrations in the lake water decreased for various reasons. In Figure 2 data labeled A represent a starting TOC of 128 mg/L while those labeled B contained 98 mg/L. In these experiments, incremental amounts of iron were being added to the lake water solutions while maintaining the targeted pH with additions of either sulfuric acid or caustic solutions.

The data shown for Solution A at pHs of 4.5 and 5.0 have been combined in the second order polymetric correlation curve and indicate that the iron contained in either ferric chloride or ferric sulfate is equally responsive for TOC removal on a g/L Fe basis. It is also apparent that it is difficult to differentiate between the response data at pH 4.5 and 5.0. The curves for pH 6A and 7A show that the addition of larger amounts of iron at the same pH will result in lower levels of TOC in the treated water. From this it can be concluded that operating in the lower 4.5-5 pH range reduces the amount of the ferric sulfate needed to be below the limit of 55 mg/L TOC. The curve for pH 4.5 Solution B indicates that there is a significant shift in the iron response curve depending upon the start TOC concentration.

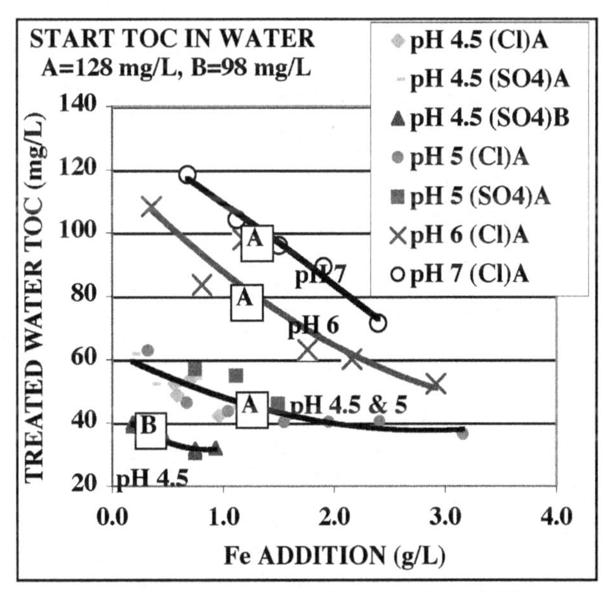

Figure 2. Chemicals Lakewater Treatment For TOC Removal With $FeCl_3$ & $Fe_2 \cdot (SO_4)_3$

3.0 Correlation Based On TOC Difference - This change in response depending upon the start TOC led to the question of whether the TOC • (difference from start to finish) versus g/L Fe addition might be a better way of representing data for different TOC start water concentrations. Figure 3 shows a re-plot of the pH 4.5 and 5.0 data for the two waters and indicates that there is still a shift in the response curve for the two starting TOCs. This shift can be explained by the fact that the response curves are not linear particularly at increments of very low iron addition.

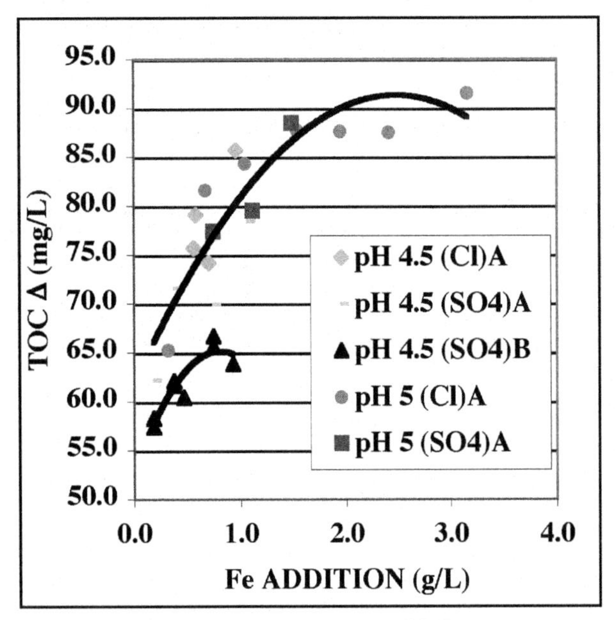

Figure 3. Chemicals Lakewater TOC Treatment

4.0 Flocculation of Iron Precipitates - Several flocculants from Cytec Industries were examined for use in settling the iron precipitate formed in the stirred tank reactor at the water treatment facility. Criteria for success included identifying a flocculant that would effectively settle the iron precipitate in the lamella clarifier, thus providing an overflow with minimal suspended solids while not adding organic carbon to the treated water concentration. Additionally, the flocculants had to be nontoxic in a marine environment. A third aspect for flocculant selection was the ability to develop a feasible system for floc addition given the remote location of the water treatment facility. Preliminary laboratory screening of flocculants included cationic and anionic charges as well as products in the form of a dry powder and water-in-oil emulsion.

Flocculants were screened at a process pH range of 4.5 to 5.5. Results from this work showed that an anionic emulsion based product, SUPERFLOC® A-1849RS, was the most effective. Testing done between emulsions and dry powder flocculants showed that emulsions had a negligible contribution towards the TOC of the treated lake water, and therefore were suitable for use in the process.

Due to the remote location of the treatment system, there were limitations on the power available for flocculant equipment and choices of water for preparing a flocculent solution. To minimize power requirements, a decision was made to use an emulsion based product instead of a dry powder. The only options for water for the preparation of a flocculant solution were treated or untreated lake water. Use of this water was challenging due to the high level of salts, such as calcium, that would adversely affect the performance of the flocculant. The impact of these salts on the flocculant was minimized by the addition of a small amount of 50% NaOH to the lake water before it was used to invert the flocculant emulsion.

5.0 Plant Treatment Results - The results of ferric sulfate treatment at pHs of 4.5 -6.0 (depending upon start TOC) were successful in controlling TOC to less than 55 mg/L. Use of SUPERFLOC® A-1849RS successfully flocculated the iron precipitate and thus controlled TSS in the treated water. In general the concentrations of the remaining ions are below permit limits.

Bayer Lakewater Treatment

1.0 Neutralization Experiments Using Sulfuric Acid - The initial set of laboratory experiments with Bayer Lakewater tested the TOC response to pH changes caused by incremental additions of sulfuric acid. The experimental procedure used 2-Liter quantities of lake water at a starting pH of 12.2 with addition of sufficient acid to lower the pHs to various levels ranging from 9 to 4.5. Figure 4's top curve illustrates the TOC response from the initial concentration of 280 mg/L to a low point at pH 5, where 66% of the TOC has been removed. TOC is reduced over this pH range by the mechanism of adsorption of organic carbon on the precipitated alumina originally present in the lake water.

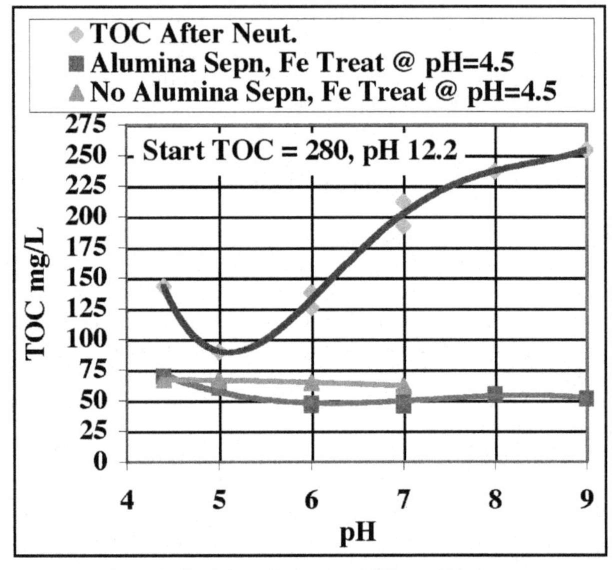

Figure 4. Bayer Lakewater Neutralization

2.0 Ferric Sulfate Treatment Of Neutralized Bayer Lakewater - The neutralized lake waters from the above experiments were further treated with ferric sulfate to a pH of 4.5 in an attempt to lower the remaining TOC to the target level of less than 55 mg/L. The water from each experiment was divided into two aliquots. In the first set of aliquots, the solids from the neutralization process were removed using vacuum filtration. The solids were left in the second set. As the two lower curves in Figure 4 demonstrate, TOC concentrations below 55 mg/L could be achieved only with solids removal. Without solids removal, the TOC remained above 60 mg/L and stayed constant at any of the treatment pHs.

The precipitated alumina formed by this neutralization is a high volume gel type precipitate that does not readily settle. Addition of SUPERFLOC® A-1849RS flocculated the solids but a very high volume of settled mass still remained; about 40% of original volume.

3.0 Experiments To Avoid Alumina Gel Formation - To avoid some or all of the alumina gel formation, a series of laboratory experiments were performed with 30-minute lime addition as a first step in the water treatment process. In this series of experiments variables included:

1. The form of lime added – $CaCl_2$, dry CaO or 100 g/L slaked CaO
2. Lime addition at room temperature (RT) or to heated lake water (HT)(80-95 °C)
3. Removal (by settling or filtration) or no removal of the lime solids prior to neutralization
4. Neutralization with sulfuric acid or CO_2
5. Removal (by filtration) or no removal of any alumina solids prior to iron treatment
6. Treatment at pH 5 with ferric sulfate

Figure 5 summarizes the results of the lime experiments. The legend sequence means:

1. RT or HT refers to whether the lake water was at room temperature or heated during the lime reaction.
2. The first Y or N refers to whether the lime solids were removed before neutralization.
3. The second Y or N refers to whether the alumina solids were removed prior to lime treatment.

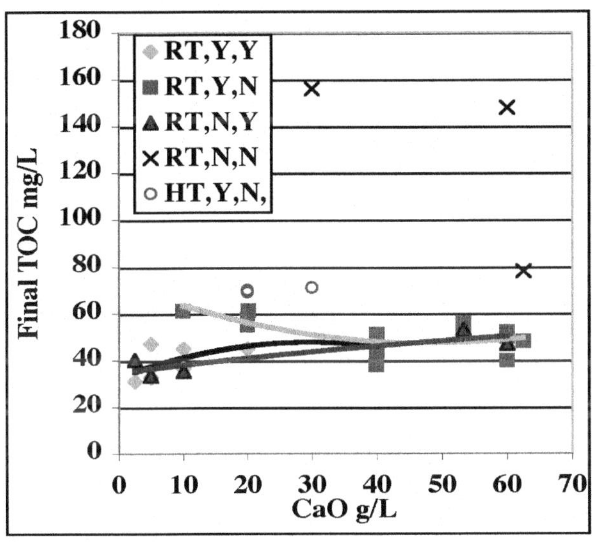

Figure 5. Bayer Lakewater Lime Pre-Treat Data (12 Fe g/L Addition)

Conclusions drawn from these and additional experiments were:

- On a g/L CaO basis, the form of the lime addition had no impact on the final results.
- Heating the water for the lime reaction (HT,Y,N) was not beneficial from a TOC removal standpoint.
- If the lime solids and the alumina solids were not removed prior to ferric treatment (RT,N,N), than the TOC values were unacceptably high.
- Between 20 and 40 g/L CaO additions all of the alumina reacts with the lime solids. Above 40 g/L, the trendlines merge. Below 40 g/L the trendlines diverge depending on whether the alumina was removed (RT,Y,Y; RT,N,Y) or not (RT,Y,N) with the TOC results going above 55 mg/L for those experiments with no alumina removal. These results are similar to the previous ones obtained during no lime addition experiments.
- Neutralization with either sulfuric acid or CO_2 did not change the final TOC results. A problem with CO_2 is that all of the CO_2 is released with vigorous bubbling and foaming when the pH of CO_2 neutralized solutions are driven down into the 4.5-6 pH range for the final ferric sulfate treatment.

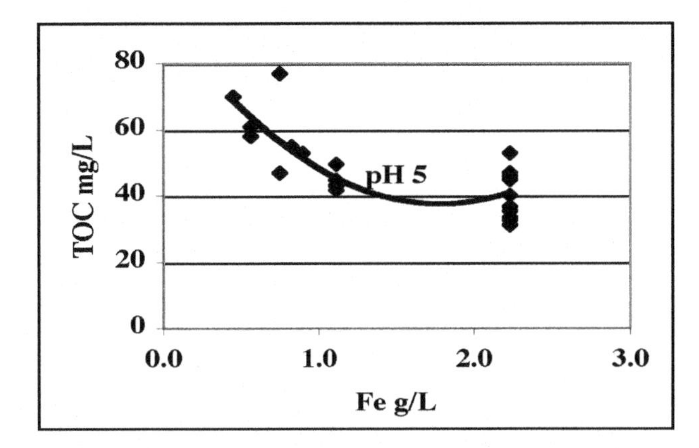

Figure 6. Bayer Lakewater (All solids removed prior to ferric treatment)

Figure 6 illustrates the results from a subset of experiments where all solids were removed prior to ferric treatment. Variable amounts of iron were added as well as sulfuric acid or sodium hydroxide to maintain a constant pH of 5. The curve in Figure 6 is very similar to the curve in Figure 2 for Solution A, the high TOC Chemicals Lakewater.

An overall conclusion reached from the above data is that the addition of lime was not effective enough to overcome the problem of alumina solids. The addition of lime also did not reduce the amount of iron needed to achieve less than 55 mg/L TOC.

4.0 Experiments To Change Alumina Settling Characteristics

A. Change In Physical Form - Additional laboratory experiments were made to attempt to change the high volume nature of the alumina gel. Fine alumina trihydrate seed was added prior to Bayer water neutralization to determine if the preformed seed would reduce the precipitate volume or act as a settling aid, with no positive results. Additionally, 50 g/L on a dry solids basis of last stage washer underflow slurry from the Bayer Plant was added to the lake water. The resultant solution was separated into two aliquots, with the sets neutralized to below pH 10 by either sulfuric acid or CO_2 addition At this pH, essentially all of the alumina has precipitated. These experiments were moderately successful because the resultant bauxite residue/alumina gel formed in the pH 9 to 10 range would settle to a volume of 200 mL per Liter of solution after synthetic flocculant addition. After solids separation and ferric sulfate treatment, the TOC vs. Fe response curve was identical to Figure 6 above.

B. Laboratory Flocculation Studies - Synthetic flocculants from Cytec Industries were tested to improve the settling of the alumina gel precipitate. In treatment strategies that consisted of neutralizing the Bayer Lakewater with sulfuric acid or CO_2 to a pH of 7.0-9.0, a highly charged anionic flocculant, such as SUPERFLOC® 1227, or a highly charged cationic, such as

SUPERFLOC® 4518 were found to be the most effective in timed settling tests.

Synthetic flocculants were also tested in the treatment option involving the addition of last stage washer underflow to Bayer Lakewater prior to neutralization of the water. Testing focused on identifying a flocculant that would be effective as the pH of the water was reduced to a range of 9 to 10. Most of the flocculants tested were anionic, with varying degrees of charge. Test results indicated that the lower charged anionic flocculants were sensitive to pH and did not perform well as the pH was gradually reduced. SUPERFLOC® 1227 outperformed all the other flocculants tested.

Following solids separation and ferric sulfate treatment of the various options discussed above for Bayer Lakewater, it was found that SUPERFLOC® A-1849RS is also effective in settling the iron precipitate formed during this process.

5.0 Removal of Arsenic and Selenium - Table III portrays the resulting ion concentrations for arsenic and selenium after various laboratory treatments discussed above followed by ferric sulfate addition for TOC removal. Ferric sulfate reduces As to less than 0.01 mg/L but the Se ion is not reduced to the permit limit of 0.1 mg/L except with lime addition. Data from a limited number of additional experiments with a combination of ferrous and ferric sulfate are found in the right hand columns in Table III. These show that the ferrous addition will further reduce Se to well below 0.1 mg/L in all cases tested.

Table III

Treated Bayer Lakewater Ions Of Concern

	Ferric Sulfate Alone		Both Ferric & Ferrous Sulfate	
	As mg/L	Se mg/L	g/L $FeSO_4$	Se mg/L
Neutralization to pH = 6	<0.01	0.07		
Neutralization to pH = 7	<0.01	0.17	1	<0.01
Lime Addition Neut to pH =7	<0.01	0.07	0.5	0.05
Last Stage Washer U'flow Residue to pH =9	<0.01	0.11	0.2	0.06

6.0 Laboratory Treatment Results - The Bayer Lakewater tested in these experiments could be treated to reduce TOC and other metallic ions to meet outfall criteria. The presence of alumina in this lake water, however, was a challenge because of the very high volume of precipitate that forms as the water is neutralized. Lime pre-treatment to remove the alumina was not promising and the use of flocculants or flocculating aids, such as bauxite residue, were of limited success because of the high volume of the remaining solids. SUPERFLOC® 1227 was the most effective of flocculants tested for the alumina/residue mixtures. SUPERFLOC® A-1849RS successfully flocculated the iron precipitate as in the case of the Chemicals Lakewater.

Conclusions

Laboratory work and installation of a treatment plant have demonstrated that the Point Comfort Chemicals Lakewater can be successfully treated and discharged while meeting all permit ion parameters. Additional laboratory experiments also show that Bayer Lakewater can be treated to meet these same outfall standards. Future work includes laboratory and field testing of mechanical separation techniques for the alumina precipitate.

References

1. *Report on Field Pilot Plant Treatability Study for Removal of Excess TOC and Suspended Solids from Lake #3 Water at the Alcoa Point Comfort Operations, Point Comfort, TX.* (Rubel Engineering, 1990).

2. Walter J. Weber, Jr. and Edward H. Smith, "Removing Dissolved Organic Contaminants from Water," *Environmental Science Technology*, 20 (1986), 970-979.

3. Wei-Chi Ying, James J. Duffy and Michael E. Tucker, "Removal of Humic Acid and Toxic Organic Compounds by Iron Precipitation," *Environmental Progress*, (7)(3)(1988), 262-269.

4. *Literature Review of Environmental Toxicity of Mercury, Cadmium, Selenium and Antimony in Metal Mining Effluents*, (A Time Network Project by Beak International Incorporated, 2002).

5. D. T. Merrill et al., "Field Evaluation of Arsenic and Selenium Removal by Iron Coprecipitation," *Journal WPCF*, 58 (1), (1986), 18-26.

6. L. Tidwell et al., "Potential Technologies for Removing Selenium from Process and Mine Wastewater," *Minor Elements 2000 Processing and Environmental Aspects of As, Sb, Se, Te, And Bi,"* Ed. C. Young, (Society of Mining, Metallurgy and Exploration, Inc., Littleton Co., 2000), 53-66.

7. Dennis Clifford, Suresh Subramonian and Thomas J. Sorg, "Removing Dissolved Inorganic Contaminants from Water," *Environmental Science Technology*, 20 (1986), 1072-1080.

8. Precha Yodnane, "Removal of Arsenic and Selenium from Fly Ash Leachate Using Iron Coprecipitation and Activated Alumina Adsorption," (#AC91-026-002, WPCF 64th Annual Conference & Exposition, Toronto 1991).

ALUMINA & BAUXITE

Bauxite and Bauxite Characterization

SESSION CHAIR
Milind V. Chaubal
Sherwin Alumina Co.
Corpus Christi, TX, USA

Light Metals 2006 *Edited by Travis J. Galloway* ***TMS (The Minerals, Metals & Materials Society), 2006***

CROSS-COUNTRY BAUXITE SLURRY TRANSPORTATION

Ramesh L. Gandhi, Jay Norwood, and Yueguang Che
Pipeline Systems Incorporated, 5099 Commercial Circle, Suite 101, Concord, California USA

Keywords: Bauxite, Hydrotransport, Rheology

Abstract

Over the last four decades, slurry pipelines have proven themselves to be a safe, reliable, and cost effective way to transport large tonnages of minerals over long distances. A variety of minerals have been successfully transported over distances ranging from a few kilometers (km) to 400 km. However, no long-distance bauxite pipelines have been built to date. Bauxite has developed a reputation for being difficult to hydraulically transport. This is about to change with the first long-distance slurry pipeline. The 250-km-long, 10 MT/y capacity pipeline was designed by PSI. Several other pipelines are currently being considered for transportation of bauxite. It will be shown that a properly designed hydrotransport system is an economically viable and reliable alternative for the long-distance transport of bauxite ore, as has been demonstrated for numerous other minerals. The focus of this paper is on how to integrate a pipeline into an existing facility.

1. Introduction

During the 1960's and 1970's a number of alumina refineries were built around the world, often near large deposits of bauxite in close to port facilities. After decades of mining, many of these nearby bauxite reserves are becoming depleted. In order to continue operation of existing refineries or, in some cases, to expand their capacity, new ore deposits farther away from the refineries will have to be exploited. Over the last 40 years slurry pipelines have moved from a status of being an intriguing, but rather risky possibility to it's present status as a cost-effective, highly reliable alternative to the conventional transportation modes. Most of the slurry pipelines transport material whose flow behavior does not change significantly during transport. Bauxite ore tends to generate fine particles (minus 45 microns) when sheared. The effective viscosity of slurry increases with an increase in fines in the slurry. Changes to the slurry viscosity during transport make any material more difficult to transport using slurry pipelines. The extent of the particle breakage and impact can be assessed such that the pipeline system can be designed to account for the change.

2. Design Considerations

2.1 Transportation System

The bauxite transportation system can be divided into slurry preparation, slurry pipeline system, and dewatering system. The bauxite ore is mined and washed to remove impurities. It is then crushed and ground to a size that is suitable for pipeline transportation. A source of water is needed at the mine to prepare slurry.

The prepared slurry is fed to slurry storage tanks that typically store about 8 hours of plant production providing a buffer between beneficiation and pipeline operations. The slurry from these tanks is fed to the slurry pipeline. Pumping facilities at the mine site, as well as at intermediate locations, may be needed to transport the slurry to the refinery. If the mine is located at a higher elevation compared to the refinery, then it may be possible to transport the slurry without using pumping facilities.

The slurry is delivered into slurry storage tanks near the refinery. The slurry will require dewatering prior to delivery to the refinery. Depending on the moisture content requirements of the filter cake, vacuum ceramic filters or pressure filters may be used to dewater the slurry. The filtrate is clarified and treated prior to disposal, or may be used as process water at the refinery. The clarified water can be recycled back to the mine if needed, substantially reducing the amount of fresh water required for preparing the slurry.

2.2 Pipeline Throughput

The pipeline throughput depends upon the refinery requirements. The operating availability of the mine, beneficiation plant, pipeline, and the dewatering plant can be different. Storage facilities are provided to accommodate differences in availability of different components.

2.3 Density of Solids

The following table presents the specific gravities of different mineral concentrate and tailings.

Table I: Typical Solids' Specific Gravity

Concentrate type	Specific Gravity of Solids
Iron	4.5 – 5.0
Copper	4.2 – 4.8
Zinc	4.0 – 4.3
Nickel Laterite	3.3 – 4.0
Phosphate	2.8 – 2.9
Bauxite	2.5 – 3.0
Coal	1.4 – 1.5
Gold Tailings	2.7 – 2.9
Copper Tailings	2.5 – 3.0

The specific gravity of bauxite solids is similar to that of sand. A solids specific gravity of 2.5 to 3.0 is normally encountered

2.4 Grind size

Typical bauxite ore is coarser than suitable for long-distance transport by pipeline — therefore, it must be ground finer than the typical process requirement to be transported in the pipeline. The optimum particle size distribution (PSD) for hydrotransport is a balance between: grinding, pumping, pipe steel, and thickening or filtering costs, as well as the impact on refinery operations. Within limits, the finer the grind, the easier it is to transport solids by pipeline. In particular, it can be transported at lower velocities without dragging a bed or sanding out. This reduces the pumping power required and the erosion rate of the pipeline. However, this comes at a price. Finer slurries tend to be transported at lower solids contents, increasing the volume to be transported and stored. The lower slurry velocity requires a larger diameter pipe for any given flow rate, and that may increase capital costs. A finer grind requires a larger mill and consumes more power and grinding media. Filtering becomes more difficult and more expensive as the particles get finer. Table II presents the typical PSDs for conventional slurry pipelines.

Table II: Typical Particle Size Distribution of Pipeline Slurry

Concentrate Type	% - 65 Mesh (210 μ)	% - 325 mesh (44 μ)
Iron	100	60 – 80
Copper	97 – 100	55 – 85
Zinc	100	60 – 80
Phosphate	65 – 85	35 – 50
Nickel Laterite	>95	60 - 90
Bauxite	95	40
Coal	50 – 60	20 – 30
Gold Tailings	100	60 – 80
Copper Tailings	70 – 90	30 – 50

2.5 Slurry Concentration

The slurry concentration should be as high as possible in order to reduce the amount of water transported with the solids. Slurry concentration is selected such that the friction loss in the pipe is not very high. For bauxite slurry the slurry concentration is expected to vary between 45, for fine slurry PSD, and 55 weight percent (wt%) solids for coarser slurry PSD.

2.6 Slurry Flow Rate

The slurry flow rate depends upon solids throughput, solids density, and slurry concentration. The mass flow rate of slurry is obtained by dividing the solids throughput by the slurry concentration in wt%. The mass flow rate of liquid (water) is obtained as a difference between the slurry mass flow rate and the solids mass flow rate. The volumetric flow rate of slurry is obtained by adding the volumetric flow rates of solids and water. The volumetric flow rate is obtained by dividing the mass flow rate by the density of the material. For example, if 5 million tons per year of bauxite solids are to be transported, and assuming an operating availability of 8322 hours per year, the hourly throughput of solids will be 600 tons per hour (tph). Assuming a slurry concentration of 50 wt% solids, the mass flow rate of slurry will be 1200 tph. Assuming a solids density of 2.65 tons/m^3, and a water density of 1 ton/m^3, the slurry flow rate will be 826 m^3/h.

2.7 Slurry Flow Properties

Bauxite slurry behaves like a Bingham plastic fluid. The flow behavior is defined by plastic viscosity and yield stress.

$$\tau = \tau_y + \eta\gamma \quad (1)$$

Where: τ = shear stress, Pa
τ_y = yield stress, Pa
η = Plastic viscosity, Pa.s
γ = rate of shear, 1/s

The yield stress of slurry is a strong function of particle surface area. For a given volume of solids, particle surface area increases when the particle size is reduced. Bauxite slurry particles tend to break down and generate fines when subjected to shear. Thus the yield stress of slurry traveling via a long-distance pipeline can be significantly higher than that of a fresh slurry sample.

The particle attrition occurs by the breaking off of the edges of larger particles, creating very small size particles. These particles tend to form flocs that immobilize water and act like a part of the particle volume.

The friction loss in a pipeline depends upon the flow regime. In laminar flow regime, friction loss depends upon yield stress as well as plastic viscosity. In turbulent flow regime, the friction loss is governed by plastic viscosity.

3. Water supply

Most existing long-distance concentrate pipelines are operated as "open" systems. Recently, some slurry pipeline systems include a return water pipeline due to a water shortage at the mine site. Fresh water (or process water from the beneficiation plant) is used to dilute the concentrate. At the terminal, the dilution water and filtrate is separated from the concentrate and used in the downstream process or is treated and discharged into the environment. Because the concentrate (e.g., copper or zinc) is a small percentage of the whole ore tonnage, and the slurry is transported at a high concentration, the total amount of excess water is small. Treatment and discharge into the environment is the economic alternative.

Bauxite is transported as a whole ore, at high tonnage rates, and can only be transported in a pipeline at a relatively low solids concentration, so a large volume of liquid is required. As a result, installing an "open" system would require finding and obtaining rights to a reliable source of fresh water near the mining area. Wells and pump stations would need to be built, powered, and operated. The water would need to be pipelined to the pump station. Large water storage ponds would be needed at the mine to ensure a continuous supply.

At the refinery, only a portion of the dilution water could be used in the process without causing water balance problems. The current operation is balanced for the moist bauxite from the mine (i.e., 82% solids by weight) and a small amount of make-up. Any filtrate that cannot be used in the process would need to be treated for discharge into a nearby water body, or in ponds designed to store water. Alternately, the water could be used for irrigation in any agricultural/pasture areas surrounding the plant, which would also require treatment and a considerable amount of distribution piping.

Shorter tailings pipelines are usually "closed" systems. For example, red mud (slurry) from the refinery is sent to the tailings impoundment area and the decanted pond water is returned to the refinery. The need for treatment of the discharge water is eliminated, as is the need for most of the make-up water. Clarification of the filtrate is not required because the small percentage of ultra-fines returning to the mine will have a negligible impact on the pipeline operation.

3.1 Carrier Fluid Options

Once a system is closed, it is no longer necessary to dilute the solids in water. In the Bayer process, the bauxite ore is ground and slurried in spent liquor before being pumped to the digesters. This spent liquor could be delivered to the pipeline beneficiation plant at the new mine. The main advantage of this is that slurry coming out of the pipeline can be directly fed into the refinery with little or no dewatering required. Slurry diluted in water needs to be put through expensive filters to reach the high solids content required to maintain refinery water balance.

4 System Design

4.1 Velocity of Flow

A key design parameter in any slurry pipeline system is the operating velocity range. In a conventional (turbulent) system, the line velocity must be high enough to maintain turbulence and to prevent deposition of the coarse particles. However, if the operating velocity is too high, pressure losses and pipeline wear rates can become too high for the system to be economic.

For a long-distance, cross-country pipeline the minimum operating velocity should be higher than the laminar-turbulent transition velocity and the deposition velocity. Methods for determining these values are described in detail elsewhere [2]. In general, viscous slurries (i.e., ones with a higher solids concentration and/or fine particles) will be transition velocity limited. Dilute slurries, and/or coarse slurries, will be deposition velocity limited. As the solids concentration of any given slurry is increased, the transition velocity will increase and the deposition velocity will decrease. As a rule of thumb, the most economical concentration with which to run a long-distance pipeline is where the two values are equal.

The particle size distribution and slurry concentration are selected such that the deposition velocity and laminar turbulent transition velocity are less than 1.5 m/s. This allows the long-distance pipeline to be designed at a velocity in the range of 1.5 to 2 m/s.

4.2 Pipe Diameter

The pipe diameter is selected such that the desired velocity of flow is obtained at the design throughput.

4.3 Throughput Variation

Normally the refinery throughput may change over a range of throughputs. This can be accommodated to some extent by a change in flow rate, as well as by a change in slurry concentration. The pipelines are designed to operate within a narrow concentration, as well as a specific flow rate range. If the plant requirement drops below the minimum continuous throughput capacity of the pipeline, then the pipeline is operated in batch mode. In a batch mode operation, alternating batches of water and slurry are transported through the pipeline.

4.4 Pipeline route

A buried cross-country pipeline is normally used for long-distance transport. If a recycle water line is needed, it will also be buried and installed in the same right of way as the slurry line. Selection of the pipeline route must consider: system hydraulics, constructability, existing land usage, access rights, and future mining plans.

The primary issue with system hydraulics is to optimize capital costs (pipe diameter, pump size, etc.) and operating costs (the amount of pumping pressure/power required). In a flat area, this is relatively straightforward: keep the line as short as possible. In mountainous areas it becomes more complicated. Pressure requirements are primarily set by high points along the route. This may require extra pumps and pump stations, and include chokes after the high points to dissipate excess pressure on the downhill runs. The total cost of a system can be reduced substantially if high points are avoided, even if the total length is much longer than a straight line. Also of hydraulic concern are keeping the hydraulic profile line close to the ground to reduce operating pressure, and therefore, the required wall thickness of the pipes. The profile of the route will also affect the magnitude of transient pressures in the system, particularly when running water batches through a slurry line. These issues are described in detail elsewhere [1].

Constructability (including existing land usage and access rights) can strongly affect the overall costs of a pipeline. Rocky or swampy areas are hard to build in. Side hill cuts, river crossings (either under or over), and tunnels are expensive. Land acquisition in developed areas can be problematic. Pipelines can be, and have been, run through all of these areas, but it may be more economical to extend the pipeline to by-pass them. Having an experienced pipeline construction engineer involved in the route selection process is usually money well spent.

4.5 Pumping requirements

The pumping requirements depend upon the friction loss through the pipeline, as well as the static head required due to changes in elevation between the mine site and the refinery site.

Bauxite slurry is expected to flow as a homogeneous slurry. The friction loss of homogenous slurry mainly depends upon the viscosity, density, velocity of flow, pipe diameter, and pipe roughness.

4.6 Number of Pump Stations

In a long-distance pipeline the pumping pressure may be such that more than one pump station is needed. The total cost of the pipeline system is equal to the sum of the cost of pump stations and the pipe steel tonnage. The maximum pressure in the pipeline is reduced when more than one pump station is used. The pipe wall thickness requirement depends upon the pressure in the pipe. Therefore, the cost of pipe will decrease, but the cost of pump stations will increase with an increase in the number of pump stations. The selection of the number of pumps stations required depends upon an economic analysis using a various number of pump stations along the pipe.

4.7 Pipe Steel Requirements

The pipe wall thickness requirement depends upon the operating pressure in the pipeline. Long-distance pipelines normally use more than one wall thickness in order to reduce steel requirements. Slurry pipelines can be designed to have a maximum allowable stress equal to 80% of the specified minimum yield stress (SMYS) of the pipe steel. Pipe steel in various grades is available. Recently, pipes have been designed using pipe steel having a SMYS equal to 4827 MPa (70,000 psi).

The wall thickness computed, based on allowable operating pressure, should be increased to account for anticipated corrosion rate during the economic life of the pipeline.

Minimum pipe wall thickness depends upon the pipe diameter. If the wall thickness is too thin, it could give rise to handling problems during construction.

Maximum wall thickness should be limited based on the capability of the pipe mill, as well as welding considerations.

5 Major Equipment

Most mine operators are familiar with slurry preparation and dewatering equipment. However, the use of a slurry pipeline requires additional equipment that may not be familiar to many mine operators, and these are briefly discussed in this section.

5.1 Pumps

The discharge pressure, controls the basic decision of whether to use centrifugal pumps or positive-displacement (PD) pumps.

Centrifugal slurry pumps can handle large flow rates, but are limited to about 40 m head per pump. These pumps are less expensive and easy to maintain. When higher pumping heads are needed, the pumps can be arranged in series. Normally the maximum pumping pressure is limited to about 5 MPa (725 psi).

PD pumps havea limited flow capability (about 300 m^3/h) per pump, but can develop very high pressures. Pumps capable of producing 25 MPa (3650 psi) have been used in long-distance slurry pipelines. Pumps can be arranged in parallel to achieve higher flow rates.

For discharge pressures below 5 MPa (725 psi), a lower capital cost with installation of centrifugal pumps provides an economic advantage over PD pumps, especially when large flow rates are encountered. However, the total cost of a centrifugal pump system can sometimes be more expensive compared to PD pump system, as a spare line of pumps is needed as spares in order to achieve a high system availability.

Two types of impellers and liners are used for centrifugal slurry pumps. Wear-resistant, metal-lined pumps are used when handling coarse particles. Rubber-lined pumps are used when handling fine slurries.

The following types of PD pumps have been used for high-pressure pumping systems.

- Piston Pumps
- Plunger Pumps
- Piston-diaphragm Pumps
- Hydrohoist (Lock-hopper) Systems

The piston pump has a piston, which is driven by a crankshaft, and is in constant contact with the cylinder wall during each stroke. This action would result in high wear if the pump were used for pumping abrasive materials such as iron ore. A modified plunger pump was therefore introduced for such service. It has a plunger, which is continuously flushed with clear water during the suction stroke, thus greatly reducing internal wear.

The parts that most often experience wear in a piston pump system are: valves, valve seats, plunger or piston packing, plunger sleeves or cylinder liners, and brass bushings.

Piston-diaphragm pumps use a diaphragm that is pushed back and forth using a piston pump that pumps hydraulic fluid from a pump reservoir that is in closed loop. The only parts experiencing wear are valve, valve seats, and the diaphragm. This feature greatly reduces the maintenance of the pump. Piston-diaphragm pumps have been used for abrasive slurries with maximum pump pressures up to 25 MPa (3650 psi).

Pump flow rates and maximum discharge pressures are a function of the piston or plunger diameter, and power end horsepower capability. As the piston or plunger diameter increases, the flow rate increases and the maximum working pressure decreases.

PD pumps have a higher efficiency rating compared to centrifugal slurry pumps. PD pumps require more skilled maintenance labor than that of centrifugal slurry pumps.

A hydrohoist system consists of two or more pipes arranged in parallel. Slurry is fed into one of the pipe chambers by opening a valve to admit slurry in that pipe chamber. Each chamber has inlet valves for low-pressure slurry feed and high-pressure water feed. On the other end of the chamber, there are outlet valves to pump out slurry that is pressurized with high-pressure water, or to pump out water using low-pressure slurry. The pipe chambers are alternatively filled with slurry and pumped out, using water under high pressure, by synchronizing the opening and closing of valves.

6 Conclusions

Slurry pipelines have been successfully used for transporting different types of concentrates and other materials.

Bauxite ore can be economically transported using slurry pipelines.

For long-distance transport of bauxite, the ore will have to be more finely ground than for current refinery feed systems.

Bauxite particles tend to break up and generate fines that increase the yield stress of slurry. The pipeline system design should be based on the rheological properties of the bauxite slurry, allowing for expected attrition during pipeline transportation.

Bauxite slurry can be transported at a slurry concentration in the range of 45 to 55 wt% solids.

Positive-displacement pumps may be needed to develop the high pressures that might be needed for long-distance transportation.

7 References

1. Hallbom, D.J., *"Over the hills and far away" – Long-distance slurry pipelines in mountainous regions.* 38th Canadian Mineral Processors Operators Conference, Ottawa, Canada (2005)

2. Wasp, E.J.; Kenney, J.P.; Gandhi, R.L., *Solid-liquid flow – Slurry pipeline transportation.* Trans Tech Publications series on Bulk Material Handling No. 4 (1977).

OCCURRENCE AND CHARACTERIZATION OF Zn AND Mn IN BAUXITE

Frank R. Feret and Jeannette See

Alcan International Limited, Arvida R&D Centre, 1955, Mellon Blvd., P.O. Box 1250, Jonquière, Québec, G7S 4K8, Canada

Keywords: Bauxite, Zn and Mn minerals, caustic solubility of Zn, XRF, XRD

Abstract

Zinc is one of the secondary elemental constituents occurring in Caribbean bauxite and the associated non-bauxitic material. It is generally believed that Zn in bauxite could either occur in gahnite or sphalerite, or substitute for Fe in goethite. Manganese represents an appreciable impurity in Caribbean bauxites and is identified on diffractograms as lithiophorite (Li,Al)$MnO_2(OH)_2$. As Zn has been observed to increase with the MnO content, the objective of this work was to better understand the mineralogical nature of Zn and Mn compounds. Data representative of 340 bauxite samples of different origins was assembled. It was found that Zn in bauxite could not possibly substitute for Fe in goethite or hematite. Strong evidence was obtained that Zn occurs in the same compound as Mn. The application of the Rietveld-XRD method to the quantification and characterization of a new mineral called zincophorite $Al(Zn_xMn_{1-x})O_2(OH)_2$ is discussed. Caustic solubility of ZnO in bauxite is also assessed.

Introduction

Minerals typically identified in bauxites include gibbsite, boehmite, kaolinite, goethite and crandallite - which are the primary sources of Al_2O_3. Other identifiable minerals include anatase, rutile, ilmenite, hematite, lithiophorite, quartz, illite and zircon [1-4]. Most secondary mineral phases are almost impossible to confirm by direct XRD analysis due to their low content and consequently low intensity of the respective diffraction peaks [5-9]. Other methods must be employed (extraction, pre-concentration, Scanning Electron Microscopy (SEM), synchrotron-based methodologies, etc.) in order to positively identify phases in which secondary elements are present. In routine bauxite analysis, the secondary phases are also important for a reliable solution of the mass balance using analytical tools such as BQUANT [10] and XDB software [11].

Unfortunately, a large number of secondary phases, which are believed to exist, have never been confirmed in bauxites processed by the aluminum industry and the resulting residue or red muds. Possible occurrence of these secondary phases is suggested by elemental composition determination using modern analytical methodology. The understanding of distribution of certain minor and trace elements (e.g., F, Na, Mg, P, S, K, Ca, V, Cr, Mn, Zn, Sr, Zr, etc.) and related mineral phases within the grains is of primary importance in the context of their extraction during Bayer digestion [12,13]. Some mineral phases are soluble under Bayer process digestion conditions. Several other phases are known to be inert or partially soluble in the digestion process. In order to understand the basic "building blocks" and the corresponding elements of typical bauxite material, the secondary mineral phases must be known. Consequently, one would be able to better predict the outcome of bauxite digestion and further optimize the Bayer process. Moreover, the accumulation of certain elements in the final product (metal) could be influenced and thus avoided.

Zinc is one of the secondary elemental constituents occurring in the bauxite matrix that is very undesirable in the production of aluminum. Concentration of ZnO is low (<0.002%) in several bauxite deposits, such as Ely and Gove (Australia), Boké (Guinea), or Trombetas (Brazil). However, in some Caribbean bauxite and the associated non-bauxitic material, the ZnO concentration may reach 0.1%, or even more and causes important production concerns. During the Bayer process, a certain amount of Zn present in bauxite is soluble and finds itself in the Bayer liquor. Consequently, Zn is found in alumina at a later stage of the process. During the electrolysis process, Zn enters with alumina and dissolves in liquid metal. Occurrence of Zn in Al and its alloys is detrimental as it affects material performance.

Goethite and hematite have a great potential for isomorphic substitution of Fe by other metals. The incorporation of Al into goethite is well known [14-15]. In addition, Mn, Cr V, Cd, Co, Cu, Ge, Ni, Pb, Ti and Zn can also be substituted for Fe [16]. It is generally believed that Zn in bauxite could either occur in gahnite or sphalerite, or substitute for Fe in goethite. However, if it occurs in association with goethite, then it should not be partially soluble under low-temperature digestion conditions. Moreover, Zn was observed to increase with the MnO content while analyzing some Jamaica bauxite and related non-bauxitic material.

Manganese represents an appreciable impurity in Caribbean bauxites. Concentrations of up to 2-3% MnO or more have been confirmed in localized sections at the bottom of bauxite deposits immediately above the calcite layer. The manganiferous mineral that has been identified in bauxite from Jamaica or Dominican Republic is lithiophorite (Li,Al)$MnO_2(OH)_2$ [17]. The presence of manganite [MnO(OH)], hausmanite [Mn_3O_4], pyrolusite [MnO_2], rhodochrosite [$MnCO_3$] or a possibility of Mn-substituted goethite [$(Fe_{1-x}Mn_x)OOH$] cannot be excluded [18].

Hence, the objective of this work was to better understand the mineralogical nature of compounds in which Zn and Mn are found.

Experimental

In order to investigate potential correlations between Zn, Mn and other major elements, the concentration data representative of 340 bauxite samples of different origins was assembled. The concentrations of element oxides were determined using X-ray fluorescence (XRF) and a fusion sample preparation technique [19]. The sample analytical portion was fused with a flux and cast into a glass disk. The flux is composed of granular lithium tetraborate ($Li_2B_4O_7$) and lithium metaborate ($LiBO_2$) pre-mixed in the 12:22 proportion. Estimated accuracy of ZnO and MnO determination is ±0.001% and 0.007%, respectively. The highest Zn concentration in the studied group of samples was 1.14%. Concentrations determined for MnO, ZnO, Cr_2O_3, ZrO_2, V_2O_5 and Li in three samples in which the ZnO concentrations was the highest are given in Table I. Li concentration was determined using an ICP spectrometer and samples fused with caustic.

Table I. Concentrations (%) of Selected Elements in Three Bauxites Samples

Constituent	Sample		
	A (%)	B (%)	C (%)
MnO	9.7	30.7	42.1
ZnO	0.32	0.61	1.14
Cr_2O_3	0.28	0.04	0.05
ZrO_2	0.056	0.02	0.06
V_2O_5	0.17	0.03	0.03
Li	<0.01	0.025	0.065

Both Table I and Figure 1 indicate that the most interesting correlation exists between Zn and Mn. The correlation coefficient obtained for the group (R^2 = 0.93) strongly suggests that Zn may occur in the same compound as Mn. By contrast, the same Figure 1 also demonstrates that Zn and Fe are anti-correlated and Zn cannot substitute for Fe in goethite or hematite as was previously presumed [10,16].

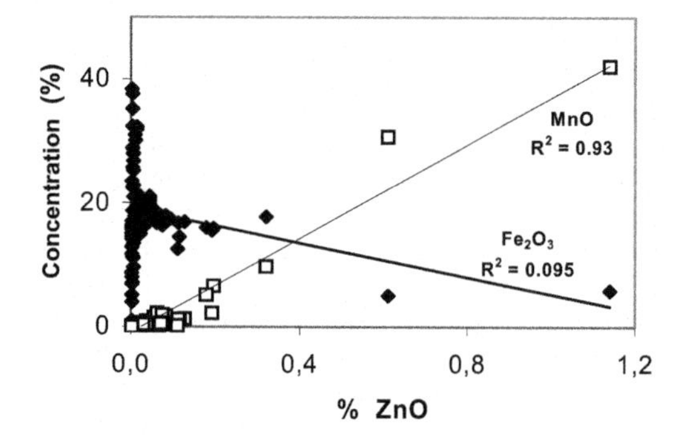

Figure 1. Correlation of ZnO versus MnO and Fe_2O_3 in Bauxite

The highest MnO and ZnO concentrations occur in the bauxite samples in which the only crystalline Mn compound identified on diffractograms is lithiophorite $(Al,Li)MnO_2(OH)_2$. Figure 2 illustrates a diffractogram corresponding to 10 – 60 ° 2θ with the four most intense peaks of lithiophorite marked by means of numbers. The peaks of gibbsite are also marked with dotted vertical sticks, but they are not numbered.

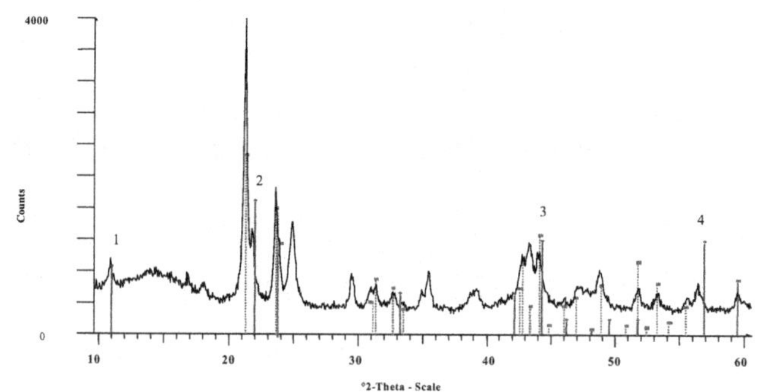

Figure 2. Diffractogram of a Jamaican Bauxite

To help identify a possible Zn-containing mineral, two samples with highest ZnO content were examined using the Jeol scanning electron microscope.

The micrographs in Figure 3 confirm a positive correlation between Zn and Mn, but certainly not all of Zn is associated with Mn. As no other Zn-bearing compound than lithiophorite was identified on diffractograms, it needs to be assumed that Zn is a part of lithiophorite. The chemical formula of lithiophorite can also be represented as $MnAl_{0.5}O_4Li_{0.5}H_2$, hence the effective composition is 63.04% MnO_2, 18.48% Al_2O_3, 5.42% Li_2O and 13.06% H_2O. It is not difficult to notice in Table I that Jamaica lithiophorite is far from this composition.

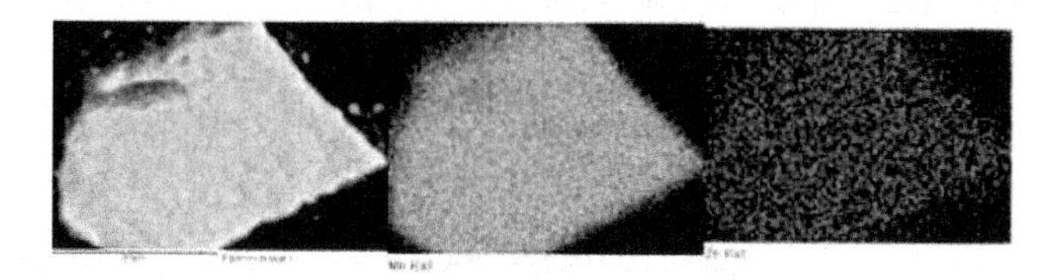

Figure 3. SEM Micrographs Indicating from Left: a) Back-scattered Image of a Grain and Elemental Maps for Mn (b) and Zn (c)

The 2003 ICDD database (ICDD – International Centre for Diffraction Data) was checked for all documented lithiophorites and eight different compounds were found. The unit cell parameters of the ICDD lithiophorites [20] are listed in Table II.

Table II. Unit Cell Parameters of Lithiophorites Listed in the ICDD Data Base

ICDD Ident.	Formula	Unit Cell				System
		a (Å)	b (Å)	C (Å)	Vol. (Å)	
82-1216	$(Li_{0.33}Al_{0.65})MnO_2(OH)_2$	2.925	2.925	28.169	241.0	R
76-0180	$Li_6Al_{14}O_{42}Mn_3Mn_{18}(OH)_{42}$	13.37	13.37	28.2	5040.9	H
73-2109	$Li_{0.32}Al_{0.68}MnO_2(OH)_2$	5.06	2.91	9.55	140.6	M
41-1378	$(Li,Al)MnO_8(OH)_2$	2.906	2.906	28.28	238.8	R
16-0364	$(Li,Al)MnO_2(OH)_2$	5.06	2.91	9.55	140.6	M
12-0717	$LiAl_2Mn_3O_6(OH)_6$	2.925	2.925	28.10	240.4	H
12-0647	$(Co,Mn)O(OH)$	3.725	12.38	9.455	436.0	O
04-0272	$(Li,Al)MnO_2(OH)_2$	2.92	2.92	9.39	80.1	H

It is clear from Table II that the crystallographic parameters of known lithiophorites are very diverse. In order to be able to compare known crystallographic parameters with those that are representative of Jamaica lithiophorite, a Rietveld-XRD analysis was employed.

Rietveld-XRD Analysis of Jamaica Bauxites

The mineralogical analysis by the Rietveld-XRD method uses all the intensity data in a diffraction pattern rather than a few of the most intense reflections. In the quantification process, the Rietveld based method involves structure data of each phase (atomic coordinates, Lorentz, scattering, polarization factors, physical and chemical constants, etc.) and does not depend on any calibration standards [21-26]. The least-square refinements are carried out until the best fit is obtained between the entire observed diffraction pattern taken as a whole, and the entire calculated pattern.

Rietveld-XRD quantification and structure refinement of phases at a low concentration level is known to be difficult, if possible. Therefore, for the Rietveld analysis, only samples believed to contain measurable quantity of lithiophorite (based on the MnO and ZnO concentrations) were selected. A Siemens D5000 diffractometer equipped with Kevex Peltier cooled detector and PANalytical X'Pert Pro software was used for the collection and interpretation of the diffraction data. The unit cell parameters obtained for lithiophorite (ICDD 73-2109) in Jamaica bauxites using the Rietveld-XRD method are listed in Table III. The last two columns give corresponding XRF concentrations for ZnO and MnO.

Table III. Unit cell Parameters Obtained for Lithiophorite in Jamaica Bauxites

Sample	Unit Cell (Å)			Volume (Å)^3	ZnO (%)	MnO (%)
	a	b	c			
1	5.07	2.904	9.87	145.3	0.082	1.7
2	5.00	2.904	9.68	140.5	0.195	6.58
3	4.99	2.902	9.80	142.0	0.178	5.15
4	5.035	2.869	9.57	138.3	0.074	0.41
5	4.945	2.943	9.868	143.6	0.059	0.3
6	4.898	2.957	9.888	143.2	0.076	0.32
7	5.198	2.778	9.718	143.2	0.077	0.58
8	5.017	2.914	9.607	140.5	0.32	9.7
9	5.023	2.909	9.65	141.0	0.61	30.7
10	5.045	2.893	9.597	140.1	1.14	42.1
11	4.924	2.905	9.88	141.4	0.026	0.32
12	4.948	2.870	9.838	139.7	0.023	0.27
13	4.556	3.042	9.887	137.0	0.029	0.38

The average a, b, c parameter value for the analyzed group of samples was 5.0, 2.9 and 9.8 Å, respectively. The closest compound among those listed in Table II resembling these unit cell parameters is $(Li,Al)MnO_2(OH)_2$. Theoretically, the lithiophorite known in the literature does not contain Zn. By contrast, lithiophorite present in Jamaica bauxite contains little Li, which is confirmed by ICP. Because there is very little Li in the Jamaica samples, it seems that Li does not replace Al in the compound. Zn cannot take the role of Li nor can it replace Al due to ionic radii differences. However, Zn could replace Mn. Under the circumstances the name lithiophorite is not justified and the formula corresponding to Jamaica lithiophorite should probably be altered. Hence, the name should be zincophorite rather than lithiophorite and the most realistic chemical formula could be $Al(Zn_xMn_{1-x})O_2(OH)_2$. Based on the data obtained, the x parameter might vary from 0.02 to 0.24. The concentration data corresponding to the investigated group of samples is given in Figure 4. In Figure 4, concentrations of ZnO are plotted against the concentrations of MnO and zincophorite. The correlation between the ZnO and MnO content is strong. Some points representative of zincophorite are placed much below those of MnO at high ZnO concentrations. This is highly suggestive that a portion of lithiophorite is X-ray amorphous. That could also explain why a hump next to the first peak of lithiophorite at 9.45 Å (Fig. 2) is so strongly pronounced. Occurrence of amorphous Zn-containing minerals is possible too.

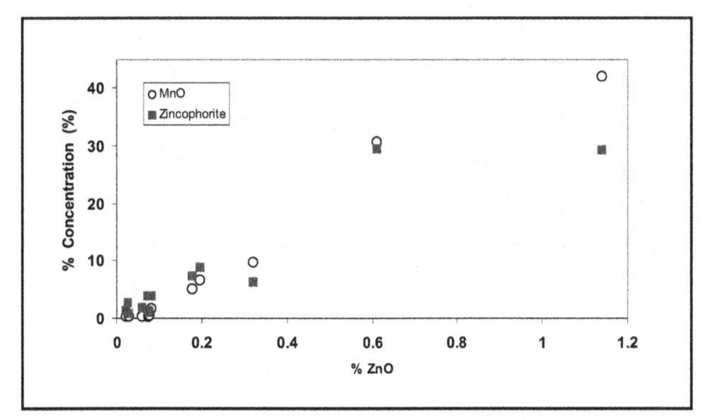

Figure 4. Concentration Data Obtained for Jamaica Zincophorite

Mettler Toledo Star TGA system was also employed to study thermal decomposition characteristics of the samples, but no clear peak corresponding to zincophorite was identified. As the determined LOM concentration cannot all be attributed to other known minerals, zincophorite in a few studied samples must have decomposed over a wide temperature range. There are several known compounds containing Zn and Mn, and a few that are hydrated. Among them, one can mention chacophanite ($ZnMn_3O_7{\cdot}3H_2O$), woodruffite $(Zn,Mn)_2Mn_5O_{12}{\cdot}4H_2O$ and hydrohetaerolite ($Zn_2Mn_4O_8{\cdot}H_2O$). However, none of these are tenable with regards to the crystal unit cell parameters estimated for Jamaica zincophorite using Rietveld-XRD analysis.

Caustic Solubility of Zn

Solubility of Zn present in bauxite during the Bayer digestion process is a known fact. Bomb digests of bauxite carried out at 135°C over 45 minutes extract a certain amount of Zn from bauxite. Atomic Absorption Spectroscopy (AAS) or Inductively Plasma Spectroscopy (ICP) may analyze the digest solution for ZnO. A similar procedure can be applied for estimation of ZnO soluble at 240 °C. Figure 5 shows that there is no correlation between the ZnO content in bauxite and the amount of caustic soluble ZnO at 135°C.

By contrast, Figure 6 confirms a clear correlation between the MnO content and caustic soluble ZnO at 240°C.

It is not certain, at present, whether one or a few different minerals contribute soluble ZnO.

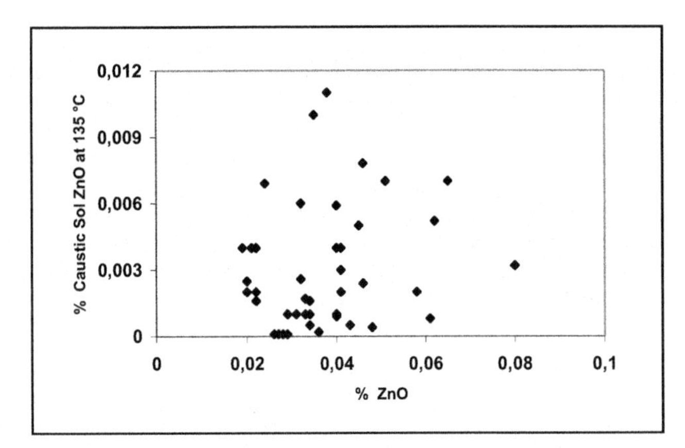

Figure 5. Relationship between ZnO in Bauxite and the Amount of Caustic Soluble ZnO at 135°C

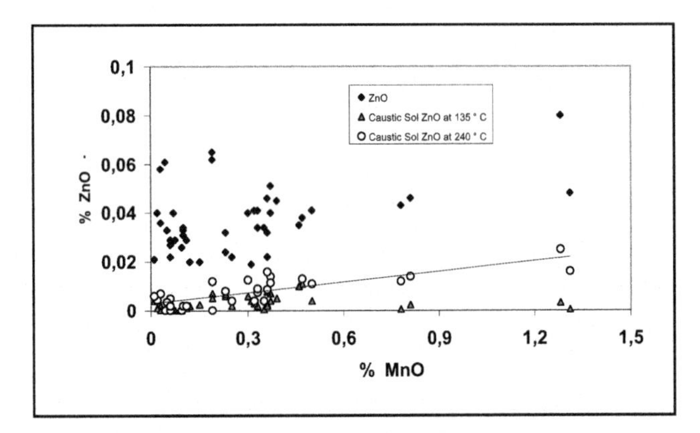

Figure 6. Correlations Between MnO and Caustic Soluble ZnO

Discussion

The secondary elemental constituents of most mineral ores play a very important role as they may affect material's economic value and/or its performance in the industrial process. Some constituents enhance ore properties (for example, Mn in iron ore) while others are more or less undesired (for example P or Zn in bauxite).

As Mn is known to be present in Bayer alumina at a very low concentration, it is usually assumed that Mn minerals in bauxite are inert under Bayer digestion conditions and Mn passes on almost entirely to the mud. Hence, a good portion of Al_2O_3 associated with these minerals is also lost. If for example MnO is at 0.58%, then based on the lithiophorite stoichiometry, Al_2O_3 and H_2O associated with it account for 0.20% and 0.14%, respectively.

Evidence points to the existence of a mineral such as zincophorite, which could explain a part of Zn occurring in bauxite. It is believed that future micro-XRD examinations could further clarify the situation.

A conventional, X-ray tube based diffraction offers limited possibilities of phase identification and quantification. For most secondary and trace phase constituents, the method's limit of detection is too high to allow any characterization. Moreover, up to 15% of bauxite material from certain deposits and up to 40% of red mud matrix is X-ray amorphous and does not contribute in the conventional XRD process. Only synchrotron-generated radiation would allow obtaining a significantly enhanced quality of diffraction leading to a more detailed study. Much better signal-to-noise ratio would allow diffraction details to be more easily discerned and located. Small volume specimens, which can be used with synchrotron radiation, would largely overcome a problem of sensitivity and preferred orientation hampering bauxite quantitative phase analysis.

References

1. H.R. Black, 1953. "Analysis of Bauxite Exploration Samples". Analytical Chem. 25: 743-748.

2. G.I. Bushinskii, "The Geology of Bauxites", Nedra, Moscow, 1975.

3. G. Bárdossy, Karst Bauxites, Elsevier Scientific Publishing Company, Amsterdam, 1982.

4. F. Feret, D. Roy, "Determination of Quartz in Bauxite by a Combined X-ray Diffraction and X-ray Fluorescence Method", Spectrochimica Acta, Part B, 57(3), p. 551-559, 2002.

5. G. Bárdossy, L. Bottyán, P. Gadó, A. Griger, and J. Sasvari, "Quantitative Phase Analysis of Crystalline Multi-component Powder Mixtures by the Diffraction Method", Bányászati és Kohászati Lapok Kohasz., 112 (9) (1979), 130 -137.

6. G. Bárdossy, G. Pantó, I.C.S.O.B.A. – "Trace Mineral and Element Investigation in Bauxites" by Electron-Probe, 3e Congrès International, SEDAL, Nice 1973.

7. F. Feret, G.F. Giasson, "Quantitative Phase Analysis of Sangaredi Bauxites (Guinea) Based on their Chemical Compositions" - Light Metals, The Minerals, Metals & Materials Society, 1991, 187.

8. F. Feret, M. Authier-Martin, István Sajó, "Quantitative Phase Analysis of Bidi-Koum Bauxites (Guinea)", Clays and Clay Minerals, 45 (3) (1977), 418-427.

9. Ni L.P., Khalyapina, O.B.,"Physical-Chemical Properties of The Raw Materials and Products of Alumina Production", Izdatielstvo "Nauka" Kaz.S.S.R., Alma-Ata, 1978.

10. F. Feret, F.M. Kimmerle, B. Feret, "BQuant: Cost-effective Calculations of Bauxite Mineralogy" - Light Metals, The Minerals, Metals & Materials Society, 1997, Vol. 9.

11. Sajó, "Powder Diffraction Phase Analytical System", Version 1.7, Users Guide, Aluterv-FKI, Ltd., Budapest, 1994.

12. S. Ostap, "Effect of Bauxite Mineralogy on its Processing Characteristics", Bauxite, Proceedings of the 1984 Bauxite Symposium, AIME, Los Angeles, California, 1984, Feb. 27- March 1.

13. M. Authier-Martin, G. Forté, S. Ostap, J. See, "The Mineralogy of Bauxite for Producing Smelter-Grade Alumina", JOM, December 2001, 36-40.

14. Thiel R., "Goethite-Diaspore System", Z Anorg U Allg Chem. 326: 70-78, 1963.

15. Norrish K, Taylor RM., "The Isomorphous Replacement of Iron by Aluminum in Soil Goethites", J. Soil Sci 12: 294, 1961

16. U. Schwertmann, R.M. Cornell, "Iron Oxides in the Laboratory", VCH, Verlagsgesellschaft mbH, Weinheim (Germany), 1991.

17. A King WR., "The Iron Minerals in Jamaica Bauxites", Light Metals 1971, AIME: 3-18.

18. U. Schwertmann, P. Cambier, and E. Murad, Properties of Goethites of Varying Crystallinity, Clays and Clay Minerals, 33 (5) (1985), 369.

19. F. Feret "Application of XRF in the Aluminum Industry, Advances in X-Ray Analysis" 36: 121-137, 1993.

20. PDF-2 Database, International Centre for Diffraction Data, 12 Campus Boulevard, Newton Square, PA 19073-3273, U.S.A., 2002.

21. A R.A. Young, The Rietveld Method, Oxford University Press, New York, 1993.

22. R.J. Hill, "Applications of Rietveld Analysis to Materials Characterization in Solid-state Chemistry, Physics and Mineralogy, Advances in X-ray Analysis", Vol. 35 (1992), 25-37.

23. J.C. Taylor, "Computer Programs for Standardless Quantitative Analysis of Minerals Using the Full Powder Diffraction Profile", Powder Diffraction, Vol. 5, No 1 (1991) 2-9.

24. D.L. Bish, S.A. Howard, "Quantitative Phase Analysis Using the Rietveld Method", J. Appl. Crystallogr., Vol. 21 (1987) 86-91.

25. D.L. Bish, J.E. Post, "Quantitative Mineralogical Analysis Using the Rietveld Full-Pattern Fitting Method", Am. Mineral., Vol. 78, (1993) 932-940.

26. M.G. Aylmore, G.S. Walker, "The Quantification of Lateritic Bauxite Minerals Using X-ray Powder Diffraction by the Rietveld Method", Powder Diffraction, 13 (3), (1998), 136-143.

BENEFICIATION OF HIGH QUARTZ CONTENT BAUXITE FROM LOS PIJIGUAOS

Jean-Marc Rousseaux, Hans Verschuur *, Pedro Flores **, Stephan Buntenbach, Fred Donhauser ***
* ALCAN Bauxite and Alumina ; ** CVG-BAUXILUM ; *** AKW Apparate + Verfahren

Keywords: bauxite beneficiation, trommeling, attrition scrubbing, wet-screening, shaking table, hydrocyclone

Abstract

At Los Pijiguaos, the economic bauxite horizon attains an average thickness of 7.6 m.
Below this layer, bauxite containing High Quartz (HQ Bauxite) can be found.
The aim of this study was to investigate the possibility to remove quartz from HQ Bauxite of Los Pijiguaos, with a minimum of Al_2O_3 losses.
The beneficiation process studied in this work was carried out in several steps. The first step was an elutriation, operated in a washing drum, aiming at liberating quartz particles. The second step was a classification process, aiming at obtaining different grain sizes according to the quartz content.
Part of the study was to investigate if the finest fractions could further be beneficiated using such technologies as cyclones and shaking tables or spirals.
This paper describes tests carried out and results obtained. Characteristics of beneficiated bauxite obtained, along with Al_2O_3 recovery are also detailed. Conclusions delineate pilot scale test conditions.

Introduction

The Los Pijiguaos bauxite mine deposit was discovered in 1976, a feasibility study was conducted by Alusuisse in 1979, and mining commenced in 1987. The mine is operated by the state-owned CVG-Bauxilum-Operadore de Bauxita. The mine production increased from less than 1 million tonnes in the late eighties to more than 5 million tonnes in the late nineties. Today production level reaches 5.8 million tonnes a year.

The Los Pijiguaos bauxite deposit is situated at the north-western section of the Proterzoic Guayana Shield. Geologically, the area is underlain mainly by the Paaguaza Granit of the Suapure Group. The Rapakivi-type granite also represents the parental rock to the ''plateau-type'' bauxite horizon that is developed on an erosional surface at an altitude between 600 and 700 m. The planation surface is deeply dissected by fluvial erosion channels, and thus, a large proportion of the mine site is occupied by deep valleys in a discontinuous bauxite blanket with the orebody being dismembered into nine isolated blocks (Escalona et coll., 1986).

At Los Pijiguaos, the economic bauxite horizon reaches an average thickness of 7.6 m. The soil cover is rather thin (< 1 m) consisting mainly of vegetal matter. A complete gibbsitic bauxite profile typically exhibits an internal stratigraphy characterized by four distinct horizons that include from top to bottom :
- a hard concretionary layer
- an earthy bauxite with loose pisolites
- a bauxite crust with pisolitic, partly porous to spongy textures
- an earthy bauxite with loose pisolites and loose spongy to cellular concretions.

Grade distribution is rather homogeneous but varies laterally and vertically within the deposit (Happel et coll., 1999).

The average grade of the bauxite is 49 % Al_2O_3 and ≈ 10% Total Silica.

The shallow bauxite layer is extracted without blasting by hydraulic shovel extractors in a conventional open-pit mining operation. Overburden is removed by dozers and stockpiled for rehabilitation. The ore is hauled to an inpit crusher by trucks. From the crusher, the bauxite is transported by a conveyor belt to the stockpile in the Suapure valley.

Presently the bottom layer is not extracted, due to its very high quartz content. This bauxite is called High Quartz Bauxite (HQ Bauxite) of Los Pijiguaos (LP).

An important feasibility project was undertaken to evaluate if recovery of this unexploited bauxite would be possible and if the obtained bauxite would be suitable for the plant.

The most important objective of this project is the possibility to increase the recoverable bauxite while obtaining an improved bauxite quality, compared to the present standard grade.
Another important aspect is the possibility to reduce the overall amount of organics in the bauxite feed to the plant.

The aim of this test work is twofold. Firstly, to determine a relevant process for the reduction of the silica content of the HQ Bauxite of LP with a minimum of Al_2O_3 losses. Secondly, to evaluate the characteristics of the beneficiated bauxite.

The process that is envisaged and presented in this study is based on an old and well-known concept of AKW. The bauxite is elutriated in a washing drum in order to liberate the coarse quartz particles, and then classified in different grain sizes according to the quartz content by wet-screening and the use of hydrocyclones.

Another aspect of this work is to investigate the possibility to recover the very fine size ranges.

It was decided to conduct this project with the following sequence of actions :
1. sampling of relevant HQ Bauxite in LP mine
2. test work in AKW A+V-Hirschau Technical Center
3. chemical analysis of all bauxite fractions and interpretation of results in Alcan-Gardanne Technical Center
4. elaborate a strategy to carry out relevant pilot tests in the LP mine based on conclusions obtained

Test Material

The bauxite sample was taken in the year 2003 by the geotechnical department of CVG-Bauxilum-Operadore de Bauxita with the assistance of Alcan.
In total, 4 big-bags of approximately 750 kg of HQ Bauxite were sampled in chosen locations of LP mine and this representative sample of around 3000 kg arrived in AKW A+V-Hirschau Technical Center in the middle of the year 2003.
This sample was divided in representative sub-samples according to international standards.
A sub-sample of approximately 55 kg was sent to the Alcan-Gardanne Technical Center for analysis and characterization.

Characteristics of HQ Bauxite of LP

Global chemical composition, PSD and Al_2O_3, Fe_2O_3 and SiO_2 contents in the various size ranges of typical HQ Bauxite of LP are presented in this paragraph.

Global Chemical Composition of HQ Bauxite of LP

Bauxite sample	Global content
Al2O3	44.5 %
Fe2O3	8.5 %
SiO2 Total	21.62 %
SiO2 Quartz	21.56 %
TiO2	1.1 %
CaO	< 0.05 %
Na2O	< 0.03 %
LOI	23.6 %
Ctotal	0.072 %
Total FX	99.5 %
Al2O3	100 % Gibbsite

Particle Size Distribution and Al_2O_3, Fe_2O_3 and SiO_2 content in size ranges

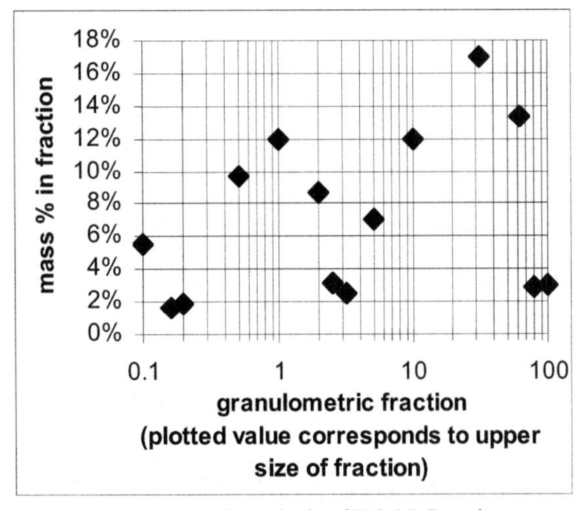

Figure 1. Granulometric analysis of HQ LP Bauxite (without preliminary treatment).

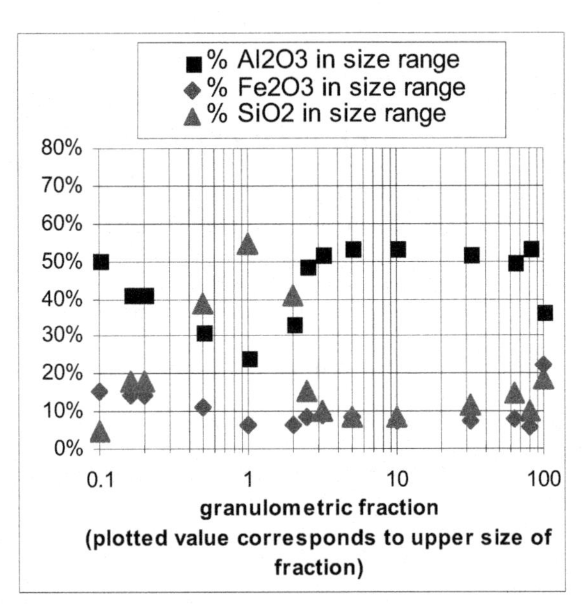

Figure 2. Al_2O_3, Fe_2O_3 and SiO_2 contents in size ranges.

Comments

HQ Bauxite from Los Pijiguaos is characterized by an accumulation of quartz in the 0.1 – 3.0 mm size range.
Finer quartz particles of HQ bauxite from LP typically are in the range of 0.2 to 2.0 mm.
For these reasons, the particle sizes cited above are to be removed for beneficiation purposes (elimination of quartz of raw bauxite).
It should also be highlighted that the Ctotal content in this bauxite (approx 0.07 %) is low, compared to standard grade bauxite, which is presently fed to the plant (Ctotal = approx. 0.20 – 0.22 %). This is in accordance with the fact that HQ Bauxite of LP is a bottom layer bauxite.

Beneficiation Tests with Washing Drum

It is well known that in some particular cases some attrition scrubbing can help favorable liberation of the quartz particles, thus enhancing the quality of the beneficiation process.

The washing drum, in which both a washing and a trommeling step are carried out, is typically used for this purpose (attrition scrubbing). During the pilot tests, the Washing Drum was operated at 320 rpm and a bauxite solid concentration of 900 g/l, during 5 min, which corresponds to the residence time in the Washing Drum. This test will be referred to as "WD Test without Gravel".

The effect of increasing the attrition intensity was investigated by adding a certain amount of gravel to the Washing Drum. During the pilot tests, the Washing Drum was operated at 24 rpm, with an addition of 8 kg of gravel (size range : 25-40 mm) and a bauxite solid concentration of 500 g/l, during 5 min, which amounts to the residence time in the Washing Drum. This test will be referred to as "WD Test with Gravel".

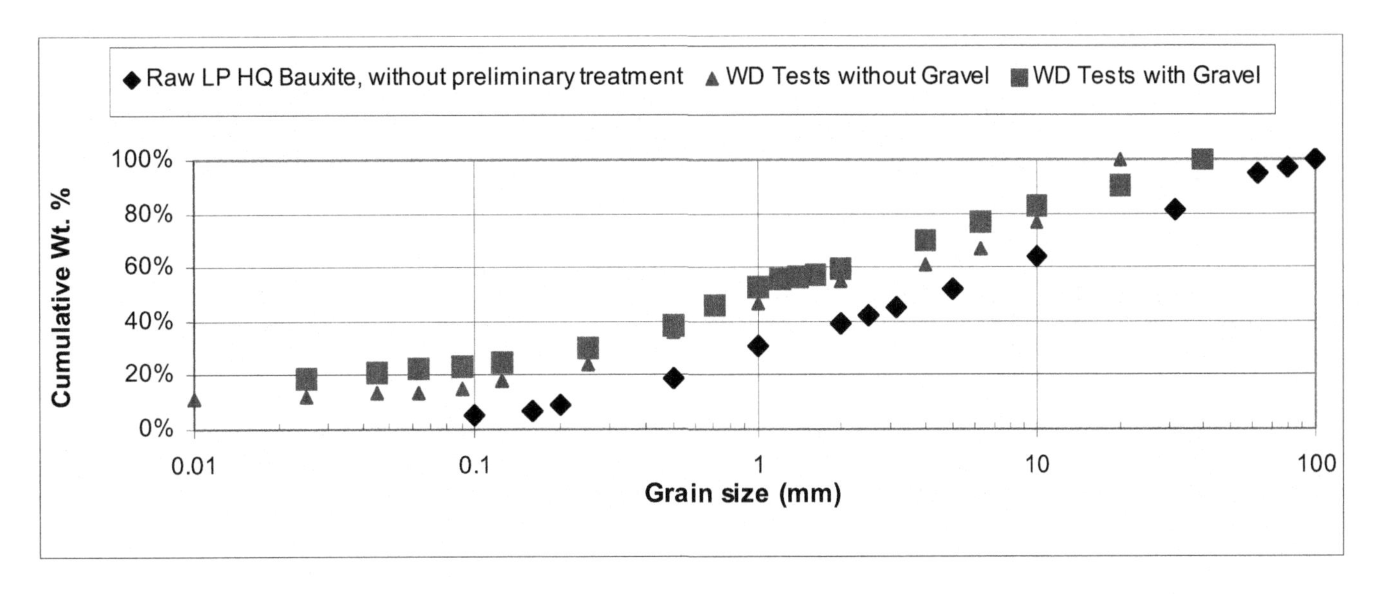

Figure 3. Grain Size Distribution without preliminary treatment and after treatment in Washing Drum (with and without gravel).

A considerable effect of the washing drum on the particle size distribution is observed.
In particular, compared to raw LP HQ Bauxite without treatment, it can be seen that grain sizes are much finer after treatment in washing drum.

Moreover, the attrition effect, introduced by the milling effect of the gravels in the ''WD Test with Gravel'' is further affecting the Particle Size Distribution, as can be seen from comparison with the ''WD Tests without Gravel''. More precisely, more fine particles are obtained in the ''WD Test with Gravel'' compared to the ''WD Test without Gravel''.

Noteworthy is that in the WD Tests with Gravel, mass fractions in following size ranges : [2mm – 1.6mm], [1.6mm – 1.4mm] and [1.4 mm – 1.2 mm] are very small.

It should be emphasized that it seems that attrition effect (in the present case, WD Test with Gravel) tends to :

- decrease the mass fraction of coarse grains
- increase significantly the mass fraction of the very fine particles (< 63 µm).

This is illustrated in Table 1, in which some major results obtained are presented.

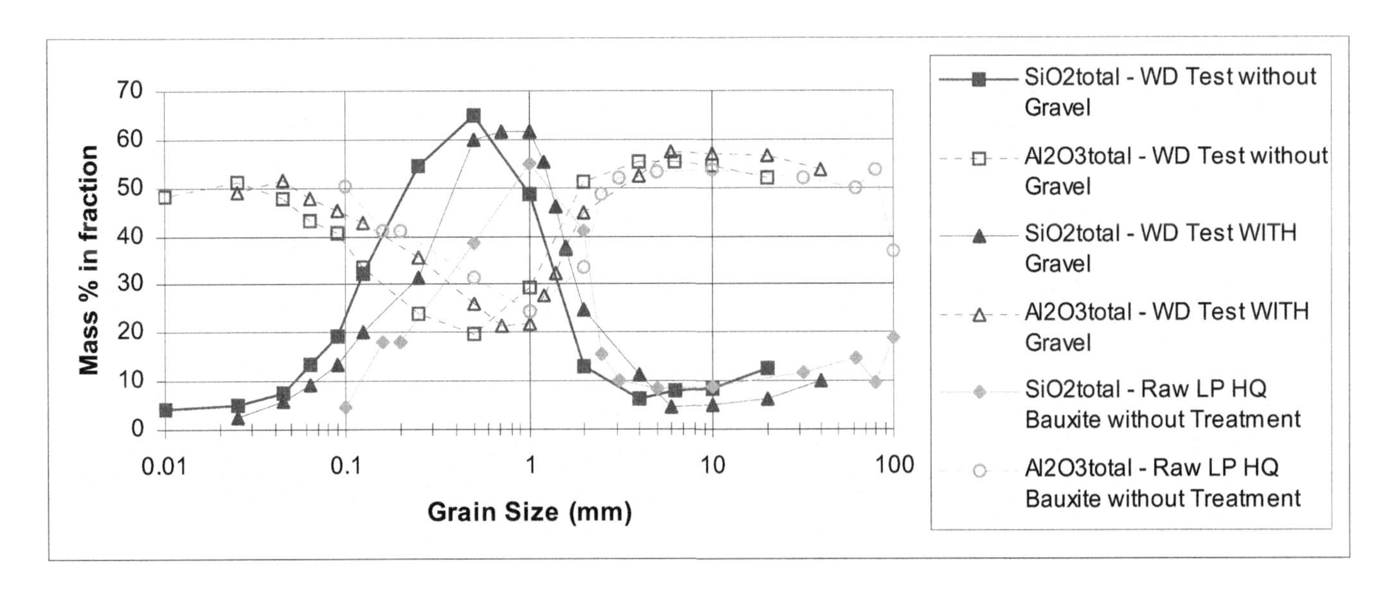

Figure 4. Total SiO_2 and Total Al_2O_3 contents in Size Ranges of samples investigated.

Table 1. Effect of preliminary treatment on Particle Size Distribution.			
	Raw LP HQ Bauxite, without preliminary treatment	Washing Drum Test WITHOUT Gravel	Washing Drum Test WITH Gravel
Mass fraction > 10 mm	48 %	33 %	17 %
Mass fraction > 2 mm	61 %	53 %	41 %
Mass fraction < 63 µm	6 %	14 %	23 %

In this context, it should be noted that almost all quartz is accumulated in intermediate fractions, typically the 0.063 mm – 2 mm fraction, as can be seen from Figure 4.
Hence, the aim is to recover the coarse grains (> 2 mm) and the very fine particles (< 100 µm).

But attention should be paid to the sum of the mass of these 2 fractions (coarse + fine grain sizes). Indeed, from analysis of results obtained on samples of following tests :
- Raw LP HQ Bauxite without preliminary treatment
- WD Test without gravel
- WD Test with gravel

it appears that :
1. if only the grain size fraction > 2 mm is recovered : more bauxite is recovered without treatment, or when the effect of attrition during pre-treatment is not too strong.
2. if all coarse and fine grain size fraction are recovered (> 2 mm and < 100 µm) : the intensity of attrition during pre-treatment will almost not affect quantities recovered, but will significantly affect the quality of beneficiation. This last point will be described in detail in the next paragraphs.

Result of WD Test with Gravel

From results given in Tables 2, 3 and 4, it appears that an optimal cut size is 2.0 mm. Indeed :
→ increasing this Cut Size to values higher than 2 mm decreases sharply % of mass recovered, without improving significantly the recovered bauxite quality (i.e. quality of beneficiation = Al2O3/SiO2total ratio).
→ decreasing the cut size below 2.0 mm has a significant detrimental effect on recovered bauxite quality.

Taking into account the analysis result of grain fractions of the global sample, it may be concluded that an optimal cut size for the wet screening process is approximately 2.0 mm, probably 1.8 mm.

The 2 other samples used for characterization and beneficiation tests (Raw LP HQC Bauxite without preliminary treatment ; WD Test without Gravel) confirms this finding.

Table 2. Determination of Recovered Quantities as a Function of Separated Granulometric Fraction (WD Test with Gravel).							
	Separation of fraction < 4 mm	Separation of fraction < 2 mm	Separation of fraction < 1.6 mm	Separation of fraction < 1.4 mm	Separation of fraction < 1.2 mm	Separation of fraction < 1.0 mm	Separation of fractions < 1.6 mm > 90 µm
% mass recovered	41.0%	42.5%	43.6%	44.0%	47.6%	54.1%	67.0%
% Al2O3 recovered	51.2%	52.8%	53.7%	54.0%	56.2%	59.5%	79.9%
% TiO2 recovered	24.5%	25.3%	25.7%	25.9%	27.2%	29.3%	66.7%
% Fe2O3 recovered	29.6%	30.5%	31.1%	31.2%	32.9%	35.4%	78.3%
% SiO2 total recovered	14.6%	16.2%	18.0%	18.8%	27.6%	45.3%	22.0%

Table 3. Characteristics of Beneficiated Bauxite as a function of separated granulometric fraction (WD Test with Gravel).								
	Global original bauxite sample	Beneficiated bauxite after separation of following fractions from original bauxite						
	global	<4mm	<2mm	<1.6mm	<1.4mm	<1.2mm	<1mm	<2mm >90µm
LOI (%)	23.9%	29.7%	29.5%	29.3%	29.2%	28.1%	26.1%	28.6%
Al2O3 (%)	44.0%	54.9%	54.6%	54.1%	53.9%	51.9%	48.3%	52.4%
TiO2 (%)	1.1%	0.7%	0.7%	0.7%	0.7%	0.6%	0.6%	1.1%
Fe2O3 (%)	8.5%	6.1%	6.1%	6.0%	6.0%	5.8%	5.5%	9.9%
SiO2 total (%)	22.6%	8.0%	8.6%	9.4%	9.7%	13.1%	19.0%	7.4%
TOTAL (%)	100.0%	99.5%	99.5%	99.5%	99.5%	99.6%	99.5%	99.5%
Al2O3/SiO2total	1.9	6.8	6.3	5.8	5.6	4.0	2.5	7.0

Results for Samples of Fine Grain Fractions (at indicated sizes) after Tests in HYDROCYLONES and GRAVITY SORTING EQUIPMENT (SHAKING TABLES, SPIRALS)

Objective of these tests

The aim of these additional tests was to evaluate the possibility to recover the fine grain fractions by an alternative beneficiation process using basic but robust equipment (Hydro-cyclone and/or Shaking Table, Spirals)

Test work carried out

- Selection of a representative sample from the originally homogenized sample of HQ Bauxite from LP
- WD Test with gravel
- Screening of the suspension on order to obtain grain sizes specified below
- Producing a fraction "0.09 mm to 0.5 mm" for a Shaking Table test
- Producing a fraction "0.5 mm to 1.0 mm" for a Shaking Table test
- Hydro-cyclone test (type RWS 105 IV at 2,5 bar) with fraction < 0,125mm
- Drying of all produced fractions for chemical analysis at Alcan-Gardanne Technical Center.

Interpretation of results

All results obtained are presented in Tables 4, 5, 6 and 7.

Results can be summarized as follows :

1. Both fractions ([0.5 – 1.0 mm] and [0.09 – 0.5 mm]) tested on Shaking Table are not interesting (bauxite without economical value because : [Al2O3total] < 35 % and [SiO2total] > 34 %).
2. Within these fractions, with the application of a sorting process (here : Shaking Tables) it proves to be impossible to extract a valuable fraction.
3. In Hydro-cyclone process of fraction < 125 µm, the overflow of the cyclone can be considered as bauxite of economical value, but given the fact that this overflow is necessarily very diluted, the economic validity of the process is questionable (given the high dewatering costs) and should be studied carefully.
4. Noteworthy is the chemical composition of the very fine particles recovered in the overflow of the hydro-cyclone : [Al2O3total] > 46 % , [SiO2total] < 3 %, and [Fe2O3] = 22 %; in other words a very low silica content and a very high iron content is obtained (compared to "conventional Los Pijiguaos bauxite").
5. It should be stated that the feed of the hydro-cyclone (fraction < 125 µm) can be considered as a bauxite of good quality (compared to bauxite quality treated at Bauxilum Puerto Ordaz refinery).

The hydro-cyclone test was carried out with AKW's RWS 105 IV Apex 4mm equipment using as a feed slurry bauxite of fraction < 125 µm. The feed to the hydro-cyclone was diluted significantly for classification purposes (so as to obtain a sharp cut). Table 5 presents results obtained and Table 6 presents chemical characteristics of feed, underflow and overflow.

Table 4. Hydrocyclone test with AKW's RWS 105 IV Apex 4mm using fraction < 125 µm.

	Concentration = (g/l)	Yield (%)	Volume split (%)
Feed	47,7		
Overflow	25,3	48,6	96,8
Underflow	732,0		

Table 5. Chemical characteristics of feed, underflow and overflow.

		LOI (%)	Al2O3 (%)	TiO2 (%)	Fe2O3 (%)	SiO2 t (%)
Overflow		26.2	46.1	2.38	22.4	2.67
Underflow		28.1	50.9	1.51	9.9	8.01
Feed		27.2	48.8	1.98	16.3	5.18

Table 6. Shaking table test : with fraction 0.09 – 0.5 mm.

	% mass in fraction	LOI (%)	Al2O3 (%)	TiO2 (%)	Fe2O3 (%)	SiO2 t (%)
Feed	100	11.9	22.9	0.48	4.5	60.92
T 1-3	22.8	15.3	29.1	0.39	4.4	50.3
T 4-5	56,9	11.1	21.8	0.37	4.1	62.11
T 6	13,2	7.9	15.5	0.38	3.9	72.22
T 7	7,1	6.6	13.3	1.79	6.8	71.18

Table 7. Shaking table test : with fraction 0.5 – 1.0 mm.

	% mass in fraction	LOI (%)	Al2O3 (%)	TiO2 (%)	Fe2O3 (%)	SiO2 t (%)
Feed	100	18.5	34.1	2.15	8.4	33.89
T 1-3	30	22.1	38.6	0.52	5.3	33.44
T 4-5	52,3	19.2	34.8	0.51	5.1	38.34
T 6	12,3	16	29.5	0.75	7.2	43.81
T 7	5,4	4	9.1	18.82	39	27.52

General Conclusions

1. With all 3 samples used for the characterization and beneficiation tests (Raw LP HQ Bauxite without preliminary treatment ; WD Test without Gravel ; WD Test with Gravel), it has been found that almost all quartz is accumulated in intermediate fractions, typically the [0.063 mm – 2 mm] fraction.

2. A Cut Size of 2 mm for the screen placed after the Washing Drum appears to be the optimal dimension to carry out the wet-screening step.

3. A simple screening (recovery of all particles > 2 mm) makes it possible to recover 65 to 70 % of Al_2O_3 and to separate more than 75 % of quartz of initial HQ Bauxite of LP ; the quantity of Al_2O_3 recovered strongly depends on the pre-treatment process.

4. More precisely, if the pre-treatment process (Washing Drum) is not too attritive, more coarse grains and less very fine particles are obtained, making it possible to recover only particles > 2 mm with a correct quality of beneficiation (Al2O3total/SiO2total ratio around 5.0) and with a Al_2O_3 recovery higher than 65 %.

5. In order to reach a Yield of Al_2O_3 greater than 80 %, it is necessary to recover the fine fraction of particles < 100 μm.

6. Points 4 and 5 are clearly demonstrated in Tables 8 and 9 given below.

These Tables summarize all results regarding the beneficiation tests obtained until now.

It can be concluded that the ''intensity of attrition'' (= tuning of grinding) introduced during pre-treatment process in the Washing Drum will have a very significant effect, both with respect to the quality of beneficiation as well as the % Al_2O_3 recovered if only the fraction > 2 mm is recovered.

More precisely, 2 cases are to be envisaged :

i. If the economic study proves that only the fraction > 2 mm is to be recovered, then the ''intensity of attrition'' undergone by the raw bauxite in the Washing Drum has to be tuned carefully. Indeed, when this intensity increases, the quality of beneficiation increases but the total mass of Al_2O_3 (or bauxite) recovered decreases.

ii. If the economic study proves that both fractions > 2 mm and < 100 μm are to be recovered, the pre-treatment in the Washing Drum seems to have a beneficial effect both on the quality of beneficiation as well as on the % Al_2O_3 (or bauxite) recovered. In addition, if ''intensity of attrition'' increases, it has been found that quality of beneficiation is improved, for identical recoveries.

7. In intermediate fractions comprised in the [0.063 mm – 2 mm] range, the various sorting tests (with the Shaking Tables) have proved that it is not possible to extract a valuable grain size fraction.

Table 8. Summary of the 3 tests. Result of beneficiation by separation of fraction < 2 mm

	Separation of fraction < 2 mm		
	WD Test with gravel	WD Test without gravel	Without Treatment
% mass recovered	42.5 %	53.4 %	61.0 %
% Al2O3 recovered	52.8 %	65.3 %	70.1 %
Al2O3	54.6 %	53.1 %	51.2 %
SiO2	8.6 %	10.6 %	11.8 %
Fe2O3	6.1 %	6.7 %	8.4 %
Quality of beneficiation : Al2O3/SiO2total	6.3	5.0	4.3

Table 9. Summary of the 3 tests. Result of beneficiation by separation of fraction < 2 mm and > 0.09 mm

	Separation of fraction < 2 mm and > 0.09 mm		
	WD Test with gravel	WD Test without gravel	Without Treatment
% mass recovered	67.0 %	67.3 %	66.5 %
% Al2O3 recovered	79.9 %	80.8 %	76.4 %
Al2O3	52.4 %	52.1 %	51.1 %
SiO2	7.4 %	9.4 %	11.2 %
Fe2O3	9.9 %	8.9 %	9.0 %
Quality of beneficiation : Al2O3/SiO2total	7.1	5.5	4.6

References

"The Variation in the Ore Grade Distribution in the Los Pijiguaos Bauxite Deposit, Venezuela : a Computer Aided Model", U. Happel, J. Hausberg, F.M. Meyer and N. Mariño, ICSOBA, Vol 26, 1999, No. 30, 12th International Symposium of ICSOBA.

"Pijiguaos Bauxite Treatment in a Low Temperature Digestion Plant", S. Escalona, C.E. Suarez, A. Sarmentero, J.M. .Gomez and A. Monge, Light Metals 1986.

Light Metals 2006 *Edited by Travis J. Galloway* ***TMS (The Minerals, Metals & Materials Society), 2006***

THIN LAYER CHROMATOGRAPHY OF BAUXITE: INSTRUMENTAL CHARACTERISATION AND QUANTIFICATION BY SCANNING DENSITOMETRY

Mohamed Najar P.A and K.V.Ramana Rao
Jawaharlal Nehru Aluminium Research Development and Design Centre
Nagpur - 440023, India

Key Words: TLC, Scanning densitometry, Characterisation, Bauxite

Abstract

Thin layer chromatography (TLC) principles were successfully used for the characterisation of various elements present in bauxites. The chromatographic characteristics have been evaluated on silica gel and cellulose layers developed with double distilled water and aqueous sodium chloride solutions. The densitometric evaluations of chromatograms of Al, Fe, Ti and Si in bauxites have shown a percentage recovery of 98 ±2. The method developed was successfully extended to the characterisation of organic impurities (< 0.2) in bauxites by coupling TLC procedures with FTIR spectroscopy. The characteristic functional groups for some organic compounds were identified in the IR spectra. The advantages of these methods over the conventional methods of characterisation were also presented.

Introduction

Alumina refineries consider bauxite as the most economical ore for the commercial production of aluminium metal. Alumina (Al_2O_3), iron oxide (Fe_2O_3), titania (TiO_2), silica (SiO_2) and loss on ignition (LOI) are the five major constituents of bauxite that determine the quality and its metallurgical and non-metallurgical applications. Also, the determination of minor constituents like CaO, P_2O_5, V_2O_5, SO_3, $C_{\text{-organics}}$ and fluorine in certain bauxites is some times important. Since the range of variation of major chemical constituents of bauxite received from various sources are different its determination and technological evaluation is necessary to use such deposits in Bayer plants and accordingly for the implementation of suitable technology. Conventional wet chemical analysis and instrumental methods such as XRD/XRF are commonly used for the qualitative and quantitative determination of bauxite constituents. This study has focused in particular on the practical application of thin layer chromatography for the detection, separation and determination of major constituents in bauxite.

Two recent studies on the TLC analysis of bauxite revealed the possibility of separating the co-existing Al-Fe-Ti in bauxite on silica gel G layers using micellar mobile phase [1] and on silica gel H layers with mixture of sodium chloride and formic acid containing mobile phases [2]. Since the chemical composition of bauxite varies with their geological origin, a basic chromatographic study has become necessary for realizing an ideal chromatographic system capable of achieving reproducible separation of Al-Fe-Ti and subsequent quantitative analysis of these constituents for rapid analysis of bauxite.

Scanning densitometry is the most preferred and accurate quantitative tool for evaluating thin layer chromatograms [3,4]. The chromatograms of Al^{3+}, Fe^{2+} Ti^{4+} and Si^{4+} were obtained at the optimimised chromatographic conditions and quantitative estimation of individual constituents is evaluated from the corresponding peak area measurements. The repeated trials indicated a percentage error of < 2% in comparison with the wet analytical data of the constituents. The Bayer liquor samples were directly loaded on the TLC plates coated with silica gel H and Silica gel H_{254} and the constituents were separated with selected mobile phases. The separated constituents were subjected for FTIR analysis.

This communication aims two fold applications i.e. development of a reliable chromatographic route for the rapid analysis of major constituents of bauxite with acceptable error limits, and exploring the possibility of using combination of TLC with FTIR for the characterization of organic species in bauxite matrix.

Experimental:

The thin layer chromatographic analysis of bauxite provides unsurpassed visual clarity through simultaneous evaluation of the three major metallic elements and silicon in bauxite. The method provides separation of aluminium, iron, and titanium on silica gel H and the detection of silicon on glass plates coated with microcrystalline cellulose.

Chromatographic trials were performed with glass plates (*7 x 2.5cm², 10 x 10cm², 15 x 3cm²*) coated with suitable adsorbents. Systronics (India) pH meter model 335 was used for all pH measurements. The photometric measurements were carried out with a Dual Wavelength Scanning Densitometer equipped with Quanta Scan Software (Model CS- 9301PC). IR spectra were recorded with FTIR spectrometer supported by EZ Omnic software (Model AVATAR 320, Thermo Nicolet Corporation, USA) for characterisation. UV detection of spots was achieved by fluorescence viewing cabinet and UV lamp (Spectroline, USA, Model M-10 and ENF-280/FE).

1% test samples of bauxite were prepared in acid mixture or by alkaline fusion according to the requirements. 1% alkali and other salt solutions were prepared in double distilled water.1% reference solutions of aluminium, iron, titanium, silicon, were prepared in acid medium.

In order to evaluate the colour characterization of various constituents of bauxite, reference standards of Al^{3+}, Fe^{2+} Ti^{4+} and Si^{4+} were prepared in aqueous (salt solutions) medium. The retardation factors (R_F) of the metallic constituents in bauxite were confirmed with the R_F obtained for reference samples. Approximately 1-2µl test solutions were applied on each chromatographic plate by using micropipettes. The coloured zones of different constituents in bauxite were identified by spraying chromogenic reagents. Al^{3+} was detected by aqueous solution of 0.1% aluminon (tri-ammonium aurin tri-carboxilate, $C_{22}H_{23}N_3O_9$). Saturated potassium ferrocyanide ($[K_4Fe(CN)_6]$) and 0.5% tiron (Yoes reagent,$C_6H_4Na_2O_8S_2$) were used for the detection of Fe^{2+} and Ti^{4+}. Si^{4+} was detected by spraying solutions of 1% aqueous sodium molybdate ($Na_2MoO_4.2H_2O$). Appropriate dilution of reagents is made according to the requirement.

Organic species were self detected on wet silica gel H plates as faint brown spot and appeared fluorescent yellow while exposure on UV light on both silica gel H and silica gel H_{254}(Silica gel H containing fluorescent indicator).

Chromatographic separation of organic species was achieved by loading 3µL of Bayer liquor sample on TLC plates made of silica gel H and silica gel H_{254}.The plates were developed separately with 1% sodium hydroxide and double distilled water. The chromatogram prepared in duplicate viz. analyte and trial plates, for the detection of organic constituents was compared with that obtained for a mixture comprising of sodium salts of carboxylic acids, amines and alcohols. The qualitative detection and spot identification was achieved by comparing R_F values of separated species. The position of spots was marked on the trial plate for identifying corresponding location of spots on analyte plates for FTIR analysis. A portion of adsorbent containing respective spots viz. SGHF COMP'1, COMP'2,COMP'3 of organic constituents were scraped out from the analyte plate and a conventional KBr pellet was prepared for recording FTIR spectra.

A schematic of the chromatographic procedure involved in bauxite analysis is shown in Fig.1.

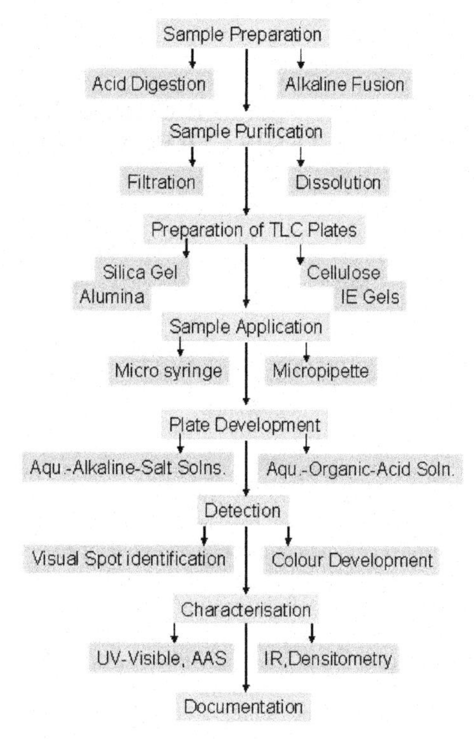

Fig.1. Schematic representation of experimental procedure

Result and Discussions

The retardation factor R_F (R_L+R_T/2, where R_L is the R_F value of leading front and R_T is the R_F value of trailing front of the spot) of bauxite constituents in specific chromatographic systems were studied by trial and error on cellulose and silica gel layers. The migration trend of Al^{3+}, Fe^{2+} and Ti^{4+} on silica gel H layers developed with mobile phases containing mixture of alkali salt solutions and formic acid (HCOOH) has been studied as function of pH for realizing optimum mobile phase combinations. Binary separations of Ti-Al and Ti-Fe are achieved in these systems and their migration trend from the point of sample application is shown in Fig.2.

The selected chromatographic systems are further investigated in terms accuracy and reproducibility by changing chromatographic parameters such as sample concentration, buffer addition in mobile phase and impregnation of stationary phases. These studies resulted the identification of a set of mobile phases capable of resulting ternary separation of Al-Fe-Ti and detection of Si^{4+} in bauxite matrix as shown in Table -1.

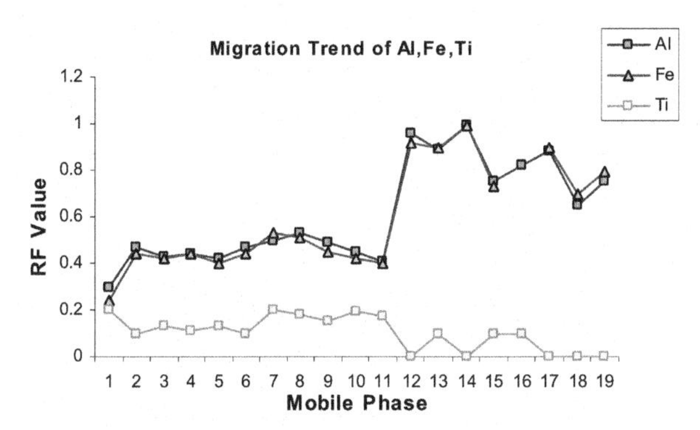

(1:1v/v Mixture of alkali salt solutions 10% + 1% HCOOH (1-11) and 10% HCOOH (12-19)

Fig.2 Variation of R_F with respect to change in concentration of HCOOH

Silicon in bauxite is detected separately on microcrystalline cellulose layers developed with dilute hydrochloric acid (1-5% v/v) and appeared highly compact and stable yellow spot on exposure to sodium molybdate.

Tabl-1 Mobile phase combinations selected for the study

Silica gel H plain or impregnated with 1% sodium formate	
10% Formic acid + 10% Sodium chloride	(3:7,2:8) v/v
10% Formic acid + 10% Barium chloride	(1:9,3:7) v/v
10% Formic acid + 10% Potassium bromide	(1:9,1:1) v/v
10% Formic acid + 10% Sodium sulphate	(1:9,1:1) v/v
10% Formic acid + 10% Potassium chloride	(1:9,1:1) v/v
10% Formic acid + 10% Sodium nitrate	(1:9,3:7) v/v
10%Formic acid + 10% sodium acetate	(1:9,3:7) v/v
1% Formic acid + 1% Ammonium chloride	1:9 v/v
1% Formic acid + 1% Sodium hydroxide	1:9 v/v
2.5% Formic acid + Saturated Sodium chloride	1:9 v/v
Hydrochloric acid	1-5% v/v
Double distilled water	

Quantitative Studies

TLC- Scanning densitometric evaluation of Al^{3+}, Fe^{2+}, Ti^{4+} in bauxite was carried out by spotting bauxite of known composition on TLC plates (10 x 10 cm^2) coated with silica gel H (0.3mm thickness) and developed with double distilled water containing and mixture of 10% solutions of sodium chloride and formic acid (2:8v/v, neutral pH maintained with acetate buffer) in 1:1v/v. For Si^{4+}, microcrystalline cellulose coated plates were used and developed with 5% HCl. Trial plates for each samples loaded were run simultaneously to locate the exact position of analyte spots. After chromatographic development, the plates were partially dried in room temperature and the spots were detected by spraying 0.01% aluminon for Al^{3+}, 0.10% potassium ferrocyanide for Fe^{2+} and 0.5% tiron for Ti^{4+}. Si^{4+} was detected on microcrystalline cellulose developed with 5% hydrochloric acid by spraying 1% sodium molybdate. The individual ions (viz. Al^{3+}, Fe^{2+}, Ti^{4+}, Si^{4+}) were detected on separate plates and the spots were subjected for densitometry at the respective λ_{max} values (530-520 nm for Al, 628-520 nm for Fe, 410-400 nm for Ti and Si) in ascending and descending concentrations. The recovery percentage was estimated to be 98% ± 2% for Al^{3+}, Fe^{2+}, Ti^{4+}, Si^{4+} with reference to the data recorded by wet chemical analysis.

Table-2 Densitometric analysis of bauxite constituents

Densitometric Evaluation of Al, Fe, Ti and Si					
Bauxite Constituents	%	Error	Mean Error	Sample Details λmax (nm)	
Al_2O_3	65.33	1.80		Al	527nm
	65.09	0.50	0.86	Fe	625nm
	65.66	0.30		Ti	405nm
Fe_2O_3	4.68	3.50		Si	410nm
	4.51	0.22	1.46	Concentration : 1% Aqueous Sample Volume: 2μl for Al, Fe and Ti and 3μl for Si	
	4.49	0.66			
TiO_2	2.80	1.06			
	2.78	1.70	1.15		
	2.81	0.70			
SiO_2	1.01	0.98			
	0.99	2.94	1.94		
	1.04	1.90			
Wet analytical data of representative bauxite					
Constituents	Al_2O_3	Fe_2O_3	TiO_2	SiO_2	LOI
%	65.45	4.52	2.83	1.02	24.96

Twenty five bauxite samples of different geological origin has been tested for error determination

TLC-FTIR Studies:

The TLC-FTIR analysis of organic constituents in Bayer liquor provides some insight on the possible characterisation of selected organic functional groups in Bayer liquor samples. The FTIR spectra recorded (Fig.3) revealed the TLC separation of some important organic constituents in Bayer liquor on silica gel H layers. Efforts have been made for the identification of functional groups from their characteristic peak values and subsequently the separated constituents were tried to identify from the list of common organic species in bauxite [5].

Preliminary studies indicated the presence of compounds with O-H and N-H stretching bands (3400-3100cm^{-1}) in the liquor. The alkene and alkyne C-H (3100-3000cm^{-1}) compounds were also observed and the triple bond stretching (2400-2200cm^{-1}) and carbonyl bands (1800-1700cm^{-1}) become prominent after the TLC separation. C=C stretching bands and C-O bond stretch at 1200-1100cm^{-1} were detected for the spots of separated constituents at 1650 -1600cm^{-1}. The detection of different peaks for the constituent spots (SGHF`COMP`2S and SGHF LIQ`S) separated by TLC methods have indicated the possibilities of some organic species.

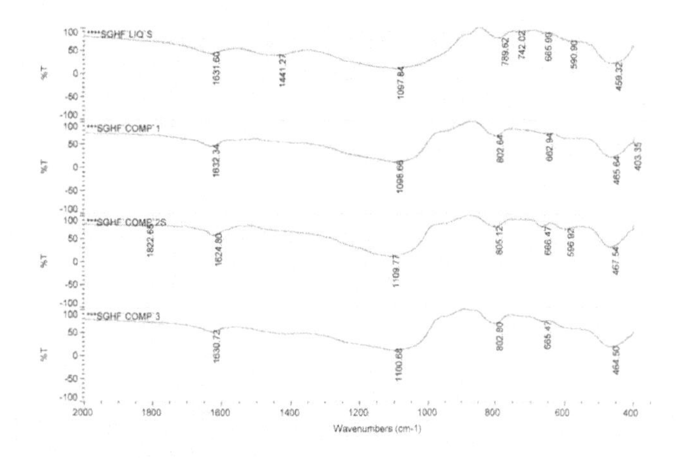

Fig -3 FTIR Spectra of organic species in Bayer liquor separated on Silica Gel H_{254}

Conclusions:

The present chromatographic studies on bauxite revealed the possibility of a rapid quantitative methodology for both qualitative and quantitative estimations of major constituents and trace elements in bauxite mineral and Bayer liquors with less than 2% measurement error. The investigations also showed that TLC procedure can be successfully modified and coupled with suitable instrumentation methods for the qualitative and quantitative determination of trace elements in ore samples. The fingerprint region of the FTIR spectrum requires a thorough investigation for more precise and accurate judging of functional groups and presence of various organic species in Bayer liquor.

Further studies are in progress for the identification of ideal chromatographic systems for the selective separation of the organic species in Bayer liquor samples and specific chromogenic reagents for compound identification prior to FTIR analysis.

Acknowledgements:

The authors thank Dr. J. Mukhopadhyay, Director JNARDDC for his constant encouragement, useful suggestions and permission for publishing the work. Science & Technology Wing, Ministry of Mines, GOI is acknowledged for financial support to the project on "*Development of rapid analytical procedures for bauxite and semi-quantitative analysis of scrap aluminium*".

References:

1. A. Mohammad and S. Hena, "Inducement of a New Micellar Mobile Phase for Thin-Layer Separation and Quantitative Estimation of Aluminium (III) in Bauxite with Preliminary Separation from Iron(III) and Titanium(IV)", *Separation Science and Technology*, 39 (2004)2731-2750.

2. M. Najar, J. U. Jeurkar and K.V.R. Rao, "Thin Layer Chromatographic Study of Bauxite and Quantitative Estimation of Co-Existing Al^{3+}, Fe^{2+} and Ti^{4+},*Chinese Journal of Chromatography*, 23(5) (2005)555-561.

3. J.Sherma and B. Fried, (Eds.) *Hand Book of Thin Layer Chromatography* (New York, NY, Marcel Dekker, Inc.1991) 55.

4. J.C.Touchstone, *"Practice of Thin Layer Chromatography"* (Wiley and Sons, Inc. New York NY, 1992).

5. K.V. Ramana Rao, P.A. Mohamed Najar, R.N. Goyal and K.S. Raju, "Spectroscopic Detection of Hydrated Phases of Alumina in Bauxite Mineral" Proceedings *of the International Seminar on Mineral Processing Technology* (MPT-2004) 214-220, 2004.

ALUMINA & BAUXITE

Bayer Digestion Technology

SESSION CHAIR
Steve Healy
Alcan Queensland R&D Centre
Brisbane, QLD, Australia

New Technology for Digestion of Bauxites

Robert Kelly[1], Mark Edwards[1], Dirk Deboer[1], Peter McIntosh[1]
[1]Hatch Associates Ltd, 144 Stirling Street, Perth, Western Australia 6000, Australia

Keywords: Bayer Process, Digestion, Integrated Digestion Technology, Tube Heaters, Jacketed Pipe Heaters, Tube Digesters, Cost Comparisons

Abstract

A number of plant design technologies are available for digestion of bauxites. Typically the major conceptual differences in these technologies revolve around the design of the heating section employed to heat liquor and bauxite slurry to the desired digestion temperature, and in the design of the equipment for extraction of alumina from the bauxite. However, the energy efficiency, operating and maintenance characteristics are determined by the overall design of the unit.

This paper reviews plant design technologies commonly used for both low and high temperature digestion. It then presents new designs that utilise tubular heating technology developed by Hatch Associates Ltd (Hatch). The safety, technical, operational and cost advantages are discussed.

Introduction

This paper addresses innovations in all three of the fundamental process steps in Bayer Digestion (i.e. Heating, Digestion and Flash Cooling) for each of the key process design alternatives:

- Dual Stream versus Single Stream bauxite/spent liquor configurations.
- Low Temperature versus High Temperature Digestion.

Traditional Plant Design Technologies

Heating

Description Of Technologies The most commonly used traditional technology is conventional Shell and Tube Heat exchangers (S&T HEX). These are characterised by having several hundred relatively small diameter (25-38 mm) special thin walled tubes in bundles held in place within horizontally or vertically aligned pressure vessels by tube plates.

Less common is the use of steam jackets inside agitated autoclaves. These are characterised by having multiple registers of vertically oriented steam heating tubes within vertically oriented pressure vessels.

Application In Dual Stream or Split Flow Digestion (spent liquor heated separately), S&T HEXs can be used successfully up to approximately 150°C for spent liquor if only conventional materials are used due to risk of caustic embrittlement. If exotic materials such as Nickel, Chrome and Molybdenum alloys are used, these heaters can be used up to approximately 250°C. At temperatures above 250°C, heater-descaling operations become increasingly onerous. Existing plants operating at about 250°C and with high caustic concentrations, already find that high temperature recuperative heaters need to be cleaned in cycles that are counted in days, rather than weeks. At high digestion temperatures, these cleaning cycles would become even shorter, while the combined effects of increased vapour pressure and increased heater resistance would place increasing demands on equipment design.

In Single Stream or Mixed Flow digestion (bauxite slurry and spent liquor heated together). S&T HEXs are generally only used up to approximately 200°C due to high scaling rates and problems with tube blockages.

Autoclaves are generally only used in Single Stream applications, but can be used up to the highest digestion temperatures normally required.

Advantages The key advantage of S&T HEXs is the large heating area per unit volume and hence high heat transfer rates in a clean condition. This can translate into apparent lower capital cost before provision of sparing is taken into account for cleaning and maintenance. A secondary advantage is that the technology is well known with many suppliers of "off-the-shelf" solutions.

Disadvantages The critical disadvantages of S&T HEXs are the usually multiplying effects of high scaling rates, propensity for tube blockages and specifically difficulties with cleaning and maintenance, particularly in high temperature applications.

Autoclaves are particularly prone to high rates of fouling due to relatively low rates of shear on the shell (slurry) side resulting in low Heat Transfer Coefficients typically only 500-750 $W/m^2{}^\circ C$.

Digestion

Description of Traditional Technology The most commonly used traditional technology is Digester Pressure Vessels (either agitated or unagitated.). these are characterized by being vertically aligned thick-walled pressure vessels designed to stringent pressure vessel design codes.

Application Digester Pressure Vessels can be used in both Dual Stream and Single Stream applications up to the higher digestion temperatures normally required.

Advantages Digester Pressure Vessels have two key advantages. Firstly, the technology is well known with many suppliers of "off-the-shelf" solutions. More importantly, for digestion holding times more than 5-10 minutes, they provide the only really practical option due to their larger specific volumes (volume/shell area).

Disadvantages The key process disadvantages of Digester Pressure Vessels is often flow short-circuiting, which impacts extraction adversely, and potentially sudden blockages from scale shifting occurs.

In high-pressure plants, scale deposited at the high temperature end is usually a hard titanate, which requires removal by mechanical means. In digester vessels or autoclaves, scale removal requires vessels be taken out of service for extended periods. The impact of such de-scaling on production is significant and requires provision of spare vessels. This in turn requires extensive slurry manifolding and valving so that individual vessels can be by-passed and isolated, adding to cost and the spatial requirements of the plant.

Flash Cooling

Extensive erosion of the Flash Tank underflow piping has long been an accepted feature of digestion plant operation and maintenance. As digestion plants were de-bottlenecked and flows increased, erosion and Flash Tank scaling increased exponentially. At the same time, heater condensate quality declined which indicated that increasing quantities of slurry particles were carried over with the vapour. This was also evident in increased heater shell side fouling which is difficult to remove.

Computer analysis using advanced two-phase analyses identified three major causes as follows:

- Highly convoluted pipe work between Flash Tanks.
- Lack of Flash Tank level controls.
- Top entry of the two-phase transfer line into the downstream flash tank.

It was found that the underflow pipe work, with its convoluted configurations and extensive manifolding causes vapour formation well upstream. Without established levels in the Flash Tanks, it is not possible to determine the two-phase flow regime in the pipe work. The erosive effects of this two-phase flow on valving and pipe work are severe. Despite extensive hard-facing being applied, large sections of pipe work and valves require frequent replacement.

New Plant Design Technologies

Jacketed Pipe Heating

Description To overcome the limitations and disadvantages of S&T HEXs, particularly for high temperature applications, "tube-in-tube" heaters have been developed. The earliest designs such as that still in operation at Stade in Germany and Zhengzhou in China, used either one or two slurry tubes inside an outer tube carrying the heating medium of vapour from Flash Tank, live steam or molten salt.

A significantly improved design has been developed using three or four slurry tubes inside the outer tube or jacket carrying the heating medium – hereinafter referred to as Jacketed Pipe Heaters (JPHs). A critical feature of the design is that JPHs are unitised allowing the number of JPH tubes to be matched to associated Flash Tanks to facilitate optimization of the Flash Train profile.

JPH design utilises thick walled 150 mm pipes rather than thin walled tubes, and hence is not subject to blockages, tube ruptures, high wear or other issues associated with traditional S&T HEXs. They are constructed of standard piping making them simple and cost effective to construct.

Tube velocity is selected to provide best compromise between erosion, scaling rates and pump pressures. Acid cleaning frequency of even high temperature JPHs is reduced to typically every 2000 hours. This is possible due to the relatively slow and steady degradation of Heat Transfer Co-efficient (HTC) over time. At the point of cleaning, the overall HTC of the heaters has reduced to typically 50-70% of its original 'clean' value.

A typical digestion unit requires a multiple number of JPH tiers operating in parallel. JPHs are arranged in banks about 50 m long, along which the tubes are run back and forth. At every return bend, access is provided for high-pressure water jet cleaning. It is not necessary to clean each JPH train over its full length during each de-scaling operation. Tube sections can be accessed individually and de-scaled according to need as not all parts of a JPH Train scale at the same rate.

Application JPH technology has particular application in single stream high temperature processes where S&T HEXs are not viable due to high scaling rates and risk of tube blockages. JPHs can be used right up to the highest digestion temperatures normally required within economical piping design (Class 600) and pumping constraints, i.e. approximately 280°C with live steam and/or molten salt.

However, JPHs can also be used advantageously in both single stream applications and dual stream low temperature processes as discussed below.

Benefits Key benefits include:

- Utilise readily available standard piping materials.
- Manufacture of heaters using piping is much simpler than heaters designed as pressure vessels. In addition, code requirements for piping are less stringent than for pressure vessels.
- JPHs do not have tube sheets and large bundles of small-diameter thin-walled heater tubes that need to be fitted into the tube sheet and regularly replaced.
- While the heaters are unitised, slurry piping is continuous throughout the complete heater train, which greatly facilitates chemical and mechanical cleaning by pigging (including the installation of permanent pigging stations) and high-pressure water jet equipment.
- Reduced complexity of JPH based integrated digestion facilities (elimination of complete interconnecting slurry pipe work and a number of the valve manifolds required for S&T HEXs) reduces capital and operating costs.
- While large tubes have lower specific heat transfer areas than small tubes, they have disproportionately greater hydraulic capacity. This, together with lower pressure losses, means that for the same flow and the same

driving force, JPHs allow higher pressures (and temperatures) to be obtained at the digester.

- Thermally balanced energy streams by matching vapour consumers (JPHs) and vapour producers (Flash Vessels), eliminating the need to export flash vapour to other process areas or reject waste heat to cooling water or the atmosphere.
- JPH systems have the potential to reduce energy costs due to the ability to achieve higher exit temperatures from the recuperative heating stage.
- JPHs can be operated in parallel streams, only one of which needs to be cleaned at a time so that the impact of de-scaling operations on production is less pronounced than it is for vessels. The need for spares and by-pass facilities is eliminated.
- High mechanical availability confirms that sparing is not required to maintain plant flows.
- Significantly reduced risk of tube blockages and ease of cleaning in-situ confirms that sparing is not required to maintain heat transfer capability.
- Ease of chemical and mechanical cleaning and elimination of tube replacement significantly reduces operating costs for cleaning and maintenance.
- Flow sheet and equipment simplicity and reduced labour requirements contribute to improved health and safety performance.

Tube Digestion

Description Tube Digesters are essentially a continuation of the slurry piping downstream of the heating facilities but in larger diameter piping to accommodate the combined flows from the multiple slurry tubes exiting the last JPHs. Tube Digesters are insulated but are not jacketed for further heating.

Application The limitations to use of the Tube Digesters are essentially driven by cost and practical constraints. For digestion holding times of more than 5-10 minutes as required for extraction of alumina from a particular bauxite under nominated process conditions (primarily caustic concentration and temperature), the length of piping required becomes prohibitive for cost and plant footprint reasons.

Benefits The key process benefit results from the fact that alumina extraction from the bauxite throughout Digester Tubes (and JPHs) occurs in a flow regime that is essentially plug-flow, eliminating flow by-pass and short circuiting that occurs in digester vessels. All plant flow is therefore subject to essentially the same residence time and hence reduce risk of loss of extraction.

Plant design benefits are essentially the same as those for JPHs:

Flash Cooling

Underflow Pipe Work/Flash Tank Level Control To reduce and potentially eliminate the erosive effects of underflow piping and valving, Hatch has developed innovative designs that significantly simplify and streamline piping systems to reduce pressure losses. This can include complete or partial elimination of bypass piping and valving.

However, piping design alone is not enough. In Hatch's Integrated JPH Digestion System, it is possible to set up the system with heater areas aligned with each flash stage so that heat transfer performance of each JPH at each stage of cleaning cycles, Flash Tank level control and temperature and pressure profile within the Flash Tank train are all in balance to prevent excessive two-phase flow.

Flash Tank Entry The only way to potentially totally eliminate two-phase in interconnecting piping is to replace "top entries" with "bottom entries". In the Hatch preferred design, any two-phase flow entering the Flash Tank is intercepted by a splash plate which reduces slurry contact with the walls and existing vapour. This greatly reduces the traditional heavy scaling of Flash Tank walls and significantly improves vapour quality.

Matrix Of Technically Viable Configurations

A summary of potentially technically viable configurations combining process design options and plant design technologies is shown in Table I.

Table I. Technically Viable Configurations

Process Type	Digestion Process	Heating Process
Dual Stream High Temp	Digester Vessel	S&T HEX
Dual Stream Low Temp	Digester Vessel	JPH preferred or S&T HEX
Single Stream High Temp	Tube Digester preferred or Digester Vessel	JPH Essential
Single Stream Low Temp	Tube Digester preferred or Digester Vessel	JPH Preferred or S&T HEX

Integrated JPH Digestion Overview

The term "Integrated JPH Digestion" is used to refer to the whole digestion process comprising JPHs, Digesters (Tube Digesters or Digestion Vessels), Flash Tanks and all associated equipment. A schematic of an Integrated JPH Digestion facility using the Single Stream Digestion Process is shown in Figure I.

Comparative Cost Studies

Scope

A comparison study was performed to determine the difference in design, capital, operating and maintenance costs of the following slurry heating options for a low temperature single 1.4 Mtpa digestion heating unit:

- Dual stream using S&T HEXs (base case).
- Single stream using S&T HEXs.
- Single stream using JPHs.

The low temperature option was chosen for this first cost comparison study since there are as yet no low temperature installations of Jacketed Pipe Heaters. An equivalent study for high temperature applications is in progress.

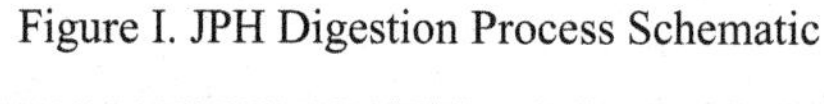

Figure I. JPH Digestion Process Schematic

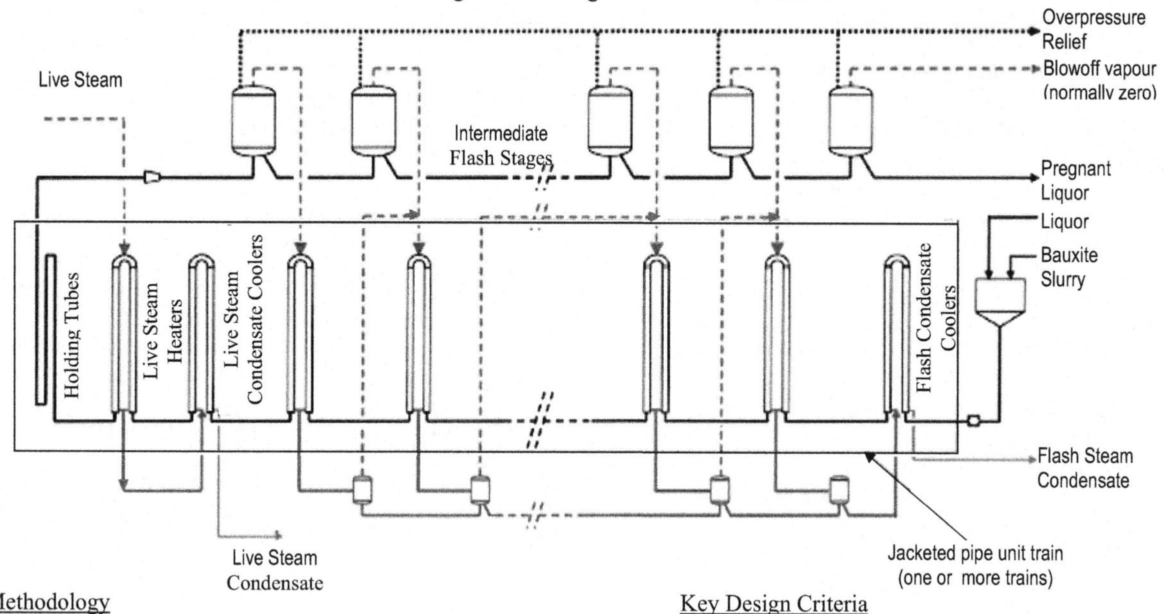

Methodology

A process model for each option was developed in SYSCAD as a full digestion model. Process data was extracted from the models to produce a Process Flow Diagram for each heating train and relevant components to simplify the comparison. This was done as the flash tank train, digesters and other equipment will be much the same for all three (3) options due to the similar slurry mass flow and temperature profile.

The same method of hydraulic and heat balance calculations, factoring, unit prices and labour rates was used for estimating each option and a consistent set of process data was used across the models to ensure the comparison was consistent.

Key Design Criteria

The following key design criteria were utilised:

- Blow-off slurry flow and properties were made common for all three (3) options (based on producing 1.4 Mtpa Alumina production).
- Single Stream options Flash Train profile was selected to give no export or blow-off steam.
- A single size 'best-fit' S&T HEX was selected for each of the S&T options.

Process & Equipment Summary

The key process data and equipment required in the options considered are summarized in Table II.

Table II. JPU Key Process Data and Equipment for Comparisons

Key Data for Heating Trains	Dual Stream - S&T HEX	Single Stream - JPH Heaters	Single Stream - S&T HEX
		Process	
Spent Liquor Temperature ex Test tank	84.7	80.2	82.2
Liquor/Slurry Flow (m^3/hr)	1063	1371	1371
Slurry/Liquor Feed Temperature (°C)	84.7	84.2	85.8
Spent Liquor Temperature to Digester (°C)	178.8	N/A	N/A
Bauxite Slurry Temperature to Digester (°C)	199	148.3	148.3
A26 LP Steam tph (180°C, 7.5 bar)	35.2	38.5	37.1
A30 LP Steam top (180°C, 7.5 bar)	120.9	195.3	204.3
MP Steam tph (194°C, 13.7 bar)	115.9	0	0
Export Steam tph	22	0	0
		Equipment	
Number of Pumps	11	12	10
Number of Condensate Pots	8	9	8
Number of Valves	440	130	440
Number of Heaters	18	76	21
Number of Tubes per Heater	1100	4	770
Weight of each Heater (t)	37	20	30
Total Power Consumption (kW)	389	331	400

Relative Capital Costs

A summary of the relative capital costs for the three (3) options is given in the Table III. The absolute values have not been included as the location and market circumstances will dictate the true value, however a relative comparison based on international prices was established.

The 1st column gives a breakdown of the approximate component that each discipline makes up of the total capital cost. The 2nd and 3rd columns is the factor in cost difference when compared to the base case. For example, 1.08 means 8% higher in cost when compared to the base case, 0.91 means 9% lower.

Overall, it is clear that the heater and piping costs have the biggest impact on the capital cost. The larger capital cost of the JPH option is almost entirely due to the cost of the JPHs themselves. However, the cost of Piping and Valves is significantly reduced due to the simplified arrangements and elimination of piping between JPHs and valve manifolds.

Table III. Relative Capital Cost Comparison Summary

	Dual Stream S&T HEX (Base Case)	**Single Stream JPHs**	**Single Stream S&T HEX**
Discipline	**Component of Total Cost**	**Cost Factor vs Base Case**	**Cost Factor vs Base Case**
Civil	1.4%	1.08	1.17
Concrete	6.2%	0.91	1.13
Structural	10.2%	1.34	1.02
Equipment	39.7%	1.43	0.76
Piping & Valves	19.8%	0.57	1.09
Insulation	4.0%	1.68	1.03
Electrical	11.9%	1.13	1.17
Instruments	7.0%	0.98	0.99
Total	100.0%	1.16	0.96

Operating and Maintenance Costs

Operating and Maintenance costs have been derived from in-house experience and knowledge of the labour and maintenance required to clean and maintain equipment. The process involved setting up a list of all equipment and associated maintenance items and allocating periodic turnaround times and associated operator, maintenance and contractor hours, complete with materials and spare part costs. A consistent approach for estimated costs was used for each option.

The three main items that impacted on operating costs were power, steam and cleaning maintenance. Power consumption was included only for equipment related to the heating train, namely the Desilication discharge pumps, Digester feed pumps and Condensate pumps. Steam consumption included LP and HP steam to both Digestion and Desilication. The assumptions for calculating operating costs for heater cleaning requirements for each option were based on Table IV.

Whole of Life Net Present Cost Analysis Results

A Whole of Life Net Present Cost (NPC) analysis was performed to take into account both capital and operating costs over a 25-year period. A spreadsheet calculation of operating and maintenance costs using in-house personnel experience and site data from various refineries based on labour hours and maintenance frequency requirements was prepared for each option for 1.4mtpa production.

Given that the inclusion of steam and power is significant and considered necessary for a realistic indication of overall costs, a separate NPC calculation was performed to exclude steam and power costs. This enabled operating and maintenance costs of the equipment to be compared separately, as shown in Table V.

Table V. NPC Cost Summary Comparisons

	Dual Stream S&T HEX (Base Case)	**Single Stream JPHs**	**Single Stream S&T HEX**
Unit Cost	**Component of Total Cost**	**Cost Factor vs Base Case**	**Cost Factor vs Base Case**
Capital	1	1.16	0.96
	Steam and Power Cost Excluded		
Operating	1	0.61	1.13
NPC	1	0.73	1.09
	Steam and Power Cost Included		
Operating	1	0.87	1.00
NPC	1	0.89	0.99

Table IV. Cleaning Cycle Assumption Data

Heater Type	**Operating Temperature (°C)**	**Acid Clean Frequency (Hours)**	**Turnaround time to Acid Clean (Hours)**	**Mechanical Clean Frequency (Hours)**	**Turnaround Time to Mechanically Clean (Hours)**	**Total Labour to Mechanically Clean (Hours)**	**Average tube Replacement (% of Tubes)**
Single Stream JPH	85-150	2000	30	Not required	N/A	N/A	Not required
Single Stream S&T HEX	85-115	500	40	2500	7	350	10-20%
	115-150	450	40	2000	10	400	20-30%
Dual Stream S&T HEX	85-115	450	40	3000	5	300	5-10%
	115-500	400	40	2500	7	350	10-20%

Cost Comparison Conclusions

The differences in other areas of the plant, such as evaporation, descilication, boiler house etc is not considered. A detailed comparison requires a full plant study. However this study is a useful comparison to identify the relative differences in the heating train, which would represent the majority of the difference in costs for complete digestion facilities.

The JPH option has an initial capital outlay disadvantage with a 16% higher capital cost than the traditional alternatives. However, when combined with the large operating and maintenance costs savings, the Whole of Life Net Present cost shows the JPH option becomes the preferred option.

It is worthwhile noting that the overall NPC cost comparison is quite sensitive to the operating and maintenance costs. Hence, it is considered that the S&T HEX options are more exposed to cost "blow-outs" due to their higher labour requirements.

It is also very important to note that the JPH option offers a more robust design with significantly lower maintenance requirements. This in turn gives a significantly lower risk to plant flow continuity and hence increased production and product quality that is difficult to quantify. This is demonstrated by the clear difference in cleaning requirements and also the significantly larger number of valves for the S&T HEX options (440) compared with the JPH option (130). These difficult to quantify risks have not been added into this analysis but provide an important further consideration for the choice of options.

Project Case Studies

Korea General Chemical Company Key Features The KGCC plant was designed by Kaiser Engineers to produce chemical quality hydrate and alumina with a capacity of approximately 220,000 tpa hydrate. Key features include:

- There is only one train of Flash Tanks feeding six independent lines of JPH heaters, rather than having a set of Flash Tanks for each set of tube heaters as at Stade. Each JPH contains four slurry tubes. Each line of heaters can be cleaned independently without the Flash Tanks having to be taken off line.
- The Flash Tanks are all arranged at the same level at grade, relying entirely on thermal drive initially applied through the live steam heaters to perform a cold start-up.
- Flash Tanks are bottom entry without entrainment separators.
- The plant utilizes Tube Digesters rather than Digestion Pressure Vessels.

Performance Experience The plant has operated successfully for thirteen years and been proven to be very operationally reliable and sturdy.

The Flash Tank arrangement has proven to be very practical with the required operating temperature of 260°C able to be reached in less than 2-2.5 hours from first start-up. In addition the plant has consistently produced high quality condensate.

Design heat transfer performance has been achieved and cleaning schedules have been maintained.

Comalco Alumina Refinery (CAR) Key Features The Comalco Alumina Refinery Integrated Digestion facility was developed by Kaiser Engineers (now Hatch) in a joint venture with Lurgi (now Outokumpu) and VAW (now Hydro).

Two identical units each of 700,000 tpa are installed at the Comalco Refinery. Each unit comprises three (3) trains of JPHs that heat bauxite slurry in 12 stages. The first 10 stages use flash vapour while the 11th and 12th stages use live steam. Once heated in the JPHs, the slurry stream flows to Digester Tubes to extract the remaining alumina from the bauxite. The slurry is then flash cooled through 10 stages of Flash Vessels.

Figure II: The Tubular Digestion Facility at CAR

Conclusions

The Integrated JPH digestion Technology as described in this paper has been demonstrated to have a number of applications that will deliver significant benefits over the use of traditional technologies. This has been proven in two successful plant installations for KGCC and Comalco.

In the Comalco Alumina Refinery, it was evident during the planned shutdowns in July and August 2005 that scaling in flash vessels was minimal after eight months of operations and vessel walls were so clean no descale was required. In addition, there was no evidence of scale in flash vessel vapour lines or on the shell side of the JPHs, highlighting the benefits of the new flash vessel design in producing high quality condensate consistently better than industry standards.

Light Metals 2006 *Edited by Travis J. Galloway* ***TMS (The Minerals, Metals & Materials Society), 2006***

OXALATE REMOVAL BY OCCLUSION IN HYDRATE

Valérie Esquerre, Philippe Clerin, Benoît Cristol
Alcan B&A Technical Assistance; Route de Biver P.O. Box 54, 13541 Gardanne Cedex, France

Keywords: Oxalate, Occlusion, Alumina trihydrate

Abstract

Occlusion in hydrate plays a significant role in the removal of oxalate at Alcan's Gardanne refinery. Up to 50% of the oxalate purge can be achieved by occlusion.

This study investigates the occlusion phenomenon and identifies the action parameters that facilitate the control of the occlusion process.

Successive precipitation cycle laboratory tests have been conducted in different operating conditions to ascertain the key factors. Having a high level of precipitated oxalate doesn't ensure a high level of occluded oxalate. The addition of CGM appears to completely inhibit oxalate occlusion, whereas an increase in caustic concentration enhances the process. An increase in temperature also tends to increase occlusion. Possible mechanisms are proposed.

A new operating instruction to maintain a high caustic concentration in the spent liquor has led to a high and lasting level of occluded oxalate.

Introduction

The removal of oxalate is a key issue in Bayer plants, where specific units, such as a side stream of evaporated liquor, are sometimes necessary to cope with this impurity.

At Alcan's Gardanne refinery, oxalate is allowed to precipitate along with part of the precipitated gibbsite (referred to below as 'hydrate'), by adapted temperature conditions in one precipitation line. The oxalate is then removed by hydrate washing on a belt filter. However, a significant amount of oxalate is eliminated by occlusion in hydrate, a no-cost purge that is innocuous for hydrate in the context of Gardanne refinery.

Only a few studies in the literature refer to occlusion. Grocott and Rosenberg [1] mention the increase in soda impurity by incorporation of sodium oxalate needles in hydrate. Occlusion is also thought as leading to additional particle breakage in calciners, which is not observed at Gardanne's plant.

This paper investigates the occlusion phenomenon in order to improve its control.
After presentation of the experimental method, the effect of different process parameters is described. An attempt towards a better understanding of the occlusion phenomenon is then proposed and the paper ends with the industrial application made of the results of this study.

Experimental

Successive precipitation recycling tests have been conducted in the laboratory to ascertain the key factors impacting occlusion.
Precipitation experiments have been performed in a rolling water bath in PVC bottles. Precipitation tests have lasted 48h typically, with a change in temperature at mid-time to reproduce a temperature profile (60°C then 50°C). Plant pregnant liquor has been seeded with oxalate-free hydrate, at a concentration of $400gAl_2O_3/L$, lower than the industrial value, in particular to accelerate the inclusion of impurities. The hydrate produced in one test has been used as seed for the following test, and so on, until a sufficient number of cycles was available to answer the question regarding occlusion.

Different operating conditions have been tested:
- Addition of CGM (Crystal Growth Modifier – Nalco 7837) to pregnant liquor. Experiments with CGM were also aimed at validating the operating mode in the laboratory, by verifying the effect of CGM to reduce occlusion, as observed in the plant. A dosage of 50ppm of CGM (volume/volume in liquor) has been introduced.
- Caustic soda concentration. The liquor has been concentrated so as to reach a caustic concentration of 180g/L (expressed as Na_2O) in the spent liquor (compared with 165g/L in the blank test).
- Temperature. The temperature has been changed from 60-50°C to 65-55°C.

Effect of process parameters on occlusion

Occlusion has to be enhanced to maximize oxalate purge of Gardanne Bayer circuit. This demands knowledge of the action parameters to better control the occlusion process.
The results are illustrated by the level of occluded oxalate as a function of the number of precipitation cycles.

Effect of CGM

The effect of CGM is shown on figure 1.
The occlusion is inhibited by the presence of CGM. The level of occluded oxalate remains stable at a low value of around $40gC_{ox}/tAl_2O_3$, even after 35 precipitation cycles.

The level of precipitated oxalate is also compared on figure 1. The precipitation of oxalate is lowered by CGM, but still reaches a significant level (only around 15% less than without CGM).
Little black balls are visible after a few cycles. The adsorption of CGM on oxalate crystals could poison the surface for further occlusion. The inhibited activity of oxalate as a seed surface for gibbsite nucleation [2] could be part of the explanation for reduced occlusion.

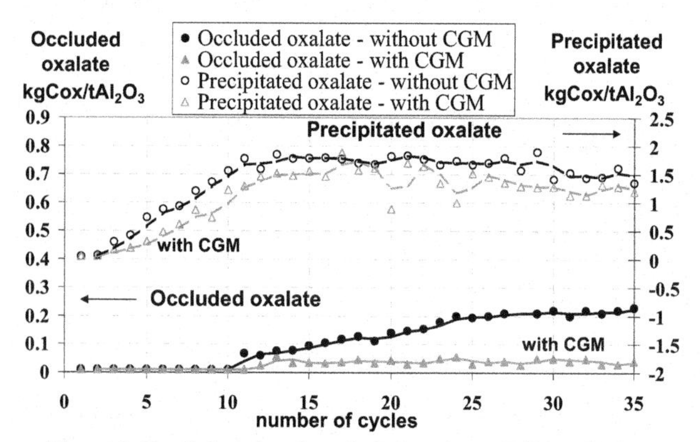

Figure 1. Precipitated and occluded oxalate - Effect of CGM

At Gardanne's plant, this detrimental effect on occlusion excludes the use of CGM to coarsen the circuit.

Effect of caustic soda concentration

The level of occluded oxalate is compared for two caustic soda concentrations (normal plant concentration and concentrated liquor) on figure 2.
Occlusion is clearly enhanced by high caustic soda concentrations: the phenomenon is faster and of greater intensity.

A higher caustic concentration leads to higher oxalate precipitation: the apparent solubility of oxalate is indeed lowered, as is the Critical Oxalate Concentration (COC).
However, a high level of precipitated oxalate is not a sufficient condition for a high level of occlusion, as it will be demonstrated in the following section.

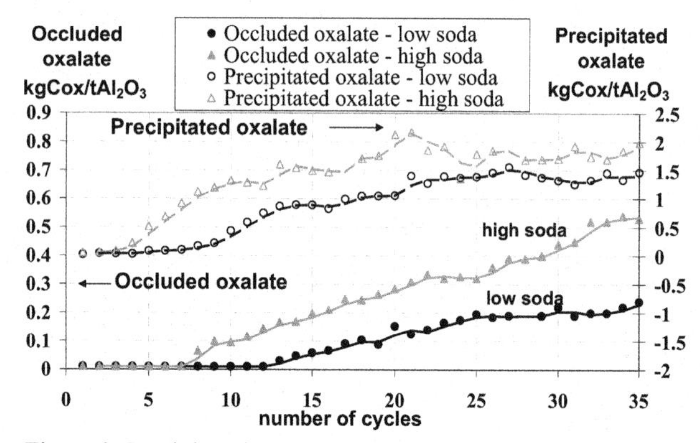

Figure 2. Precipitated and occluded oxalate - Effect of caustic soda concentration

Effect of temperature

The effect of temperature has been measured by increasing the temperature from 60-50°C to 65-55°C in cycle n°37.
The complete evolution of the occluded oxalate is reported in figure 3 for two experiments:
- "blank test" (normal plant soda concentration)
- high caustic soda concentration

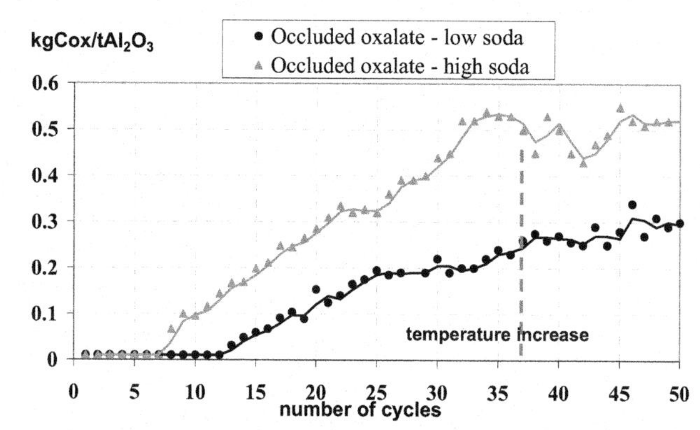

Figure 3. Occluded oxalate - Effect of temperature

When the temperature is increased, the precipitation of oxalate (ie oxalate lost by the liquor) is reduced. The occluded oxalate appears however to increase, with a clearer trend in the blank test. The change in conditions during processing of cycles implies that the effect is only visible after a few cycles, time for the seed to be renewed.

This result on the effect of temperature is also visible on the analysis of industrial data (see last paragraph).

Decorrelating the occluded oxalate from the amount of precipitated oxalate and the hydrate growth (which would both act in the opposite sense of the observed results), the effect of temperature is suggesting a type or size of oxalate crystals easier to occlude.

Comprehension of the occlusion process

Laboratory experiments enabled identification of one powerful actuator on occlusion, which is a major result for Gardanne's plant (see last paragraph). The discussion below intends to give hints for a better understanding of the occlusion phenomenon.

Link with precipitated oxalate

One test has been conducted with a higher oxalate concentration in the liquor to increase supersaturation, and then enhance precipitation. This test should answer the question about a relationship between the level of occluded oxalate and the level of precipitated oxalate.
The test liquor has been doped with a synthetic oxalate solution (average Cox=0.23%Na2Otot vs 0.205%Na2Otot).

In the oxalate-enriched liquor, an enhanced precipitation is effectively measured, as well as a higher occlusion (figure 4).

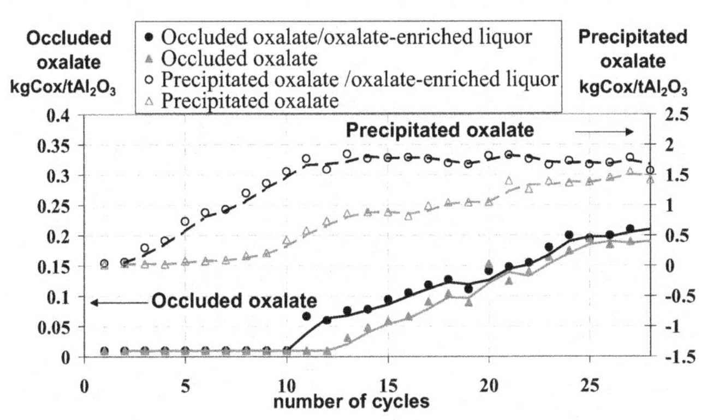

Figure 4. Precipitated and occluded oxalate - Link between precipitated and occluded oxalate (1)

When the enrichment in oxalate is stopped (21st cycle), the two curves join each other after a few cycles.

Similarity between both tests and results (caustic concentration, granulometry, hydrate production) suggest that the difference in occlusion in this case is only due to the difference in the level of precipitated oxalate.

However, this test with oxalate-enriched liquor makes it possible to further comment on the link between precipitated oxalate and occluded oxalate.
Indeed, the level of precipitated oxalate obtained in the oxalate-enriched system is similar to the one obtained in the test conducted at high caustic soda concentration. The evolution of precipitated oxalate and occluded oxalate are compared for both tests in figure 5.

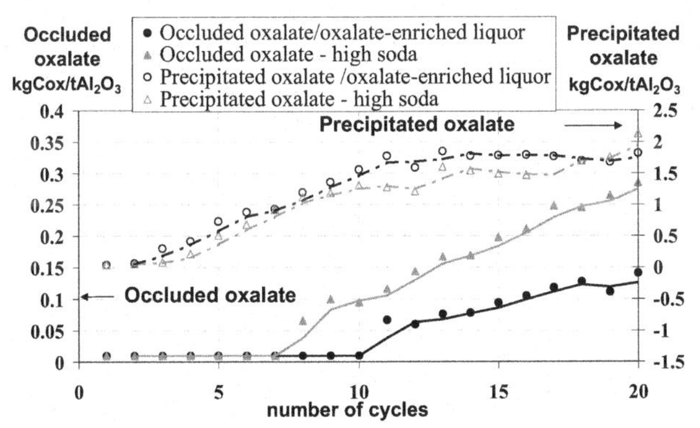

Figure 5. Precipitated and occluded oxalate - Link between precipitated and occluded oxalate (2)

Despite a similar precipitation, the occlusion is significantly different, with a ratio of two between the two conditions.
On the basis of these results, there is no direct relationship between precipitated and occluded oxalate.

Link with productivity and hydrate growth

One could suppose that a higher productivity or alumina supersaturation leads to a higher oxalate inclusion, considering the higher hydrate growth.
However, productivity is less at high caustic concentration, where the occlusion is at a maximum. The hypothesis of a higher potential to occlude related with a higher productivity is here dismissed, as had already been suggested by the study of the effect of temperature.

Link with type of oxalate crystals

Occlusion cannot be directly correlated to a high level of precipitated oxalate or to a high hydrate growth. A difference in the type or size of oxalate crystals as a function of conditions could be an explanation for the different behaviors towards occlusion.

In the presence of CGM, oxalate does not precipitate as individual fine needles, but more as "bunchs" of many needles. One of this crystal morphology is visible on figure 6. These larger crystals are more difficult to occlude.

Figure 6. Oxalate crystals on hydrate with CGM

Considering the effect of high caustic soda concentration, one could link the enhanced occlusion with an oxalate liquor concentration closer to the Critical Oxalate Concentration.
Indeed, a high caustic concentration reduces the COC.
Extra experiments following oxalate precipitation during hydrate precipitation for two caustic concentrations and comparing oxalate concentrations to COC values (obtained via models) support this idea. Indeed, in the case of concentrated liquor, oxalate liquor concentration is running closer to the values of COC and even goes above the COC at the end of precipitation (see table I).

One possible explanation to the enhanced occlusion observed in precipitation cycles at higher soda concentration could be then a finer oxalate product generated by spontaneous nucleation above the COC.

Table I. Oxalate precipitation kinetics during hydrate precipitation for two caustic concentrations - Difference between oxalate liquor concentration and COC

Time (h)		0	1	3	5	7	22	25	28	30	48
Cox liq - COC	low soda	-0.031	-0.053	-0.050	-0.041	-0.050	-0.071	-0.049	-0.013	-0.006	-0.018
(%Na2Otot)	high soda	0.003	-0.009	-0.019	-0.017	-0.020	-0.030	-0.007	**0.017**	**0.035**	**0.020**

Photographs for three conditions of precipitation cycles are visible on the following figure 7. It looks as if big oxalate needles are less numerous for the test at high soda concentration.

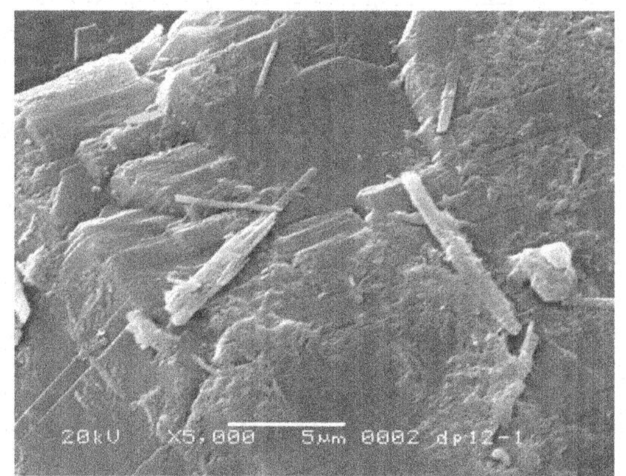

7.a) blank test (normal plant soda concentration)

7.b) oxalate-enriched liquor

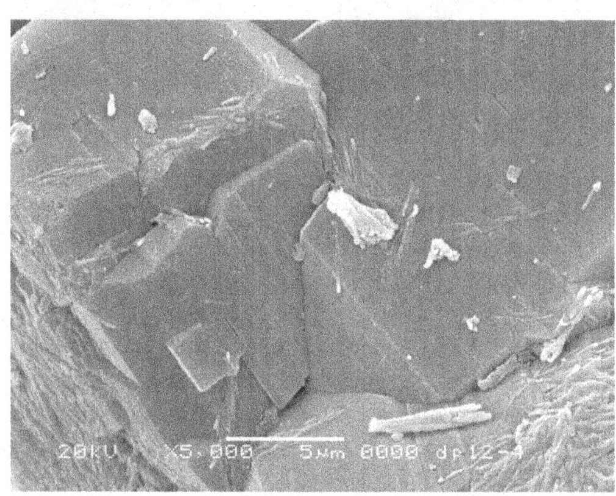

7.c) high soda concentration

Figure 7. Oxalate crystals on hydrate in different conditions

However, observations made on the effect of temperature don't support this hypothesis, since a higher temperature increases the COC, and should have lead to a decrease in occluded oxalate. The growth of oxalate crystals may also play an important role here. The reduced growth at lower temperature could lead to an easier product to occlude.

A clear mechanism is not proven yet, but proposals made here could give leads for further consideration.

Process control in the plant

The caustic concentration has been identified as a powerful control parameter on occlusion. In the plant, the operating instruction is now to maintain a high caustic concentration in the spent liquor.

Figure 8 below illustrates the good correlation between the caustic concentration and the oxalate occlusion. Each time the soda concentration is reduced, the level of occluded oxalate decreases.

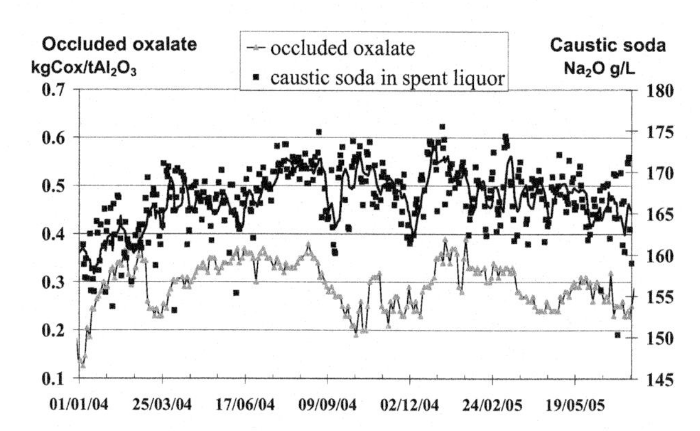

Figure 8. Industrial data: caustic soda concentration in spent liquor and level of occluded oxalate

The effect of temperature is overwhelmed by the effect of caustic concentration. However, the temperature, as illustrated in figure 9, can explain some inconsistent evolutions between caustic concentration and occlusion.

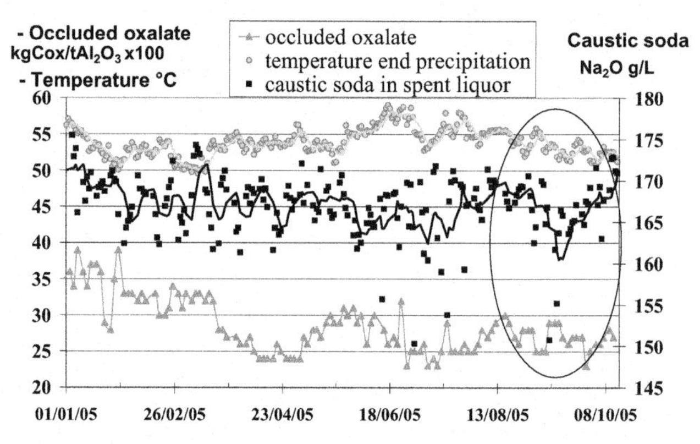

Figure 9. Industrial data: caustic soda concentration in spent liquor, temperature at the end of precipitation and level of occluded oxalate

The September 2005 period shows a hesitating level of occluded oxalate, despite a caustic soda concentration close to 170g/L. It is associated with a decreasing temperature, which is likely to counteract the effect of a high soda.

Periods maintaining a caustic concentration of around 170g/L in the spent liquor, with a temperature relatively high, ensured a good value of the occluded oxalate of $350 gC_{ox}/tAl_2O_3$.

Conclusion

The complexity of the occlusion phenomenon is underlined in this study. Although no clear mechanism for the occlusion of oxalate has been obtained, the type and size of oxalate crystals is supposed to play a major role in the occlusion, and should be considered for further study.

Moreover, the impact of different process parameters on occlusion has been evaluated, which constitutes an answer to plant concerns.

- The addition of CGM completely inhibits the occlusion of oxalate.
- The temperature is considered as a second-level action parameter, since the higher the temperature, the higher the occlusion, but the lower the productivity.
- The caustic soda concentration is a powerful actuator on occlusion; a high level of caustic soda is necessary to guarantee a high occlusion of oxalate.

Finally, this work has given the Alcan Gardanne refinery a solution for increasing the occlusion of oxalate in hydrate, thus purging oxalate at a lower cost.

Acknowledgements

Many thanks to Jean-Pierre Macario, for his significant contribution to this work.

References

1. S.C. Grocott and S.P. Rosenberg, "Soda in Alumina. Possible Mechanisms for Inclusion", *Proceedings of the First International Alumina Quality Workshop (Gladstone, Australia)*, 1988, 271-287.

2. M.M. Reyhani et al., "Gibbsite nucleation at sodium oxalate surfaces", *Proceedings of the Fifth International Alumina Quality Workshop (Bunbury, Australia),* 1999, 181-191.

Light Metals 2006 *Edited by Travis J. Galloway* ***TMS (The Minerals, Metals & Materials Society), 2006***

ORGANIC CARBON IN INDIAN BAUXITES AND ITS CONTROL IN ALUMINA PLANTS

K.V.Ramana Rao and R.N.Goyal
Jawaharlal Nehru Aluminium Research Development and Design Centre
Nagpur - 440023, India

Key words: Organic carbon, Organic controls, Indian Bauxites, Oxalates, Humates, Bayer Process

Abstract

The organic build up in alumina plant liquor is known to cause not only various plant operational problems, but also reduces liquor productivity and quality of product hydrate. The organic situation in Indian alumina plants is still not alarming for the reason that the bauxites contain low organic carbon. However, every plant feels the impact of the organic build up as it becomes old.

The rate of organic build up is plant specific and depends upon the digestion conditions, such as digestion temperature and pressure, caustic concentration, TOC values and nature of organic compounds in the mineral. In this paper the influence of digestion parameters, TOC in Indian bauxites and influence of special additives on organic build up in Indian alumina plants are discussed. The details of various control methods investigated for removal of oxalates, humates and TOC for Indian plants are also presented.

Introduction

The build up of organic carbon in alumina plants is well known to cause not only various operational problems in the alumina plant but also reduces the liquor productivity and impairs the product quality in terms of generation of more fines and contamination of product hydrate.

Its removal from the Bayer circuit or its control below the critical saturation limit is essential for managing the detrimental effects of these organic impurities and enhancing the productivity of the alumina plant. However, total removal of organic impurities from Bayer liquor is not warranted as some of these species are observed to be useful in increasing the stability of sodium oxalate in aluminate liquor, thus preventing its co-precipitation along with alumina trihydrate.

At present, Indian alumina plants do not use organic control methods on regular basis, hence the severe impact of organic build up is already being felt by older plants. Many control methods reported in the literature are plant-specific and are intended to control a specific type of organic impurity such as TOC or humates or oxalates for the reason that plant conditions differ widely. Furthermore, the nature of organics present in the Bayer liquor are seldom known, as bauxites from different origins contain different types of organic species depending upon the environmental factors such as local vegetation, rainfall, temperature, geology, permeability and microbiological degradation of the soil.

In this work the organic carbon in various Indian bauxites is determined and a case study of organic build up in Indian alumina plants is presented. Reported organic control methods were tested on plant liquors for their suitability in alumina plants.

Experimental

Analytical procedures reported earlier [1-2] were utilised in the present work. Total Organic Carbon (TOC) in Indian bauxites is determined by conventional wet chemical method in which initially the carbonate content in the bauxite is eliminated by smooth boiling with phosphoric acid solution for nearly 2 hours.

The organic carbon content in carbonate free bauxite is oxidised quantitatively to carbon dioxide which is distilled into a standardised alkaline solution. The quantity of absorbed carbon dioxide is determined by back titration. The humates were estimated in terms of TOC, and speciation of intermediate molecular weight organic species is carried out by gas chromatography. All these parameters were expressed in terms of Corg.

Results And Discussion

India is endowed with huge reserves of good quality bauxite to an extent of 3100 MT. The abundance of the bauxite reserves in India is shown in Fig.1. About 70% of these reserves are situated in the east coast region of the country, and are referred to as "East Coast bauxites".

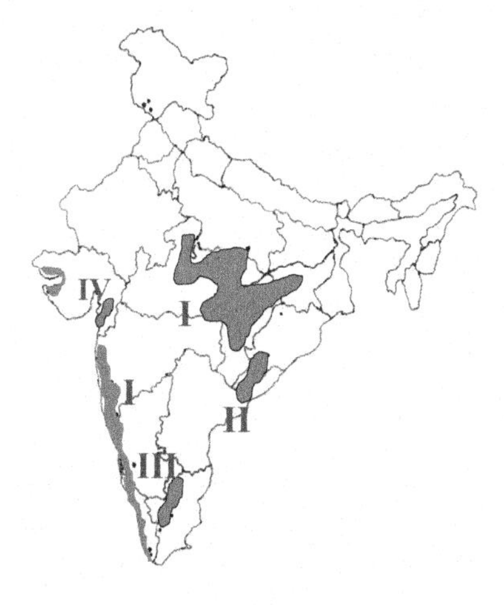

Fig.1. The abudance of bauxite in India
I. Central India II. Eastern ghats (East coast)
III. Western ghats and IV. Gujarat

The bauxite from this region is of good quality in terms of low organic carbon and low silica, but high iron content. A low bond work index of about 9-10 KWh/t is exhibited by these bauxites, hence easy to grind and consume less energy in comparison with other Indian bauxites which have a bond work index of about 15 KWh/t [3]. Mineralogy of Indian bauxites is basically gibbsitic, or mixed phases of gibbsite and boehmite, with boehmite phase in small quantities.

The organic problem has not drawn the attention of Indian alumina plants till recently for the simple reason that the total organic carbon (TOC) in Indian bauxites, as shown in Table 1, is in the low to medium range (0.045% to 0.20%) in comparison with several other bauxites of the world. The process liquors too, have not attained critical levels.

Table 1. Organic content in typical Indian bauxites

Sl. no	Bauxite source	Type of Parent Rock	Mineralogy	Range of TOC as C org (%)
1	Central India	Deccan Trap Basalt / Granitenies s	Mixed gibbsitic-boehmitic	0.110 - 0.182
2	Eastern ghats	Khondalite / Charnokite	Gibbsitic, high iron and low silica	0.045 - 0.094
3	Western Ghats	Deccan Trap Basalt / Granitenies s	Gibbsitic, mixed gibbsitic - boehmitic	0.125 - 0.240
4	Gujarat	Deccan Trap Basalt	Mixed gibbsitic – boehmitic, contains CaO	0.080 - 0.185

However, our recent measurements on some plant liquors have indicated a TOC in the range of 10-14 gpl and over a period of time more plants may reach such high organic levels. Typical organic levels in Indian plants are shown in Table 2 which indicates alarming levels of sodium oxalate (10-12%) and humates (5-6%) in process liquors of a plant.

Table 2. Typical organic levels in Indian plants

Data	Balco	Nalco	Hindalco	Indal
Digestion temperature (°C)	240	106	240	145
Caustic as Na_2O_C in liquor at digestion (gpl)	175	210	160	160
Caustic as Na_2O_C in liquor at precipitation (gpl)	140	146	105-110	110
TOC in process liquors C org (gpl)	14	6	9	10
Sodium Oxalate as C org (gpl)	1.5-1.8	0.55-0.70	0.67-0.85	0.50-0.89
Humates as C org (gpl)	0.844	0.336	0.415	0.60

Model digestion tests carried out at different temperatures have shown that the rate of dissolution is in the same range of 55-70 %. However, the Eastern ghats bauxites have shown the least dissolution rates. A gas chromatograph record showing various organic species present in a typical plant liquor is shown in Fig. 2.

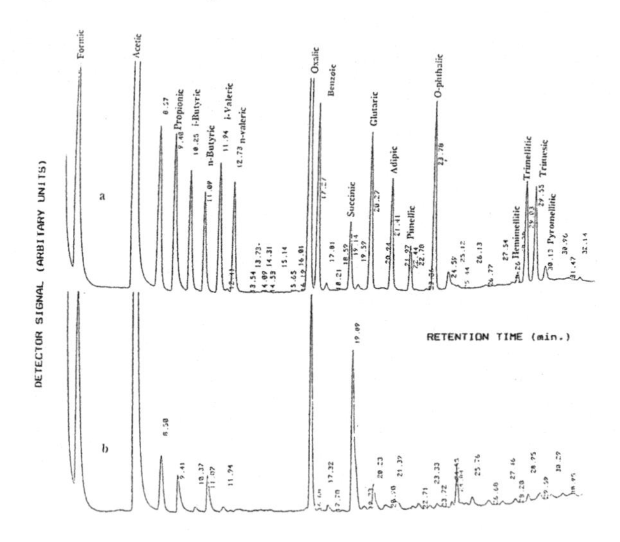

Fig.2 Typical chromatograms of n butyl esters of (a) calibration standard and (b) a typical plant liquor

The organic speciation of the plant liquors as shown in Table-3 and Table-4 has revealed that humic fraction is about 2-6 %, but the build up of oxalate to 12 % of TOC and higher in plant liquors is attributed to the continuous degradation of not only humic acids but also due to the high molecular weight other organic species under high caustic and high temperature of digestion.

Table 3. Organic speciation shown by gas chromatography of plant liquors

Species	Range (in % TOC)
Humic acids	2-6
Aromatic acids (pthalic, hemimelitic, trimellitic, trimesic, pyromellitic)	15-16
Dicarboxylic acids (succinic, glutaric, adipic)	20-25
Monocarboxylic acids (formic, acetic, propionic, butaric, valeric etc.)	20-60
Oxalic acid	5-12
Unidentified	10-15

Table 4. Humic fraction distributed in various stages of a typical plant liquor

Liquor	Humic fraction (%) in TOC of the liquor
Spent liquor	2.36
Digestion liquor	2.72
Strong liquor	2.40
Pregnant lliquor	2.37
Salt solutions	6.70

Digestion studies on a typical plant liquor have shown that the caustic concentration and digestion temperature >140°C have significant influence on the organic build up in the plant as shown in Fig.3 and Fig.4 respectively.

Table 5. Typical organic control methods adopted in alumina plants

Sl. No	Method of Control	Observations/ Remarks
1	Bauxite roasting	Carried out at > 600°C, Broad spectrum control, Efficient removal of 80-90% TOC, Energy and Cost intensive. For gibbsitic bauxites care to be taken for partial or full dehydration.
2	Liquor calcination / liquor burning	Carried out at temperature 1000-1200°C. Broad spectrum control, efficient removal of 80-90 % TOC, Energy and Cost intensive.
3	Oxidation methods	Oxidants like oxygen, oxides, peroxides and ozone are employed. Wet oxidation process at 270°C is effective in humate destruction. Known to form sodium oxalate and sodium carbonate which must be removed by subsequent causticisation.
4	Evaporative crystallisation of sodium oxalate	Process liquor evaporated and then cooled. Oxalate / hydrate seed added to concentrated liquor. 28-30% Sodium oxalate removed.
5	Salting of sodium oxalate crystals	Based on solubility criteria, caustic is added to the evaporated spent liquor. Helps in removal of 25-30 % of sodium oxalate from plant liquor.
6	Adsorption methods	Granular activated carbon (GAC) is effective in removing 80 % TOC and 20 % sodium oxalate. Carbon bed can be reactivated and reused. Reactor beds of polymers / cellulose not found beneficial.
7	Microbial degradation of oxalate	Microorganisms degrade sodium oxalate. Identification and culturing of organism a limiting factor.
8	Oxalate removal by polysulfone membranes	Though the process is effective for removal of sodium oxalate, the membranes are not only expensive, but cannot be easily reactivated.
9	Humate control by manganese ore and salts	Addition of manganese salts to digestion is effective in removal of 45-50% of humates, but some salts create filtration difficulties.
10	Addition of polymers and specialty chemicals	Cationic polymers are known to form complexes with humates and then separation from plant liquors. Anionic polymers work as CGM.
11	Seed washing followed by wash water evaporation and salt separation by treatment	Separation of salts can reduce the organic content by precipitating sodium oxalate.
12	UV-irradiation	UV irradiation for 2 hours can reduce the organic content. This method is also useful for preparation of by-products like oxalic acid. But cannot be implemented in plants.
14	Liquor causticisation	Lime addition is reported to reduce organics in liquor by adsorption on to calcium salts. However, removal is marginal.
15	Addition of barium and magnesium salts	Barium salts are effective but hazardous. Kieserite ($MnSO_4$) is useful in humate control, but may create problems in filtration circuits.

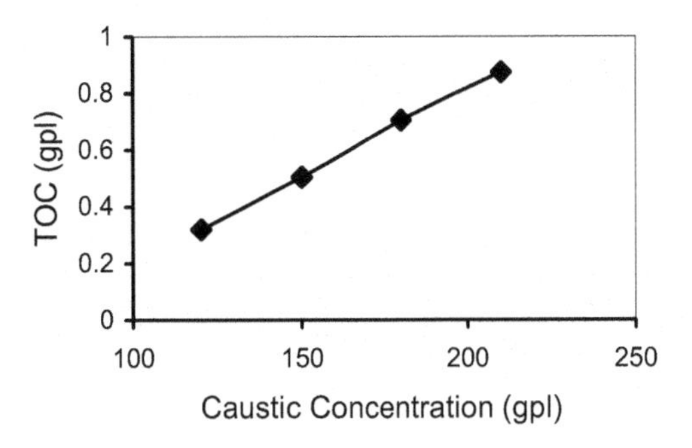

Fig. 3 Organic build up as a function of caustic concentration at digestion temperature 106°C

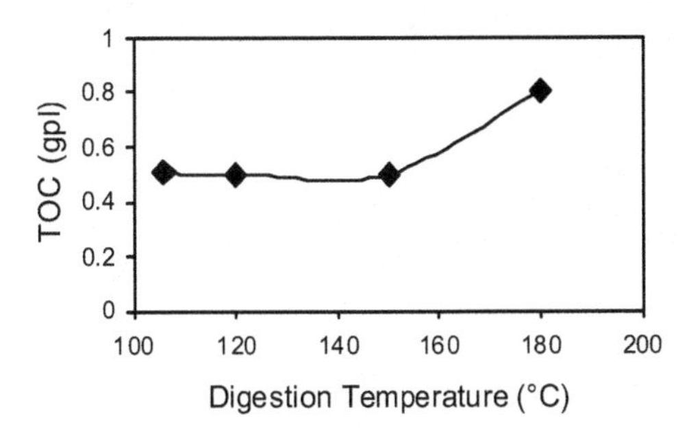

Fig.4 Organic build as a function of digestion temperature at caustic concentration of 150gpl

Most processes developed for organic control were intended to controll a specific type of organic impurity such as total organics or humates or sodium oxalate. Often plants also use more than one method to ensure the total control of organics in the plant.

The best way of controlling / minimising organic impurities in Bayer circuit is to choose low organic carbon bauxites. This can certainly prolong time taken to reach the critical levels in the plant. Bauxite beneficiation studies in our laboratory have also established that by eliminating fine fractions of bauxite (< 0.5 mm), we can not only reduce the silica content in the process bauxite but also the organic input into the plant as shown in Fig.5. But the process involves selective mining and beneficiation.

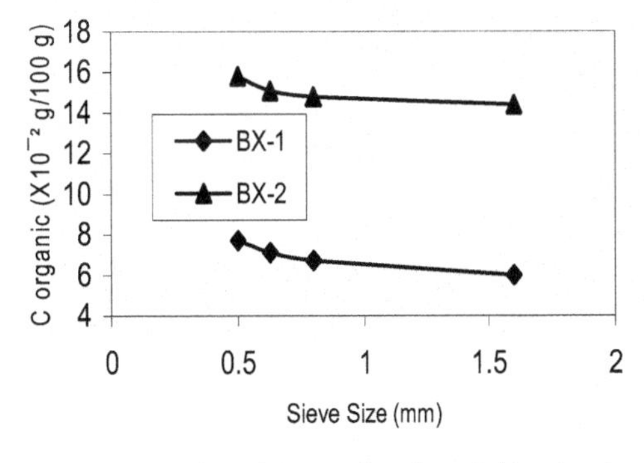

Fig.5 Organic carbon as a function of sieve fraction

The solubility of sodium oxalate in Bayer liquor is directly related to temperature and inversely related to caustic concentration. For a typical synthetic liquor, the apparent solubility of sodium oxalate at 50ºC and caustic concentration of 112 gpl as Na_2Oc is about 2.1 gpl as Corg. [3,4].

However, in plant liquors the apparent solubility of sodium oxalate may be still higher due to the presence of other ions. Sodium oxalate build up in a typical plant liquor is shown in Fig.6 and sodium oxalate removed (28-30%) from plant liquor using the above solubility criteria is shown in Fig.7.

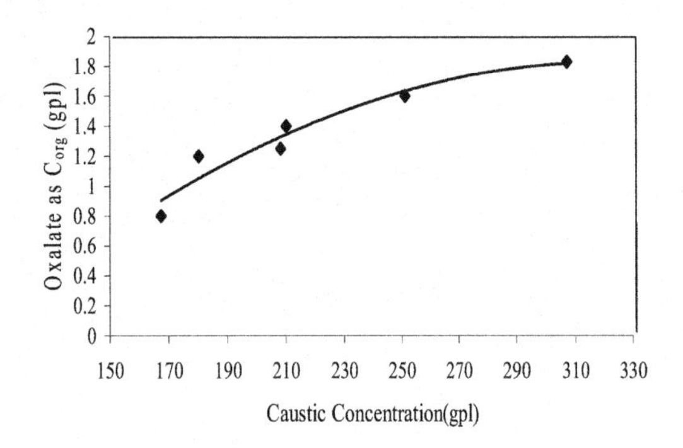

Fig.6. Oxalate build up in a saturated plant liquor

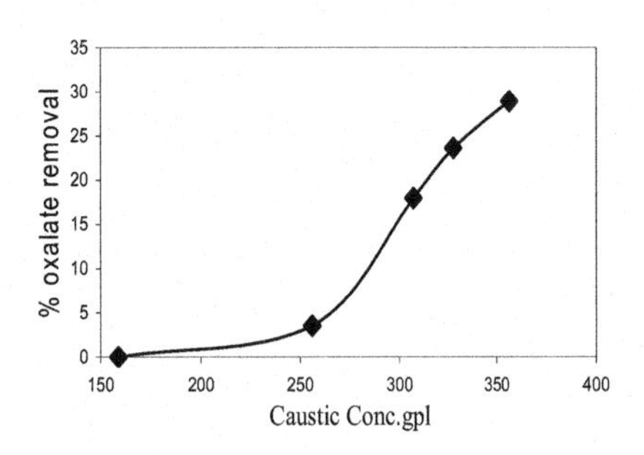

Fig.7 Percentage oxalate removal on seeding a saturated plant liquor

The micrograph depicting the needle shaped crystals precipitated along with humates in the liquor is shown in Fig.8.

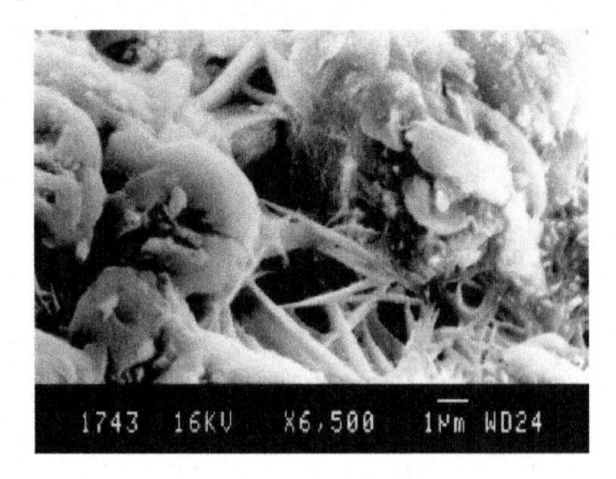

Fig.8 A micrograph depicting oxalate crystals precipitated in our trials

Some reported control methods [5,6] were also investigated for their applicability to plant liquors and the results were presented in Table-5. Some untried methods were also included for comparison. Studies on specialty chemicals like CGM and synthetic flocculants have revealed that synthetic flocculants are more stable and CGM use for longer periods may increase TOC in plant liquors. Further investigations are in progress to confirm its contribution to organic problems in alumina plants.

Conclusions

Organic levels in Indian plant liquors are still not alarming for the reason that bauxites contain low-intermediate range TOC. However, old plants using special chemicals and bauxites with higher TOC have started facing problems associated with organic build up. Several organic control methods were tested for their suitability to Indian plant conditions. Organic control methods which are not energy intensive, though may not be of broad spectrum organic controls, have only to be implemented for energy savings in Indian alumina plants.

Acknowledgements

The authors are thankful to the earlier Directors Dr. T.R.Ramachandran and Dr.V.V.Kutumbarao for many useful discussions and the present Director Dr.J.Mukhopadhay of JNARDDC for constant encouragement and granting permission to publish the work. Dr.P.G.Bhukte, Scientist is also duly acknowledged for his geological map of Indian bauxites. The authors also thank S & T Division, Ministry of Mines, Government of India for sponsoring the project on Organics in Bayer process.

References

1. I.Molnar-Perl,V.Fabian Vonsik and M.Pinter-Szakacs, Chromatographia, 18 (12).1984, pp 673-676
2. K.V.Ramana Rao and T.R.Ramachandran, Proc. Int. conf. on Aluminium INCAL-98, Feb,.1998, pp 341-346.
3. K.V.Ramana Rao and R.N.Goyal, Proc. Int. Conf. on Aluminium INCAL-03, New Delhi, April 2003, pp 291-297
4. T.Yamada,K.Hashimoto and K.Nagano, Australian Patent no. 63270/73,5 Dec 1973
5. Kwat The, Proc.11 th Int. Sym. ICSOBA on Quality control in Aluminium Industry, 1996 pp 211-218
6. Andre Lalla and Rolfe Arpe, Light Metals, 2002 pp 177-180

Light Metals 2006 *Edited by Travis J. Galloway* ***TMS (The Minerals, Metals & Materials Society), 2006***

EFFECTS OF ULTRASONIC ON BAYER DIGESTION OF DIASPORIC BAUXITE

Jianli WANG [1, 2], Qiyuan CHEN[1], Wangxing LI[2], Zhoulan YIN[1]
[1]Central South University, Changsha, Hunan, 410083, China
[2]Zhengzhou Research Institute of Chalco, Zhengzhou, Henan, 450041, China

Keywords: ultrasonic, diasporic bauxite, Bayer digestion, effect, mechanism

Abstract

The effects of ultrasonic irradiation on digestion of diasporic bauxite are studied by measuring the relative digestion ratio of diasporic bauxite. Experimental research has been carried out at different digestion temperature and caustic concentrations. The results show that the ultrasonics can improve the digestion of diasporic bauxite. In the process of Bayer digestion irradiated by ultrasonic, both the digestion ratio and digestion rate rise. Compared to a conventional Bayer digestion of diasporic bauxite, the digestion time can be shortened by 20-40 minutes, the digestion temperature can be decreased by about 20ºC, and the caustic concentration can be reduced over 20g/L. The mechanism by which ultrasonic irradiation affects the digestion of diasporic bauxite is discussed.

Introduction

The dissolution of bauxite ores, containing varied aluminum hydroxide crystalline phases such as gibbsite, bayerite, boehmite and diaspore in concentrated caustic solutions and subsequent recrystallisation of purified $Al(OH)_3$ is called the Bayer process [1]. The digestion of diasporic bauxite is a critical step in the Bayer process where diasporic bauxites are the principal feed material. Apart from red side efficiencies, it can subsequently influence precipitation, in terms of the yield and quality of product.

There are abundant resources of centrally distributed bauxite, in China, most of them located in Henan, Shanxi, Guizhou and Guangxi, which is advantageous to alumina production. But most of the bauxites are diasporic, the most difficult to digest among the aluminum hydroxide crystalline phases. During digestion, the diasporic bauxite ores cannot be easily penetrated by the hydroxide ions because the diasporic crystals are perfect and connect firmly.

The high content of ferric oxide in diasporic bauxite also hinders the digestion. Moreover, during the process of digestion, rutile or anatase in ores react with sodium hydroxide in caustic liquor to form Na_2TiO_3, which covers the surface of the diaspore and forms a dense layer to prevent the digestion. Very high digestion temperature, strong caustic liquor concentration and long digestion times are therefore employed in the process of digestion [2].

In order to improve the process of digestion, much research has been done in the world, but with limited success. Recently, ultrasound has been applied to chemical reactions in both homogeneous liquids and solid-liquid systems [3]. When ultrasonic is applied to the liquids, it can cause heating, mechanical effects, cavitation, etc.

There are two cavitation effects caused by ultrasonic near the extended liquid–solid interfaces in the liquid–solid system: micro jet impact and shock wave damage. The deformation of the cavity during its collapse sends a fast-moving stream of liquid through the cavity with velocities greater than 100 m/s [4-7]. The cavitation-induced surface micro jet attack is also accompanied by shock waves created by cavity collapse in the liquid.

The process of diasporic bauxite digestion is a complicated liquid-solid reaction system, with several chemical reactions and mass transfers taking place simultaneously. If the digestion of diasporic bauxite is irradiated by ultrasonic, both digestion efficiency and reaction rate should be increased. Based on the features of ultrasound, ultrasonic irradiation seems to be an attractive new technology for the digestion of diasporic bauxite. The aim of this paper is to study the effects of ultrasonic radiation on the Bayer digestion of diaspoic bauxite.

Experimental

2.1 Experimental materials

Caustic liquor: from alumina processing operation of Henan Branch of CHALCO, China.
Diasporic bauxite and lime: both from Henan Branch of CHALCO, China. Their chemical composition and particle size distribution are given in Table 1 and Table 2.

Table 1 Chemical composition of diasporic bauxite and lime

Ore	Al_2O_3	SiO_2	Fe_2O_3	CaO	TiO_2	K_2O	Na_2O	LOI
Bauxite	71.20	5.89	3.76	1.00	3.37	0.58	0.08	14.24
Lime	1.20	1.82	-	95.20	-	-	-	-

Table 2 Particle size distribution of experimental ores

Ore	>425 μm	>250 μm	>150 μm	>90 μm	>63 μm	<63 μm
Bauxite	1.54	3.39	10.57	19.29	38.87	61.13
Lime	0	0	1.30	11.72	47.39	52.61

2.2 Experimental apparatus

The experimental apparatus is shown in Fig.1. The frequency of the ultrasonic generator used for the experiment is 20 kHz, and the ultrasonic intensity is about 10 watt·cm^{-2}. The volume of the high-pressure reactor with ultrasonicator is 5 liters. An electric heater is used to heat the reactor. The temperature of the reactor is held constant within ±1ºC.

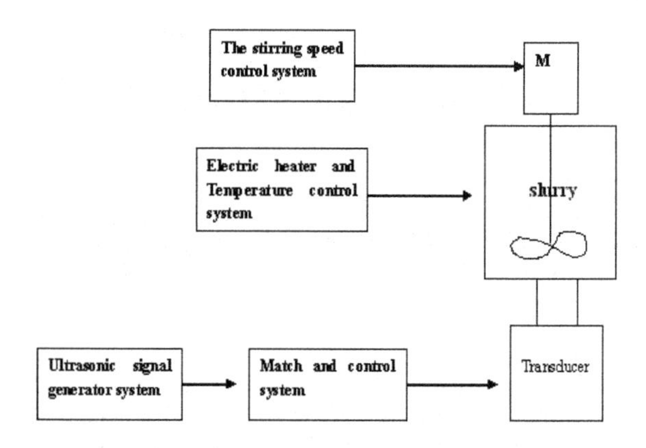

Fig. 1. Schematic diagram of experimental apparatus

2.3 Experimental methods

The feed slurry was made by mixing diasporic bauxite, diluted caustic liquor, and lime. The molecular ratio of Na_2O/Al_2O_3 in feed slurry is 1.5. The mass ratio of Al_2O_3/SiO_2 in red mud is 1.5 (in case of perfect digestion), while mass of the lime to be added is 8% of the bauxite's. The stirring rate of reactor is 380 rpm. The reactor was heated at a rate about 3°C/min. After having been digested at the given temperature for the required time, the digested slurry was taken from the reactor and filtered, before the filter cake was washed with hot distilled water. The red mud was dried and its chemical components analyzed.

The relative digestion ratio (η_R) is calculated according following formula [7]:

$$\eta_R = \frac{(A/S)_O - (A/S)_M}{(A/S)_O - 1} \times 100\%$$

Where η_R is the relative digestion ratio of diasporic bauxite, $(A/S)_O$ is the mass ratio of Al_2O_3/SiO_2 in bauxite, $(A/S)_M$ is the measured mass ratio of Al_2O_3/SiO_2 in red mud.

Results and discussion

3.1 Effect of irradiation on digestion at different digestion temperatures

The feed slurry, which was formulated using diluted caustic liquor (Na_2O in liquor is 220 $g \cdot L^{-1}$), was introduced into the reactor, and then the reactor was heated at a rate of 3°C/min. After the target temperature was reached, the digestion was carried out for 90 minutes. The digestion with ultrasonic was irradiated when the temperature reached 95°C. Fig.2 shows the results of these experiments at different temperature.

From Fig.2, it can be seen that ultrasonic irradiation can increase the digestion rate, and that the effect of irradiation on digestion is more obvious at lower temperature. The digestion rate of diasporic ores with ultrasonic irradiation at 220°C is almost equal the digestion rate without irradiation at 237 °C.

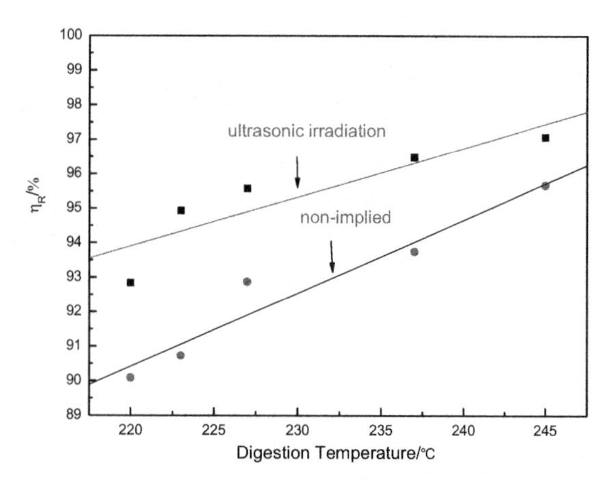

Fig. 2 Effect of irradiation on digestion at different digestion temperatures

3.2 Effect of irradiation on digestion at different caustic concentration

The evaporated caustic liquor produced by the alumina refinery was diluted with distilled water to caustic concentrations of 220g/l, 200g/l and 184g/l, as Na_2O. Then the digestion of slurry formulated using the various concentrations of caustic liquor was carried out at 245°C for 90 minutes. The ultrasonic irradiation was started when temperature reached 95°C. The experimental results are listed in Table 3.

Table 3 Effect of irradiation on digestion at different caustic concentrations

NO	Na_2O Concentration ($g \cdot l^{-1}$)	Irradiation method	Composition of red mud (%)		Digestion ratio
			Al_2O_3	SiO_2	η_R (%)
1	220	Not-applied	22.30	15.07	95.67
2	220	Ultrasonic irradiation	20.90	15.50	96.86
3	200	Not-applied	23.70	15.00	94.77
4	200	Ultrasonic irradiation	21.90	15.21	96.03
5	184	Not-applied	24.50	14.67	93.96
6	184	Ultrasonic irradiation	23.10	15.17	95.28

The experimental results show that ultrasonic irradiation can enhance the digestion ratio, and allow the use of a lower caustic concentration. The digestion ratio at a caustic of 200 $g \cdot L^{-1}$ with ultrasonic irradiation is a little larger than that at caustic concentration 220 $g \cdot L^{-1}$ without ultrasonic irradiation.

3.3 Effect of irradiation on digestion rate

The feed slurry, which was formulated using diluted caustic liquor (the concentration of Na_2O is 220 $g \cdot L^{-1}$), was introduced into the reactor, and then the reactor was heated at a heating rate of 3°C/min. When the temperature reached the required digestion temperature, about 100ml of slurry was taken for chemical analysis at 0, 20, 40, 60, 80, 100 and 120 minutes, respectively. The digestion ratios at different digestion time are shown in Figure 3, where 245N, 227N and 223N refer to digestion without ultrasonic irradiation at 245°C, 227°C and 223°C respectively, while 245Y, 227Y and 223Y refer to digestion with ultrasonic irradiation at 245°C, 227°C and 223°C, respectively.

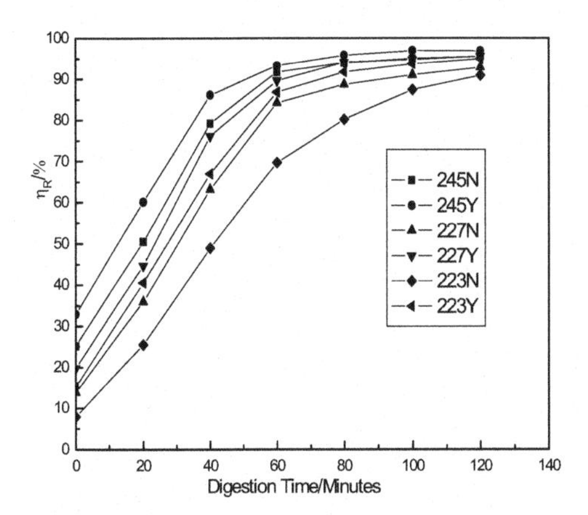

Fig. 3 Effect of ultrasonic irradiation on digestion rate

The experimental results in Fig.3 illustrate that the Bayer digestion rate of diasporic bauxite can be improved by ultrasonic irradiation. The lower the temperature, the more obvious the effect of ultrasonic irradiation. The final digestion ratio with ultrasonic treatment at 223°C can reach that without ultrasonic treating at 245°C. The experimential results show that ultrasonic can soften conditions of digestion and accelerate the rate of digestion of the diasporic bauxite.

3.4 Effect of irradiation time on digestion

The feed slurry, formulated using diluted caustic liquor (the concentration of Na_2O is 220 $g \cdot L^{-1}$), was introduced into the reactor. The reactor was heated from room temperature to 237°C and kept there for 120 minutes. The ultrasonic irradiation was applied to the digestion at different times: starting when heating of the reactor first begins, when temperature reached 100°C, and when temperature reached 237°C. These cases are expressed as Continuous Irriadited, 100 Irradiated, and 237 Irradiated, respectively. The experiment results are shown in Fig.4.

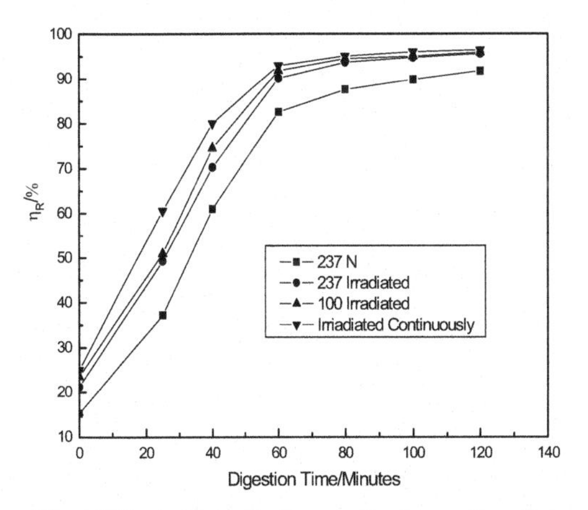

Fig.4 Effect of irradiation methods on digestion

From fig. 4 we can see that the digestion ratio at 60 minutes with ultrasonic irradiation is a little larger than that at 100 minutes without ultrasonic irradiating. The results show that the final digestion ratio of the diasporic bauxite irradiated by ultrasonic is two percent higher than that without ultrasonic at 237°C. The result with Continuous Irriadiation is the best, but the effect of irradiation methods is slight.

Proposed mechanisms

The reaction between the disporic bauxite and the caustic liquor is a complicated multi-phase reaction. The reason that ultrasonic irradiation affects the digestion of diasporic bauxite can be explained by the following mechanism. At lower temperature (<215°C), the reaction is controlled by chemical processes, while the reaction is controlled by diffusion at the higher temperatures (>215°C) [7].

When the feed slurry is irradiated with ultrasonic, the high-energy sound wave induces mechanical effects in the slurry, which can cause the materials to vibrate intensely, increasing the diffusion effect. In addition, when feed slurry is irradiated by ultrasound, acoustically induced cavitation takes place causing markedly asymmetric bubble collapse to occur, generating a high-speed jet of liquid directed at the solid-liquid surface [8–10].

The impingement of this jet and related shock waves can create localized erosion in the surface of ores [8–11], cause particle fragmentation, produce fresh reactive surfaces, and improve mass transfer [12–14]. The induced cavitation and the shock waves it creates can reduce the viscosity of the slurry and produce high-velocity interparticle collisions, which is favorable for the dispersion of diasporic bauxite particles [15].

Conclusion

Ultrasonic irradiation can improve the digestion of diasporic bauxite. In the process of Bayer digestion irradiated by ultrasonic, both the digestion ratio and digestion rate rise. Compared to a conventional Bayer digestion of diasporic bauxite, the digestion time can shorten by 20-40 minutes, the digestion temperature can decrease by about 20°C,and the caustic concentration can be reduced by over 20 $g \cdot L^{-1}$.

The proposed mechanism of ultrasonic irradiation improving the diasporic bauxite is that during ultrasoinc irradiation of digestion slurry, induced cavitation and the shock waves it creates can produce high-velocity interparticle collisions and reduce viscosity of the slurry, which is favorable for the dispersion of ore particles. The high-speed jet of liquid and shock waves can create localized erosion and refresh the reactive surface areas.

References

1. James Counter, Andrea Gerson, and John Ralston, "Caustic aluminate liquors: preparation and characterization using static light scattering and in situ X-ray diffracton",*Colloids and Surface A: Physicochemical and Engineering Aspects,* 126(1997), 103-104.
2. Xianjian Gao, Dianzuo Wang, and Weian Ding, "Micro-wave assisted digestion of diasporic bauxite", *Nonferrous Metals,* 47(1995), 55-57.
3. Timothy J. Mason, J. Phillip Lorimer, "An introduction to sonochemistry",*Endeavour*, 13(3) (1989), 23-128.
4. Aharon Gedanken, "Using sonochemistry for the fabrication of nanomaterials", *Ultrasonics Sonochemistry,* 11(2) (2004), 47-55

5. K. S. Suslick et al., "Characterization of sonochemically prepared proteinaceous microspheres",*Ultrasonics Sonochemistry,* 1(1) (1994), S65-S68
6. Yongxin Li, Baoqing Li, "Study on the ultrasonic irradiation of coal water slurry", *Fuel,* 79(2000), 235-241.
7. R. L. Hunicke, "Industrial applications of high power ultrasound for chemical reactions",*Ultrasonics*, 28(5) (1990), 291-294
8. Zhongyu Yang, *Technique on Alumina Production Process* (Beijing: Metallurgical Industry Press, 1993), 70-90.
9. K. S. Suslick et al., "Effects of high intensity ultrasound on inorganic solids",*Ultrasonics*, 25(1) (1987), 56-59
10. D. J. Flannigan, S. D. Hopkins, and K.S. Suslick, "Sonochemistry and sonoluminescence in ionic liquids, molten salts, and concentrated electrolyte solutions",*Journal of Organometallic Chemistry*, 690(15) (2005), 3513-3517
11. Gareth J. Price, "The effect of high-intensity ultrasound on diesel fuels ",*Ultrasonics Sonochemistry,* 2(2) (1995), S67-S70
12. A. Alippi, A. Galbato, and F. Cataldo, "Ultrasound cavitation in sonochemistry: Decomposition of carbon tetrachloride in aqueous solutions of potassium iodide",*Ultrasonics,* 30(3) (1992), 148-151
13. Ruo Fen, Huamao Li, *Sonochemistry and Application* (Hefei: Anhui Science and Technology Press, 1992), 10-30.

ALUMINA & BAUXITE

Joint Session of Alumina & Bauxite and Aluminum Reduction Technology

SESSION CHAIR
Renaud Santerre
Alcan Primary Metal
Chicoutimi, QC, Canada

Light Metals 2006 *Edited by Travis J. Galloway* *TMS (The Minerals, Metals & Materials Society), 2006*

FLUIDCON - A New Pneumatic Conveying System for Alumina

ANDREAS WOLF
Claudius Peters Projects GmbH,
Schanzenstr. 40, Buxtehude 21614, Germany
PETER HILGRAF
Claudius Peters Technologies GmbH,
Schanzenstr. 40, Buxtehude 21614, Germany

Summary

A description is given of a simple dense phase conveying system that combines the advantages of airslide conveying and pneumatic pipe transport. The characteristics of the FLUIDCON conveying system are an extremely low transport velocity and a low power requirement. The design of the conveying pipe, the layout of the system as well as the requirements to the bulk materials which are conveyable with FLUIDCON are discussed. The operating behaviour of FLUIDCON in the case of the transport of "sandy alumina" is represented with the aid of the results of extensive systematic measurements carried out in a test plant in the Claudius Peters Technologies research center and is applied to a real plant implementation. The results of this plant design show clearly the advantages of the new conveying process.

1 Introduction

The requirements on a system for the handling and transport of alumina are:

- no grain abrasion and no grain fracture; the particle size portion of < 45 μm, critical for the further processing procedure, must not be increased,
- no segregation according to grain size, meaning the critical portion < 45 μm must not accumulate during transport or storage etc., neither spatially nor over time,
- systems must be designed wear-resistant and must have low wear during operation.

The two points stated first essentially influence the quality of the product "sandy alumina" and therefore also the operation of the subsequent aluminium electrolysis cells. The third point influences the economic efficiency of the complete production line. For alumina storage silos and feeding systems for the electrolysis cells highly efficient and proven solutions already exist with ASS (→ Anti Segregation System) and ADS (→ Aerated Distribution System) [1].

The transport of alumina between different plant areas can be realized with either mechanical or pneumatic conveying systems. Simple plant design and the closed, i.e. environmentally-sensitive, conveying line are advantages of pneumatic conveying. Disadvantages compared to the mechanical conveying are the system-inherent higher power consumption and an increased wear sensitivity. With the presented FLUIDCON conveying procedure these disadvantages can be substantially reduced.

2 Conveying principle and application

FLUIDCON can be described as a combination of airslide conveying and pneumatic pipe conveying. It consistently uses the advantages of both processes and eliminates essential disadvantages by combining them. Airslide conveying is characterized by an extremely low energy consumption but needs an inclination of the airslide in flow direction. It therefore has a limited flexibility in the pipe routing, i.e. no vertical conveying is possible. The advantage of the pneumatic pipe conveying is the almost unlimited flexibility in the pipe routing, its disadvantage is the much higher power consumption.

Fig.1 shows the design of the FLUIDCON pipe and Fig. 2 the structure of a respective conveying plant. The total gas flow supplied by a pressure generator is divided into a fluidizing gas flow and a driving gas flow. The fluidizing gas flow quantity is adjusted by a controller and is fed to the conveying pipe, distributed along the transport route, for fluidization of the bulk material. The driving gas flow is fed at the beginning of the conveying pipe and triggers the axial solids transport. Here the pressure drop of the driving gas flow replaces the inclination of the airslide. Due to the fluidization the bulk solid is transferred into a fluid-like state with nearly no internal friction and is lifted off the pipe bottom and introduced into the driving gas flow. It therefore does not support itself on the (horizontal) pipe wall. These are optimum conveying conditions to realize the same conveying velocities as on an airslide. The fluidizing gas is fed to the conveying pipe via fluidization elements, which are adjusted in their geometry to the circular conveying pipe cross section and which can be exchanged and dismantled individually without modifications of the conveying pipe. The maximum length of these elements, independently fed with gas, is currently ΔL = 2 m. For the gas distribution normal airslide fabric is used, if necessary covered with perforated plate as wear protection. In special cases metal fabric can be used as distributor. Bends in the conveying line and vertical pipe segments are not fluidized.

Suitable and proven solid feeders for FLUIDCON are: pressure vessel, screw feeder, rotary-valve feeder and various flap-type feeders. The implementation of multi-point feeds, i.e. the more or less simultaneous feed of bulk material into one transport line through several feeders in parallel, is also possible and used for example for the fly ash removal from power plant filters. The gas supply of FLUIDCON plants can in many cases be realized by blowers. Conveying pressures of up to approx. 3 bar(g) are possible.

The bulk materials that are particularly suitable for FLUIDCON are all those that can be fluidized with low gas velocities and then expand substantially homogeneously. A high gas retention is also of advantage. Bulk materials with appropriate characteristics are to be found in the entire hatched part of the Geldart diagram shown in Fig. 3. The suitability of products outside the hatched area has to be analyzed for each individual case by fluidization and various other tests. All bulk materials plotted in Fig. 3 have already been transported successfully with FLUIDCON. Table 1 summarizes their main characteristics. Gas velocities at the inlet of the conveying pipe of $v_{F,A} \cong (1.0 - 4.0)$ m/s with specific fluidizing gas flows of $\dot{q}_{WS} \cong (0.3 - 1.0)$ $m^3/(m^2 \cdot min)$ are realized.

For all bulk materials tested a stable and pulsation-free conveying through pipes inclined upward up to $\alpha_R \cong 30°$ against the horizontal (→ tested so far) was possible without backflow. This behaviour has also been confirmed in operating plants. Restarting after conveying that has been interrupted by, for example, a power failure, i.e. starting up with a full line, is absolutely no problem with FLUIDCON. The conveying gas is fed to the conveying system at different times: when the fluidizing gas has been applied, the driving gas flow is switched on after a time delay. This picks up the bulk material that has already been transformed to a fluidized state and takes it away evenly and without significant pressure fluctuations.

A test plant with pipe diameter DN 100 is available in the CP technical center for testing of new bulk materials. A calculation model has been prepared. Details are given in [2].

3 FLUIDCON for alumina

Some of the characteristics of sandy alumina, relevant for the specific behaviour in pneumatic conveying systems, are compiled in Table 2. Sandy alumina is a bulk material at the transition of Geldart groups A/B. It can therefore be fluidized with low gas velocities; minimum fluidizing velocity $u_L \cong (0.20 - 0.35)$ m/min. The gas flow exceeding u_L flows through the bed in form of gas bubbles. Alumina runs like a liquid over airslides with an inclination of $\alpha_{WS} \cong 1.0°$. The gas retention property is low (→ de-aeration time of 2 kg fluidized solid: $\Delta\tau_{WS} \leq 20$ s), which has the effect that in case of an interruption of the gas supply the fluidized bed is vented via the bubble phase and goes abruptly back to a behaviour like a rigid body. Alumina is a free-flowing product.

Sandy alumina consists of single particles with porous structure. This becomes clear in the difference between BET surface, $A_{BET} \cong 50$ m^2/g, and Blaine surface, $A_{Blaine} \cong 0.15$ m^2/g, as well as between solid density, $\rho_S \cong 4500$ kg/m^3, and particle density, $\rho_P \cong 2200$ kg/m^3.

The wear behaviour of alumina can be characterised by its Vickers hardness, HV(alumina) $\cong 11.3$ kN/mm^2, in relation to the hardness value of the respective wall material. The comparison with normal steel St 37, HV(St37) $\cong 1.3$ kN/mm^2, makes clear, that with this material combination, (HV(alumina) / HV(St37)) $\cong 8.7 >> 1$, an extremely high level wear has to be expected. This is also valid for other metallic materials, as their hardness normally clearly lies below HV $\cong 5$ kN/mm^2.

For conventional pneumatic conveying of alumina gas velocities at the conveying pipe inlet of $v_{F,A} \geq 12$ m/s are necessary for a safe operation (→ lean phase conveying). A reduction of the gas velocity results in gas/solid separation along the conveying pipe. A moving strand at the bottom of the pipe develops over which conglobations and dunes run irregularly. These can combine uncontrolled to plugs of various lengths, which fill up the complete pipe cross section. Conveying under these conditions is strongly fluctuating with regard to solids mass flow and pressure. Plugs of excessive length result in pipe clogging due to the low gas retention property of the alumina combined with its also low gas permeability. An operation in this mode is only possible with special by-pass conveying methods. With these initial conveying gas velocities of $v_{F,A} \geq (6 - 7)$ m/s are possible [3].

With sandy alumina, but also with Al_2O_3 -hydrate, systematic FLUIDCON conveying tests were carried out. Varied parameter range, e.g.: $\dot{q}_{WS} \cong (0.4 - 1.0)$ m^3/ (m^2·min) $\cong (1.5 - 4.0) \cdot u_L$, $v_{F,A} \cong (1 - 7)$ m/s. Fig. 4 compares exemplarily two typical measuring charts. Plotted against the time axis τ_V are the conveying pressure p_R at the start of the line and the increase in weight M_S (G_S) in the receiving silo. The slope ($\Delta M_S / \Delta\tau_V$) corresponds to the solid throughput $\dot{M}_S$. Further measuring values are not shown for sake of clarity. At identical specific fluidizing gas flows, $\dot{q}_{WS} = 0.74$ m^3/(m^2·min), and nearly identical conveying pressure drops, $\Delta p_R \cong 1.4$ bar, different flow modes take place in the conveying line, depending on the selected driving gas velocity, $v_{F,A} = 2.8$ m/s or 4.7 m/s. For realization of a pulsation free conveying operation in the above case, the bulk solid mass flow $\dot{M}_S$ related to the available conveying pipe cross section A_R has to exceed a minimum mass flux $\dot{m}_S = (\dot{M}_S / A_R)$. This necessitates a higher driving gas velocity, because with the specified boundary conditions an increase in the throughput $\dot{M}_S$ is only possible by an increase of $v_{F,A}$. The specific mass flux in Fig. 4a is $\dot{m}_S \cong 1820$ t/(h·m^2) but in Fig. 4b it is $\dot{m}_S \cong 2981$ t/(h·m^2).

In detail the process is more complex. The cause for this behaviour is, inter alia, the already discussed bubble formation in alumina fluidized beds, where the surrounding suspension phase remains in the condition of minimum fluidization. This results in a strongly decreasing solid concentration over the height of the (horizontal) conveying pipe cross section, which can only be compensated by a sufficiently high mass flux $\dot{m}_S$. This is shown schematically in Fig. 5. Optimum conveying behaviour is achieved by operating conditions as shown in Fig. 5a.

In Fig. 6 and 7 the dependencies of the alumina mass flow $\dot{M}_S$ on the conveying pressure difference Δp_R and the initial conveying gas velocity $v_{F,A}$ under constant boundary conditions are shown. $\dot{M}_S$ increases, as in conventional conveying systems, slightly over-proportional with Δp_R, see Fig. 6. The fact that the dependencies $\dot{M}_S(v_{F,A})$ in Fig. 7 also increase with increasing $v_{F,A}$ shows, that the underlying measurements were taken in the dense phase conveying range, because only there the requirement $\dot{M}_S \propto v_F^x$ is fulfilled. For lean phase conveying the following is valid: $\dot{M}_S \propto (1/v_F^x)$ [3]. Further it can be seen from both figures, that alumina can be conveyed with initial gas velocities of $v_{F,A} < 2$ m/s.

Fig. 8 shows the dependencies $\dot{M}_S = \dot{M}_S(\dot{q}_{WS})$, that means the influence of the specific fluidizing gas flow $\dot{q}_{WS}$ on $\dot{M}_S$, of different bulk materials under constant boundary conditions. Stucco and cement run into a horizontal curve leg after exceeding a critical $\dot{q}_{WS}$ value. Up to now this course has been found in nearly all tested bulk solids and defines an energetic optimum operating point for these kinds of systems. If $\dot{q}_{WS}$ exceeds the critical value $(\dot{q}_{WS})_{crit}$ this does not lead to a further increase of $\dot{M}_S$, but increases the demand of compressed conveying gas and the cost for the equipment. In Fig. 8 sandy alumina and pet coke still lie above their critical fluidization. In case of pet coke a stable conveying was possible with only $\dot{q}_{WS} \cong 0.08$ m^3/(m^2·min) and $v_{F,A} < 0.5$ m/s. A comparison of the solid mass flows of the various bulk materials in Fig. 8 under consideration of the various boundary conditions shows that alumina is a bulk solid difficult to convey pneumatically.

From the above described measuring values the associated friction coefficients can be re-calculated according to the selected calculation method.

4 Case study

$\dot{M}_S$ = 135 t/h of sandy alumina has to be conveyed over a total distance of L_R = 410 m including a total height of H_R = 35 m. The conveying height is divided into three height steps: One of them with an upward inclination of $\alpha_R \cong 8°$ and two with vertical arrangement. A no. of n_R = 7 90° bends are installed in the conveying line.

Due to constructional limitation a screw feeder, type CP X-pump, size 300, was chosen as the feeder for introducing the bulk material into the FLUIDCON pipe. The X-pump is a high-speed, $n_P \cong (500 \ldots 1500)$ min^{-1}, compression screw system with two-sided bearings, whose screw duct is completely filled with bulk solid. Sealing of the conveying pressure Δp_R is carried out via the bulk material in the screw duct itself. Conveyor pipe = feeder pressure differences up to $\Delta p_R \cong 2.5$ bar can be realised. During start-up and shutdown a check flap closes the screw duct. These screw feeders can be used for bulk solids with low gas permeability. Details are given in [4]. The wear of a screw feeder system can be reduced by reducing the conveying pressure Δp_R, decreasing the screw speed n_P, selecting the correct screw type for the corresponding bulk solid and by numerous proven anti-wear measures. Sufficient experience is available [5].

For the case at hand the low construction height and the continuous operation of the screw feeder were decisive for selecting it. The screw speed was determined with n_P = 730 min^{-1} under load, the associated conveying pipe pressure difference was limited to $\Delta p_R \cong 0.95$ bar by respective design of the FLUIDCON pipe and all wear parts were protected by suitable armouring.

The conveying pipe diameter for the stated boundary conditions was calculated as D_R = 388.8 mm, pipe (∅406.4 x 8.8) mm. During operation the conveying system needs a pre-pressure of p_{vor} = 1.25 bar(g) and a total gas volume flow of $\dot{V}_F$ = 7485 m^3/h at 1.0 bar (abs) and 20°C. The compressor is designed with Δp_C = 1.50 bar. The superficial gas velocity at pipe inlet is $v_{F,A}$ = 3.0 m/s, at pipe outlet $v_{F,E}$ = 16.2 m/s. A specific fluidization gas flow of $\dot{q}_{WS} \cong 0.65$ $m^3/(m^2 \cdot min)$ is used. A summary of operation data and calculation results is given in Table 3. In Fig. 9 the course of the gas velocity along the conveying line is shown. Over approx. three quarters of the conveying distance the gas velocity remains lower than v_F = 10 m /s. The solid velocities are approx. 1.0 m/s lower than those of the gas. By staggering the FLUIDCON-pipe a further reduction of the final gas velocity is possible.

Table 3 shows, that during operation a power consumption at coupling of P_C = 254 kW for the compressor and of P_P = 65 kW for the drive motor of the screw feeder is necessary. This results in a specific energy demand of the conveying plant of

$$P_{spec} = \frac{P_C + P_P}{\dot{M}_S \cdot L_R} = \frac{(254+65)kW}{135\frac{t}{h} \cdot 4.1 \cdot 100m} = 0.576 \frac{kWh}{t \cdot 100m}.$$

This is a very low value for the product „sandy alumina". Comment: the discussed plant design is based on a real order. The complete design therefore still includes safety margins.

References

[1] Karlsen, M., Dyroy, A., Nagell, B., Enstad, G.G., Hilgraf, P.: New Aerated Distribution (ADS) and Anti Segregation (ASS) Systems for Alumina. TMS Annual Meeting 2002, Seattle, USA, 2002 February 17-21. P.5. A. (02-1000-8).

[2] Hilgraf, P.: FLUIDCON - a new pneumatic conveying system for fine-grained bulk materials. CEMENT INTERNATIONAL 2 (2004) No. 6, pp. 74 - 87.

[3] Hilgraf, P.: Review of pneumatic dense phase conveying, part 1 und 2. ZKG INTERNATIONAL 53 (2000) No. 12, pp. 657 - 662 and 54 (2001) No. 2, pp. 94 - 105.

[4] Hilgraf, P., Paepcke, J.: Introducing bulk materials into pneumatic conveying lines with screw feeders. ZKG INTERNATIONAL 46 (1993) No. 7, pp. 368 - 375.

[5] Hilgraf, P.: Wear in Pneumatic Conveying Systems. powder handling & processing 17 (2005) Nr. 5, September/October, S. 272 - 284.

Table 1: Bulk materials investigated

No.	Bulk solid	Average Particle-∅ $d_{s,50}$ [μm]	Bulk density ρ_{SS} [kg/m³]	Solids density ρ_S [kg/m³]
1	Fly ash	20	1120	2300
2	Cement 1	13	1100	3100
3	Cement 2	29	1260	3110
4	Sandy alumina	78	1020	4760
5	Al_2O_3-hydrate 1	105	1260	2690
6	Al_2O_3-hydrate 2	83	1120	2480
7	Fe-II-sulfate	180	870	1980
8	Titanium ore	19	2000	4700
9	Stucco	20	1060	2790
10	FGD gypsum	15	660	2510
11	Petcoke	36	570	1330
12	Kiln dust	<10	620	2850

Table 2: Characteristics of „sandy alumina“

Dry sandy alumina		
Geldart group	[-]	A/B
Bulk density	ρ_{SS} [kg/m^3]	950 - 1050
Vibrated bulk density	ρ_{SR} [kg/m^3]	1150 - 1250
Aerated bulk density [1]	$\rho_{S,aer}$ [kg/m^3]	775 - 875
Solids density	ρ_S [kg/m^3]	4300 - 4800
Particle density	ρ_P [kg/m^3]	approx. 2200
Average particle-∅ at R = 50 wt-%	$d_{S,50}$ [μm]	75 - 95
Particle size distribution: RRSB slope	α_{RRSB} [degree]	65 - 75
Particle shape acc. FEM 2582	[1]	mixtures of III, IV, V
Angle of repose	α_{SS} [degree]	28 - 38
Specific surface acc. Blaine	A_{Blaine} [m^2/g]	0.1 – 0.2
Specific surface acc. BET	A_{BET} [m^2/g]	$\geq$ 50
Minimum fluidization velocity	u_L [m/min]	0.20 - 0.35
Bed expansion at $v_F = 5 \cdot u_L$	H_{exp}/H_L [1]	1.24 - 1.28
De-aeration time of 2 kg solid [1,2]	$\Delta\tau_{WS}$ [s]	5 - 22
Airslide inclination at $\dot{q}_{WS}$ = 1.0 m/min	α_{asi} [degree]	$\leq$ 1.5
Internal / effective angle of friction	$\varphi_i \cong \varphi_e$ [degree]	35
Wall friction angle against steel St37	φ_W [degree]	approx. 21
Jenike flow function	$FF = \sigma_1/f_C$ [1]	>> 10
Compressibility acc. Carr	$R_C=(\rho_{SR}-\rho_{SS})/\rho_{SR}$ [1]	<< 0.2

1) fluidized with $\dot{q}_{WS} \cong$ 4.0 m/min, 2) diameter of fluidized bed: 100 mm

Table 3: Case study

Bulk solid		sandy alumina
Conveying gas		air
Type of conveying system		FLUIDCON
Type of solid feeder		screw feeder
Solids mass flow	$\dot{M}_S$ [t/h]	135
Total conveying distance	L_R [m]	410
Including: total height	H_R [m]	35
No. of height steps along pipe	[1]	3; 2 vertical, 1 at 8° in-clined above horizontal
No. of 90° bends	[1]	7
Pipe diameter	D_R [mm]	388.8 (∅406.4 x 8.8)
Total gas volume flow	$\dot{V}_F$ [m^2/h at 20°C, 1 bar]	7485
Average spec. fluidization gas flow	$\dot{q}_{WS}$ [m^3/(m^2·min)]	0.65
Gas velocity at pipe inlet	$v_{F,A}$ [m/s]	3.0
Gas velocity at pipe outlet	$v_{F,E}$ [m/s]	16.2
Pipe pressure difference	Δp_R [bar]	0.95
Total pressure difference	Δp_{vor} [bar]	1.25
Power consumpt. of compressor	P_C [kW]	254
Type of screw feeder	[-]	CP X-pump
Screw diameter	D_P [mm]	300
Screw speed	n_P [min^{-1}]	730
Power consumption of feeder	P_P [kW]	65
Solid air ratio at pipe inlet	μ_A [(kg/h)$_S$/(kg/h)$_F$]	41.2
Solid air ratio at pipe outlet	μ_E [(kg/h)$_S$/(kg/h)$_F$]	15.4
Total specific power consumption	P_{spec} [kWh/(t·100m)]	0.576

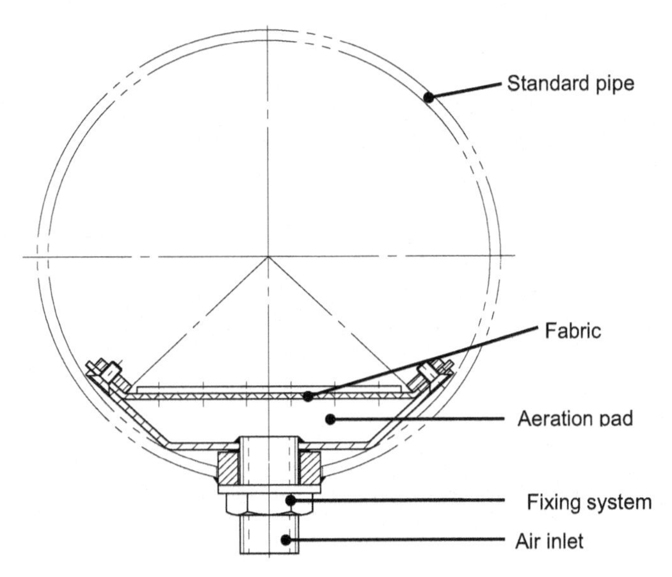

Fig. 1: Design of a FLUIDCON pipe

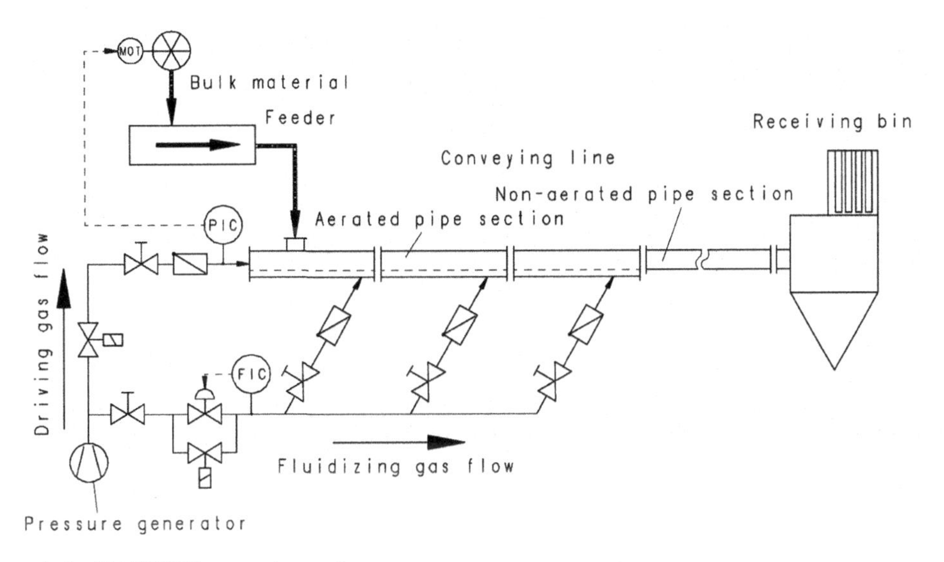

Fig. 2: Layout of a FLUIDCON conveying system

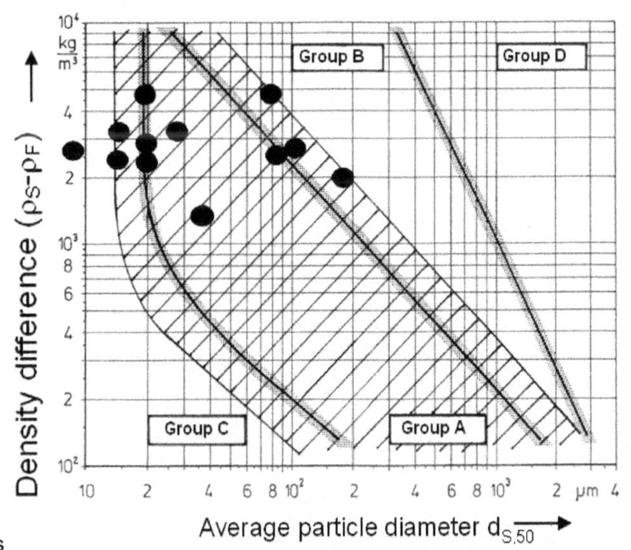

Fig. 3: Suitable bulk solids

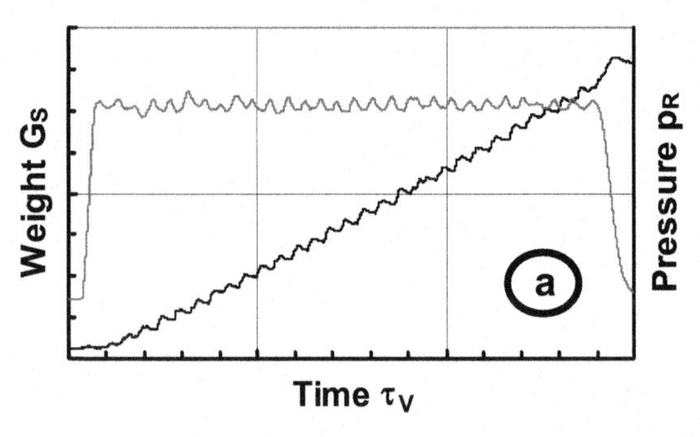

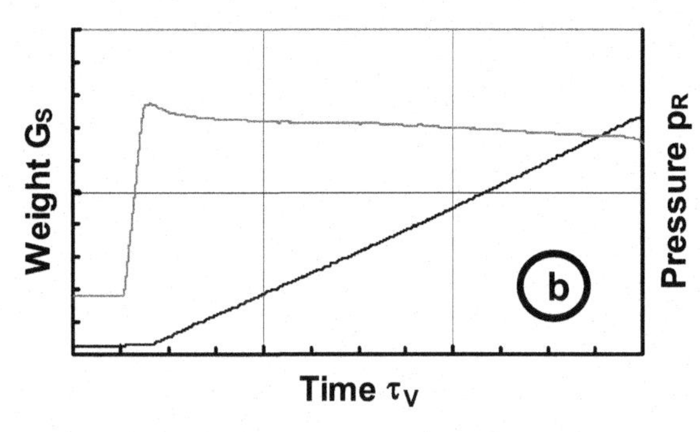

$v_{F,A}$ = 2.8 m/s, q_{WS} = 0.74 m/min, Δp_R = 1.41 bar

$v_{F,A}$ = 4.7 m/s, q_{WS} = 0.74 m/min, Δp_R = 1.38 bar

Fig. 4: FLUIDCON conveying of sandy alumina with different operational settings

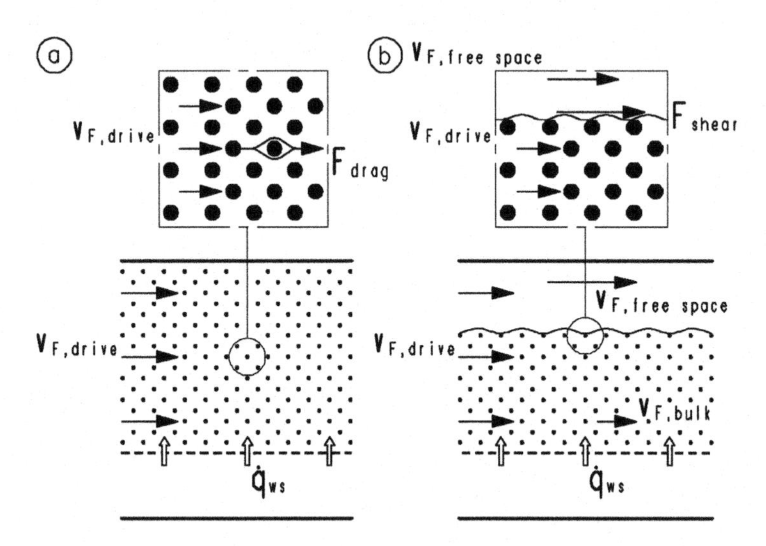

Fig. 5: Possible flow modes and forces acting

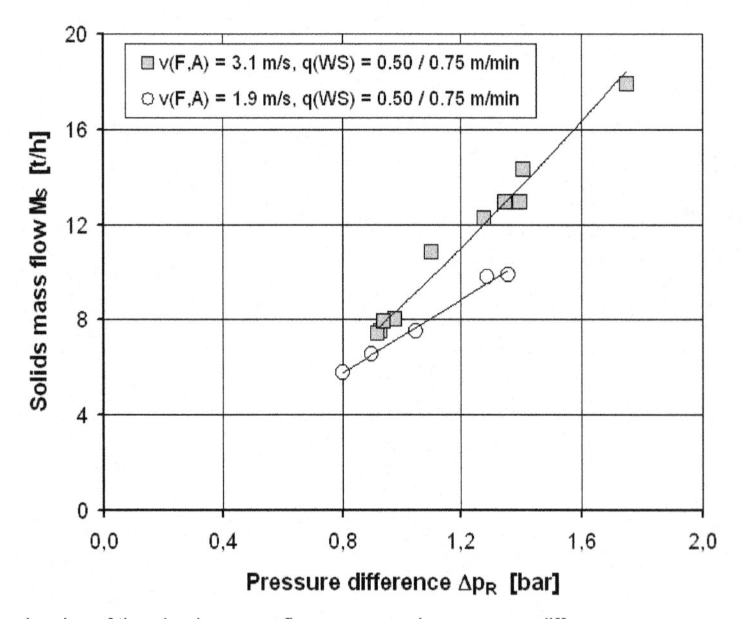

Fig. 6: Dependencies of the alumina mass flow on conveying pressure difference

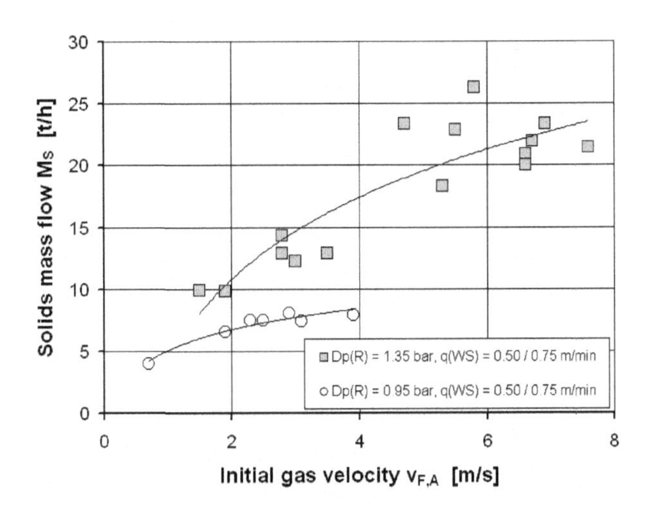

Fig. 7: Dependencies of the alumina mass flow on initial gas velocity

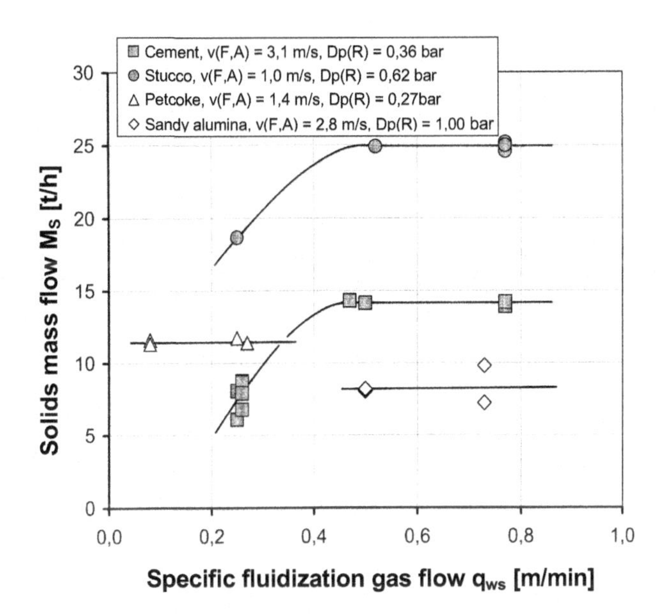

Fig. 8: Influence of the specific fluidizing gas flow on the FLUIDCON conveying characteristic

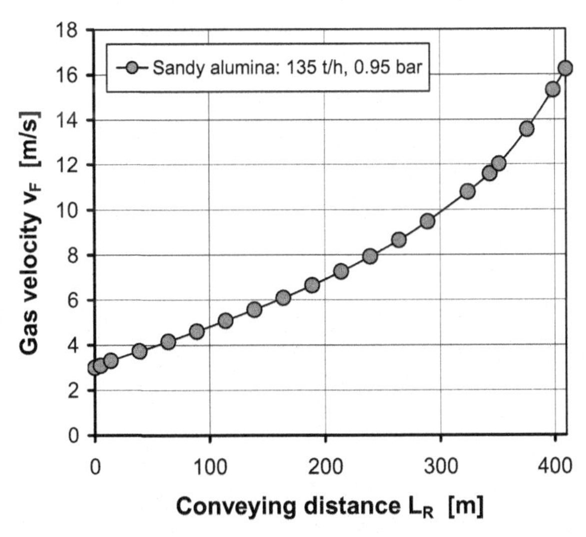

Fig. 9: Conveying gas velocities along the conveying line

EVOLUTION OF MICROSTRUCTURE AND PROPERTIES OF SGA WITH CALCINATION OF BAYER GIBBSITE

James Metson, Tania Groutso, Margaret Hyland and Scott Powell.
Light Metals Research Centre, The University of Auckland
Private Bag 92019, Auckland, New Zealand

Keywords: Gibbsite, Smelter Grade Alumina, Calcination, Microstructure,

Abstract

The microstructure of smelter grade alumina (SGA) is relevant to smelter performance in a number of critical areas. The connection between the degree of calcination, the residual structural hydroxide content and HF generation, has recently been explored, and impacts on alumina dissolution have also been reported. A critical parameter of relevance to both of the above, and particularly to dry-scrubber performance, is the evolution of surface area and its relationship to the progression of transition alumina phase. The relationship of surface area and microstructure with equilibrium heating of gibbsite has been explored in the laboratory and results compared with those from several industrial calciner technologies. For the equilibrium samples, it is possible to accurately represent the transition alumina structure with the γ, γ', θ model previously developed, and to correlate the evolution of phase with changes in surface area. The relative rates of hydroxyl loss versus loss of surface area are of critical importance in understanding the balance between HF generation at the cell and HF capture in the scrubber.

Introduction

The calcination of Bayer process Gibbsite $Al(OH)_3$ to produce smelter grade alumina is frequently referred to as the "removal of the waters from alumina trihydrate". This presents a quite misleading view of the actual transitions involved when gibbsite is calcined. Indeed two quite distinct processes are involved:

- The dehydroxylation, initially of the gibbsite, and then of the transition alumina lattices through the removal of adjacent hydroxyl groups, with the subsequent elimination of a water molecule from the lattice.
- The rearrangement of the lattice from that of Gibbsite, likely through the oxyhydroxide AlOOH, and then through a sequence of transition aluminas, and finally if heating is sustained, into the most thermodynamically stable alumina, alpha.

The sequence or pathway of this transformation is important in that it has a significant impact on the nature of the final product. For example the relationship between specific surface area and the population of structural hydroxides that remain in the lattice [1,2]. Thus SGA properties are sensitive to a range of factors, most prominently the conditions under which the Gibbsite was precipitated, and the technology by which it is calcined.

As alumina represents the single largest raw materials input to an aluminium smelter reduction cell and is the sole materials feed to the scrubber, we would expect alumina properties to be a major determinant in the effective operation of the smelter. Parameters such as SSA and residual hydroxide content (as reflected in the LOI) have significant impacts not only in the feeding and dissolution of alumina [3,4], but also in both the generation of HF [5,6] and in the performance of the dry-scrubber.

As indicated above, calcination technology has significant impacts on alumina properties. In particular the very rapid heating of gibbsite in for example gas suspension calciners, can effectively decouple the dehydroxylation and lattice rearrangement processes identified above, the structural rearrangement may simply not keep up with the rate of water removal from the lattice.

Thus it is of interest to understand how the properties of the final SGA are related to the conditions under which they are calcined. Wefers and Misra [7] still provide the most authoritative description of the interrelationship of the transition alumina phases with respect to SGA including the pathways as a function of calciner technologies. There is ongoing debate in refining and understanding these pathways, and in particular making connections to alumina properties. Unfortunately it is difficult both to accurately identify and quantify the subcrystalline transition aluminas, and to simulate the extremely rapid heating rates of current calciners in the laboratory. For example, the influence of the partial pressure of water, which is high in industrial flash calciners, is particularly difficult to simulate in the laboratory.

However several groups, most notably Davies, Ingram-Jones et.al. have sought to compare flash calcining with soak calcining of gibbsite in laboratory trials [8]. This study used XRD and ^{27}Al NMR to examine the pathway of such reactions both as a function of calcination conditions and particle size. Whittington and Ilievski [9], report on laboratory calcining of gibbsite under conditions which simulate those of fluid bed calciners. The careful control of calcination time allows the phase distribution (although perhaps not the degree of structural disorder) of the FBC products to be reproduced in the laboratory. Meinhold et. al. provide a detailed analysis of disorder in the transition aluminas based on high field MAS NMR [10], but examine only products from laboratory based slow calcinations.

Although the dehydroxylation and structural transformation can potentially be decoupled in flash calcination, under slow or soak conditions, these processes are essentially simultaneous. Thus it might be expected that the pathway and evolution of microstructure will differ for the two regimes. In this study we have examined soak calcining of Bayer Gibbsite to follow the calcination pathway and the evolution of parameters such as specific surface area and pore size distribution.

Experimental

Samples of an industrial Bayer Gibbsite were subjected to slow heating in a horizontal tube laboratory furnace to temperatures between 250 and 1000 °C. Calcination times ranged from 3 to 24 hrs and were generally carried out in air, in some cases with water injection to raise the partial pressure of water. Samples were also size fractionated into above and below 63 μm to check for particle size effects. However given the Ingham-Jones et. al. classification into coarse and fine, with median sizes of 14 and 0.5 μm respectively, significant differences were not expected in the soak calcinations.

The calcined samples were characterised by X-ray diffraction (Siemans) with Rietveld refinement using the FullProf package [11]. BET Surface Area and Pore size distribution were measured on a Tristar 3000 (Micromeritics). Loss on ignition (LOI 300-1000 °C) and Moisture on Ignition (MOI 0-300 °C) were also carried out on some samples using standard methods.

Phase Analysis and Calcination Pathways

The calcination of Bayer Gibbsite $Al(OH)_3$ proceeds by the progressive loss of the structural hydroxides through the capture of a proton from an adjacent -OH group, with the subsequent expulsion of a water molecule [12]. These water molecules diffuse out of the structure via anion vacancies. Thus the process slows markedly as calcination proceeds to the point where there are no longer adjacent hydroxyls, and where structural rearrangement is required [Freund 1965]. In parallel, the lattice undergoes rearrangement through a sequence of distinct structures, however many studies have highlighted the closely structural relationship, high degree of disorder, and thus indistinct boundaries between these forms.

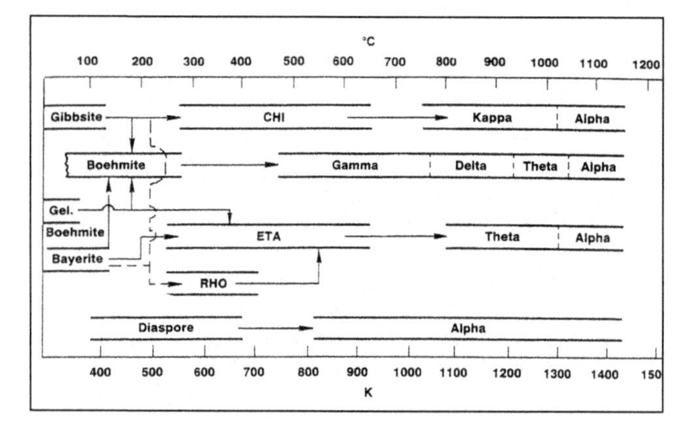

Figure 1. The thermal transformations of aluminium hydroxides and oxyhydroxides, from Wefers and Misra [7].

Figure 1 is reproduced from Wefers and Misra [7]. It outlines the sequence of phases or forms of alumina arising from the heating of the hydroxides (gibbsite, bayerite) or oxy-hydroxides (boehmite, diaspore) as a function of temperature. In the type of soak calcinations undertaken in this work, the gibbsite samples are expected to follow the accepted boehmite-gamma-delta-theta route [7,8,9].

We have previously reported extending the Rietveld refinement of the X-ray powder diffraction data for smelter grade aluminas [2,13]. This was developed to better accommodate the degree of structural disorder observed and has allowed the exploration of new models for the transition aluminas especially the gamma phase, optimized for SGA. A new phase described as γ'- Al_2O_3 by Paglia et. al. [14,15] was introduced and significantly improves the fitting particularly for SGA from fluid bed and gas suspension calciners. Indeed the SGA analyses are invariably dominated by the two γ phases, and clearly understanding the structure of these phases is key to understanding the properties of SGA. The Rietveld refinement is also significantly improved when hydrogen is present in γ-Al_2O_3 with the hydrogen incorporated into the tetragonal 8c position in the structure. The difference made by hydrogen incorporation is summarized in the improvement in the "goodness of fit" parameter in Table 1, reproduced from reference 2.

Table 1. 'Goodness of fit' for the Rietveld refinement of several SGA samples, fitted with a γ, γ', θ, α model – see reference [2] - with and without hydrogen incorporation.

Phase/ Sample	γ %	γ' %	Θ %	α %	GOF	GOF Non H
A	42	39	15	4	5.55	11.6
B	45	39	15	1	4.25	10.4
C	30	33	32	5	6.36	11.1
D	43	40	15	2	4.49	11.7

Although not specific to SGA, Sohlberg et al. [16] review the location of hydrogen in aluminas and introduce the notation $H_{3m}Al_{2-m}O_3$ {*Where* $m=2n/(n+3)$}, rather than the incorrect and misleading Al_2O_3 .nH_2O widely used in the industry. The latter incorrectly suggests that excess oxygen is associated with the loss of hydrogen and that the lattice species, rather than just the leaving species, is water. Ashida et.al. estimate a stoichiometry of $H_2Al_{10}O_{16}$ for typical SGA samples [13].

An additional factor is that although the oxygen sub-lattice is generally well ordered for SGA samples, satisfactory fitting of the powder diffraction data is only obtained when a substantial proportion of the aluminium ions of the transition aluminas are located in interstitial sites, rather than the octahedral and tetrahedral sites of the ideal spinel [13]. Ashida estimates this interstitial population at around 25% of the Aluminiums. The high field NMR data for γ alumina samples prepared under similar conditions [10] suggests these are non-spinel octahedral and tetrahedral interstices rather than 5 coordinate sites.

On the other hand, Whittington and Ilievski [11] compare x-ray diffraction patterns of laboratory and industrial samples using simulations based on well ordered standards. Although this is clearly useful in the validation of the laboratory simulation of the products of the industrial calciner, the influence of disorder expected to be significant in the performance of the alumina in the smelter – for example in dissolution and HF generation – is not accounted for by the ideal structures model.

The application of the Rietveld model previously used to analyse SGA samples [2,13] has been applied to our laboratory calcined materials. The model provides reasonable accuracy to around 800 °C (figure 2), although the goodness of fit does not reach that observed for the industrial SGA samples. Part of the mismatch is attributed to the persistent presence of some chi alumina consistent with the upper pathway in figure 1.

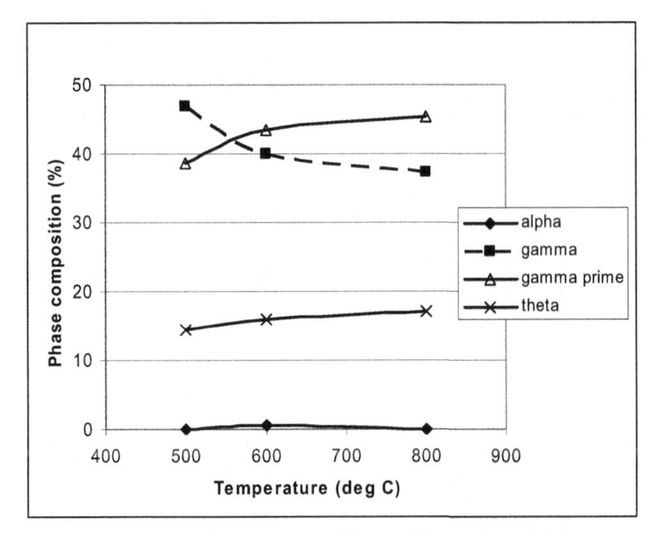

Figure 2. Phase composition with soak calcining temperatures between 500 and 800 °C. Above this range the presence of kappa alumina makes the application of this model problematic.

Figure 2 suggests that although the gamma/ gamma prime ratio changes significantly over this temperature range, little else changes. However beyond this temperature range, problems emerge with the application of the model with the appearance of a well ordered phase that causes increasing deviation in the fitting. This phase is most accurately indexed as kappa alumina (figure 3). Even at 1000 °C, alpha is only ever a minor component. In terms of figure 1, this suggests that not only does some of the gibbsite not transit through the anticipated boehmite pathway, but some of the material which does follow the boehmite-gamma pathway then crosses into the upper pathway.

The influence of partial pressure of water is discussed extensively in terms of impact on calcinations pathway [7,9,17]. Initial experiments involving water injection with an air feed into the laboratory furnace had little influence on the calcination pathway and significant kappa was always observed in 900 and 1000 °C calcinations. Thus, as suggested by other authors, there are substantial crossovers in the pathways identified in figure 1. There is considerable debate, reviewed by Whittington and Ilievski [11], over the initial steps in fluid bed calcination, partially driven by the extreme difficulty of positively identifying the earliest transition alumina phases which form under these conditions.

Clearly the pathways In the light of the results from this study and the lab simulation of flash and fluid bed calciners [8,11] it is worth considering if understanding of this process has advanced much beyond that in figure 1.

Zolotovski et.al. [18] consider the thermochemical activation of crystalline compounds using the aluminium hydroxides as the example cases. This paper examines the effect of heating rate on the decoupling of dehydroxylation/dehydration and oxygen framework reconstruction and the impacts of this coupling on creating a highly amorphous structure. Although targeting activated products in a lower temperature regime (maximum temperatures around 500°C) than SGA, Zolotovski et.al. produce materials which are essentially 100% amorphous. Importantly, in the light of our and other published results from soak calcinations the importance of fast cooling in keeping the two processes decoupled is also identified. Thus the modern, energy efficient fluid bed or flash calciner which recovers heat from the calcined material provides an ideal reactor in which to achieve this decoupling.

It remains a considerable challenge to

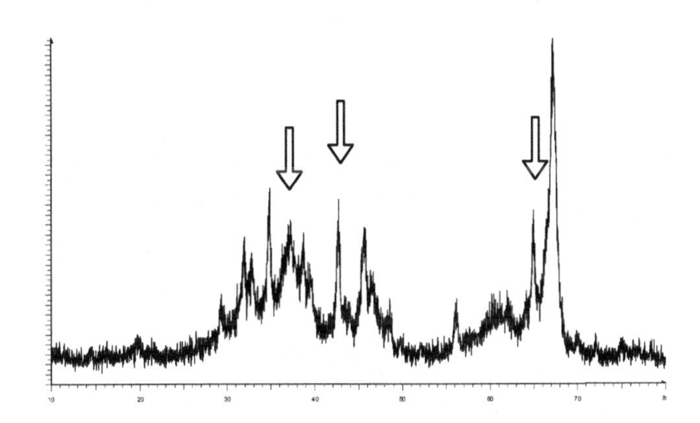

Figure 3. Powder diffraction pattern from 10 to 80 deg 2θ, of Gibbsite heated to 900 °C for 5hrs. The peaks identified with arrows are best fitted to kappa alumina, showing reversion to the upper pathway in figure 1.

Development of Surface Area and Pore Distribution

In TGA curves the gibbsite – boehmite transition is resolved between 250 and 270 °C, at which point surface area increases dramatically (figure 5) as a network of pores centered in the 2nm range develop. Surface areas beyond 300 $m^2\ g^{-1}$ can be produced in this region, Surface area is then lost steadily with increasing temperature, reaching around 50 m^2g^{-1} at 1000 °C again in line with the data of Wefers and Misra [7]. Total pore volume increases steeply until around 400 °C at which point little variation is seen up to 1000 °C (figure3). On the other hand, mean pore radius increases significantly (figure 4), accounting for the decrease in surface area.

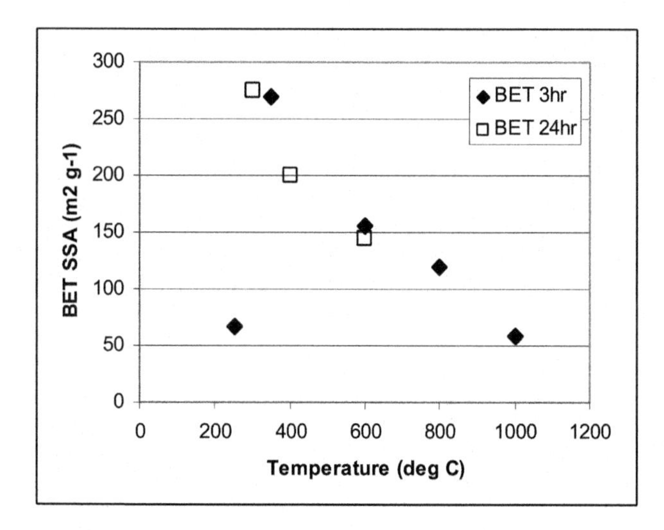

Figure 4. Development of surface area as a function of temperature for Gibbsite samples soak calcined for 3 and 24 hours. This can be compared with figure 4.4 from Wefers and Misra [7].

The structural hydroxides of the lattice are lost progressively by capture of a proton from an adjacent –OH group, with the subsequent loss of a water molecule [12]. Thus the process slows markedly as calcination proceeds to the point where there are no longer adjacent hydroxyls, and where structural rearrangement is required. The boundaries between these transition forms or phases are thus indistinct, due to their disorder and close structural relationship.

Rietveld refinement of the X-ray powder diffraction data has allowed the exploration of new models for the transition aluminas especially the gamma phase, optimized for SGA [ ]. These refinements indicate a gamma alumina dominant material with lesser amounts of the theta phases and small amounts of alpha alumina (see Table 1).

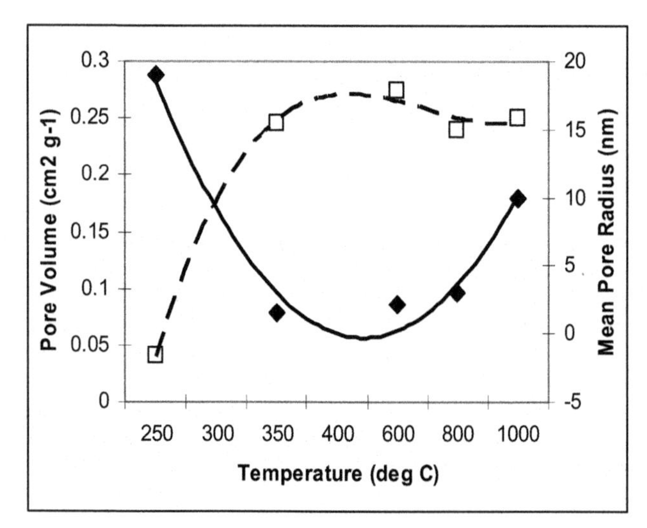

Figure5. Development of pore volume and the mean pore radius with the calcining of Gibbsite.

The extensive structural studies on transition aluminas have dealt almost exclusively with well characterised laboratory prepared samples, or with commercial catalyst support materials, however a number of the conclusions from this work are clearly relevant to SGA. To preserve the Al_2O_3 stoichiometry, the spinel structure (modelled on the structure of $MgAl_2O_4$) requires cation vacancies. Energetically these are favoured in the octahedral site [12], however in the presence of structural –OH, these vacancies tend to cluster, as near-neighbour vacancies are energetically preferred.

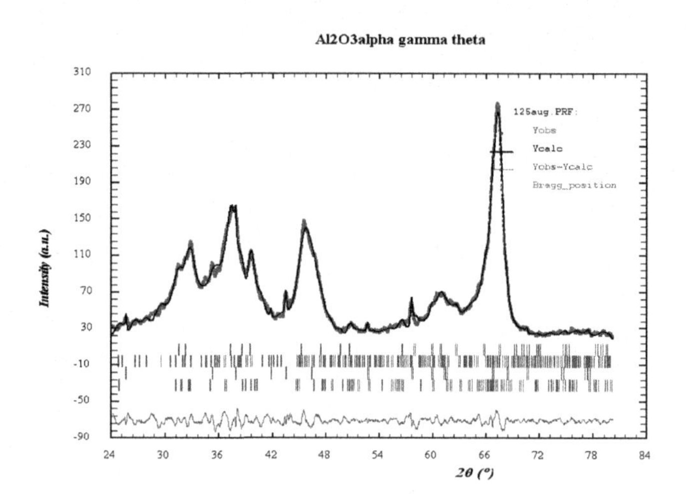

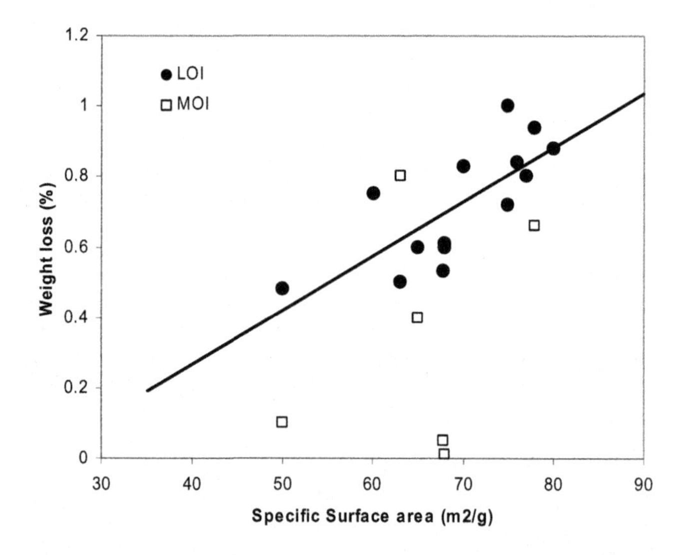

Conclusions

Acknowledgments

The authors would like to acknowledge the input of Professor Barry Welch in the emissions studies and Professor Toshi Ashida in the early Rietveld refinements. The assistance of a number of a number of smelters in providing alumina samples is also gratefully acknowledged.

References

1. I.Levin and D.Brandon. *J. Am. Ceram. Soc.* 81. (8), 1994-2012 (1998)

2. J. B. Metson, M. M. Hyland and T. Groutso. Alumina Phase Distribution, Structural Hydroxyl And Performance Of Smelter Grade Aluminas In The Reduction Cell. *Light Metals 2005 Edited by Halvor Kvande.* The Minerals, Metals & Materials Society), p. 127-131, (2005).

3. R. Haverkamp PhD thesis, University of Auckland,

4. R.G.Haverkamp, B.J.Welch and J.B.Metson. *Light Metals* 1994, pp. 365-371

5. M.M. Hyland, E. Patterson, and B.J. Welch. *Light Metals 2004* Edited A.T. Tabereaux, The Minerals, Metals & Materials Society p.361-366 (2004).

6. E. Patterson, M.M. Hyland, V. Kielland and B.J. Welch, *Light Metals* 2001, 365-370.

7. K.Wefers and C.Misra, Alcoa Laboratories, Technical paper No. 19 Revised (1987).

8. V.J.Ingram-Jones, R.T. Slade, T.W. Davies, J.C. Southern and S. Savador. *J Mat. Chem.* **6**(1), p.73-79 (1996).

9. B. Whittington and D. Ilievski. *Chem. Eng. Journal* **98**, p.89-97. (2004).

10. R.H.Meinhold, R.C.T. Slade and R.H. Newman. *Appl. Magn. Reson.* **4**, p.121-140, (1993).

11. J. Rodriguez-Carvajal, "FULLPROF: A Program for Rietveld Refinement and Pattern Matching Analysis", XV Congress of the IUCr, p. 127, Toulouse, France (1990).

12. B.C. Lippens and J.H. de Boer. *Acta. Cryst.* **17**, 1312-1321. (1964)

13. T.Ashida, J.B.Metson and M.M.Hyland. New Approaches to Phase Analysis of Smelter Grade Aluminas. *Light Metals 2004.* The Metals Minerals and Materials Society, Ed. A. Tabereaux. p.93-97, (2004).

14. G.Paglia, C.E.Buckley, A.L.Rohl, R.D.Hart, K.Winter, A.J.Studer, B.A.Hunter, and J.V. Hanna. *Chem. Mat.* **16,** 220-236, (2004)

15. G.Paglia, C.E.Buckley, A.L.Rohl, B.A.Hunter, R.D.Hart, J.V. Hanna and L.T.Byne. *Phys. Rev. B.* **68**, 144110 (2003).

16. K. Sohlberg, S.J.Pennycook, and S.T. Pantelides. *J. Am. Ceram. Soc.* 121, 7493-7499 (1999)

17. Water effect.

18. B.P. Zolotovski, E.A. Taraban and R.A. Buyanov. *In Flash Reaction Processes,* T.W.Davies ed. P.73-94.

19. B.C. Lippens and J.H. de Boer. *Acta. Cryst.* **17**, 1312-1321. (1964)

20. S.J.Wilson. *J.Solid.State Chem.* (1979), **30**, p.247-255,

21.

Light Metals 2006 *Edited by Travis J. Galloway* ***TMS (The Minerals, Metals & Materials Society), 2006***

BERYLLIUM IN POT ROOM BATH

Stephen J. Lindsay[1], Dr. Charles L. Dobbs[2]
[1]Alcoa Inc.; Primary Metals; 300 N. Hall Rd. MS-01, Alcoa, Tennessee, 37701-2516, USA
[2]Alcoa Inc.;Alcoa Technical Center; 100 Technical Drive, Alcoa Center, Pennsylvania, 15069, USA

Keywords: Beryllium, Beryllium Fluoride, Beryllium Difluoride, Sodium Fluoroberyllate, Pot Bath, Concentration, Dilution

Abstract

Concentrations of beryllium in pot room bath have become a multi-national issue of concern in recent years. In this paper the authors discuss the mass flow mechanisms that permit beryllium to become concentrated in pot bath as well as those factors that limit the reduction of beryllium concentration when the source of contamination is removed. Some general guidelines are proposed regarding decay rates in pot room bath.

Introduction

Concerns about worker exposure to materials that contain beryllium, Be, or beryllium compounds have grown substantially in recent years. Industrial health efforts have been primarily focused on the exposure of workers to Be in metal working or welding. Some attention has been turned to the aluminum industry and to pot room operations in particular due to the observation that Be concentrations in smelter bath can be elevated. However, not much is known or has been published on the nature of beryllium in pot room bath. What is commonly known is that it appears to concentrate more in some smelter operations than in others and that it takes years for concentrations to decay once the source of Be is removed.

The ability of beryllium to concentrate in pot room bath is not surprising. In the Periodic Table Be is located near lithium, sodium, magnesium and calcium and is a Group 2 metal. Beryllium is an alkaline earth metal, many compounds of which are toxic. Discovery of the element in the mineral beryl, was reported by Nicolas Louis Vauquelin in 1789. Beryllium metal was isolated by Wöhler and Bussy in 1828. Group 1 and 2 metals (e.g. lithium, sodium, magnesium, and calcium) readily form fluorides in pot bath and all have very low partition coefficients with the product metal. The loss rates of Be to bath and metal production limit its concentration in typical potroom bath.

Beryllium is lost to a number of sinks including the cathode, dust losses and especially to pot bath that exits the smelter. The science is not as complex as the details of the overall mass balance can be. Understanding the loss mechanisms primarily boils down to having a good understanding of the sodium mass balance. Some generalities can be derived from such analysis that can also serve as general guidelines for those concerned with Be in pot bath. This paper is aimed at these generalities and building upon common understanding of the Be mass balance for pot room bath.

Discussion – Beryllium Inputs

Beryllium may enter the pot room bath from many potential sources. Alumina, coke, pitch, aluminum fluoride, cryolite, purchased bath, cathode materials and others have the potential to place beryllium in contact with pot room bath where it dissolves to form Be ions. These combine to form compounds such as beryllium fluoride, also known as beryllium difluoride, BeF_2, and sodium fluoroberyllate, Na_2BeF_4, a material that is not dissimilar to atmolite (its aluminum analog). The most common sources of Be are fresh alumina, and pot room bath that has come from another smelting location.

Although beryllium may be reported as BeO in raw materials the convention is to refer to the quantity of Be, rather than BeO, that is in material inputs, pot room bath or metal.

"The source of beryllium in alumina is trace levels present in some bauxite deposits." [1] The mineral beryl, $Be_3Al_2(SiO_3)_6$, is an alumino-silicate like many other impurities that are found in bauxite. The form of beryllium in bauxite is not known and is probably a combination of minerals like beryl, bertrandite, chrysoberyl, etc.

In smelting grade alumina, SGA, the concentration of BeO, expressed as the oxide, can vary from non-detectable to as much as 20 ppm which is roughly equivalent to 7 ppm of Be. There is no commercially available method to remove low levels of beryllium from either bauxite or SGA. The level in alumina is controlled by the source of the bauxite.

Typical levels of Be in alumina are not of particular concern for the SGA itself since it is so dilute. However, due to the large amounts of alumina that pass through a smelter, and the nature of Be to concentrate in pot bath, SGA is quite often by far the major input stream of beryllium into a smelter.

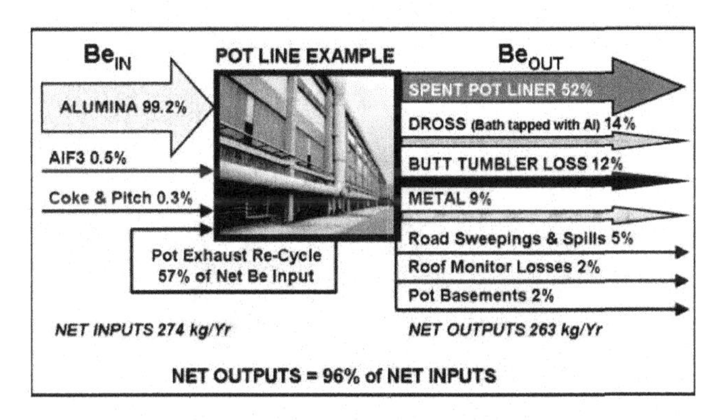

Figure 1 – Actual Beryllium Mass Balance Example

Coke, pitch, AlF3 and anthracite are also derived from materials taken from nature. While the input levels of Be from these products is typically very low or non-detectable, trace levels may be found that vary with the sources of crude oil, fluorspar, hydrate and/or coal. In cases where smelters have extremely low levels of Be in alumina the input from all sources will typically generate an equilibrium concentration of less than 10 ppm in pot bath. In cases where the major source of Be is alumina, the combined total of AlF3, coke, pitch and cathodes typically contribute less than 1% of the total input of Be to the smelter as in figure #1.

The other common source of Be to smelters is pot bath that has been purchased or transferred from a smelter that has used a raw material source containing beryllium. Since Be is concentrated in pot bath the transference of many tons can bring with it a substantial amount of Be that is generally in the form of beryllium fluorides. In some cases this transfer mechanism has been known to raise the average Be concentration to 70 ppm in liquid bath.

Discussion – Beryllium Concentration

As with some other metallic oxides BeO is miscible in Na_3AlF_6 and as with other alkaline earth metals such as Na, Li, Mg and Ca, Be will tend to stay in the bath and combine with fluoride rather than report to the metal. These Group 1 and Group 2 elements have very low partition coefficients due to their strong affinity to halogens such as fluorine. Thus only a small amount of Be will be found in aluminum production. Locations that have in excess of 150 ppm of Be in pot room bath will have less than 1 ppm of Be in the product metal.

Without other sinks for beryllium beyond the small amount that leaves with metal production it concentrates in the pot bath until the mass of Be_{IN} at low concentrations in alumina approximates the mass of Be_{OUT} in pot bath at high concentrations. One estimate indicates that Be would concentrate to levels greater than 2500 ppm in bath if metal were the only avenue to exit the system. This is on the order of 500 to 1000 times the concentration of an SGA that is considered to be relatively high in Be content. At steady-state the actual concentration of Be in pot bath is on the order of 40 to 90 times that of fresh SGA, with 60X being most typical.

While beryllium concentrates to approximately sixty times that of fresh alumina in liquid bath limited data indicates that it does not concentrate uniformly in all forms of bath. Bath in anode cover, also know as crust bath, will have beryllium concentrations that range from the proportion of bath that is mixed with alumina to as much as one half to a full order of magnitude greater that the concentration of Be in liquid bath. See figure #2.

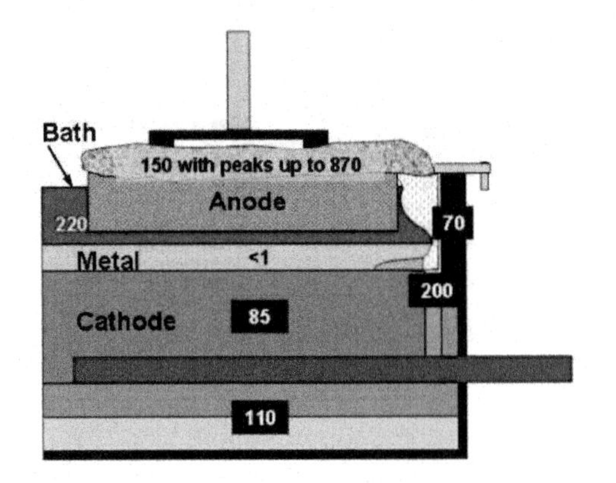

Figure 2 – Beryllium Concentration Examples

The mechanism that drives concentration of Be within the layer of anode crust also drives a similar phenomenon in the cathode. The hottest regions of the cathode will have relatively uniform concentrations of beryllium. Regions near the freeze isotherms can have Be concentrations that are three times greater than those of the hot zones. Colder zones of the cathode that are well beyond the freeze isotherms have very low concentrations of beryllium. This is shown in figure #3 below.

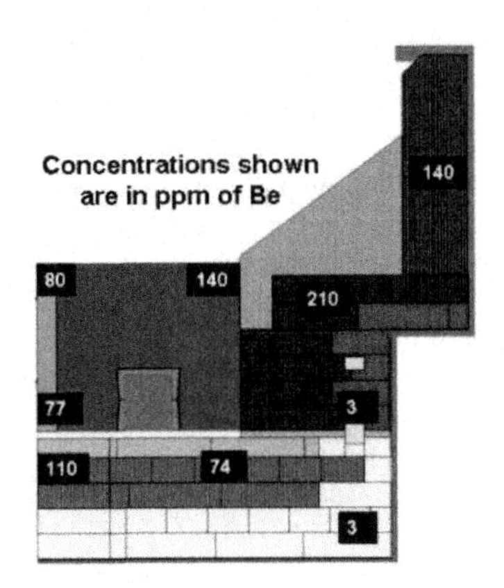

Figure 3 – Beryllium concentrations in cathodes

The vapor pressure of beryllium fluoride compounds and heat are both contributing factors that are suspected of being of primary importance to this phenomenon. This may be related to the vapor pressure of BeF_2, melting point 544^{0}C, sublimation at 800^{0}C [2] and migration or deposition of BeF_2 and Na_2BeF_4 within the layer of crust bath and within the cathode along isotherm lines.

Discussion – Beryllium Outputs

Aside from carbon dioxide, metal production accounts for the largest output by weight from reduction cells. But with very low concentrations of Be, aluminum falls far from accounting for the bulk of beryllium lost from the cell. However, metal production does carry away 5% to 10% of total Be input and thus accounts for a significant fraction of Be losses.

Other potential sinks are relatively small in terms of annual tonnage when compared to metal production. For Be_{IN} to equal Be_{OUT} this implies that at least one exit stream will have a Be concentration and mass loss that is relatively high.

This does not imply that all sinks are highly concentrated with beryllium compounds. Secondary alumina has concentrations of Be that are higher than that of fresh alumina, but much lower than that of bath and other materials. Dust losses of secondary alumina will therefore be only a minor sink for beryllium.

To understand the various exit streams for Be one should follow the paths of two other elements. The first is sodium in every form of bath exiting the smelter. The second element to follow is the fluorine that Be is so highly attracted to.

The sum of bath losses including that to cathodes will always account for the bulk of beryllium that exits any smelter. However, this involves many small and varied streams. An obvious stream to consider is that of excess tapped bath that may be crushed and sold or transferred to other smelters. Tonnages

may be relatively small, but with high concentrations a significant portion of the Be mass can flow to this exit stream, often in the range of 5 to 15% of total beryllium input.
Using a hypothetical example consider a smelter that produces just over 208,300 mt of aluminum per year. This location consumes 400,000 mt of alumina per year with a reported BeO equivalent of 8.3 ppm or 3.0 ppm of Be. There are no other significant inputs of Be to this smelter. Thus approximately 1200 kg of Be will enter this facility each year. If the concentration of Be in pot bath is 180 ppm and the plant generates 50 mt of excess bath for sale or transfer each month then roughly 108 kg or 9% of the net Be input leaves with the excess bath stream.

Closing this fraction of the Be mass flow loop helps to understand the bigger picture of beryllium losses to bath streams. Investigation into how much bath is lost each year to other streams will naturally follow in any undertaking to develop a complete mass balance. These streams will include; dust losses, dross leaving the ingot plant, accumulations of excess bath in piles, storage, material under pots, and other losses.

The key to understanding the many individual bath streams is to have good knowledge of the sodium balance for the smelter overall. It may be a surprising statistic but many smelters receive enough sodium and fluoride each year to displace between 33% and 50% of the total bath inventory. Most who work with plant operations do not realize that this much bath turns over each year. This is so since the flows of sodium are insidious rather than obvious.

Let's take a closer look at this claim using our hypothetical smelting example. There are a total of 468 pots operating at 160 kA in this smelter. The total bath inventory including liquid, solid, in anode cover and in all forms of storage is 10,250 mt, or about 20 mt/pot plus 10% in inventory.

The Na_2O content of the alumina to this smelter is a fairly typical 0.41%. Some of this goes to sodium intercalation in cathodes and some to other sinks, but if all went to pot bath it would generate 4,200 mt each year.

Consumption of soda ash is relatively high at this smelter at 0.60 kg/mt Al. If all of the sodium in this stream went to make bath it would generate 190 mt of bath each year.

This smelter also consumes 30 mt of cryolite each month to support new pot starts. This stream will generate an additional 410 mt of bath each year.

In all this smelter brings in enough sodium each year to displace 47% of the average bath inventory, or 4800 mt.

A quick look at AlF3 consumption confirms the magnitude of these calculations. Our example smelter consumes 19 kg AlF3/mt Al produced, or 3960 mt/yr. Enough to make 4830 mt of bath if all of it went to this sink.

These calculations are not offered as precise mass balances for either sodium or fluoride. They only serve to illustrate the point that the mass balance for pot bath is highly dynamic and that a large fraction of the bath inventory in a smelter turns over in any given year.

This also supports the premise that the major sink for beryllium is pot room bath loss in its various forms. Thus, net inputs of Be are primarily counter-balanced by steady dilution of pot room bath. While there are faults with these over-simplified calculations, consider one additional Be metric from our example smelter. If 1200 kg of Be enters this example smelter each year and 4800 mt of bath at 180 ppm were to leave it implies that 864 kg of Be, 72% of the input, leaves with the bath in one form or another. The balance leaves with metal, secondary alumina dust and fume, carbon dust skimmings, in forms of bath in that have higher Be concentration than tapped bath, and even a little concentrated in any hard gray scale that is accumulated and eventually discarded.

Since not all incoming sodium goes to make new pot bath with the proper amount of excess fluoride it is necessary to understand where the Na goes.

Some is accumulated and then held, sold or transferred as was illustrated in the example in which 9% of the net Be_{IN} was accounted for via bath sales. Accounting for the beryllium in this particular stream is relatively easy to do.

Much sodium and basic bath permeates the cathodes via ingress and intercalation. Following the beryllium mass flow here is not so straightforward. Since the dry weight of new vs. consumed cathodes generally increases by 30% to 50% we might expect that the average Be concentration in spent pot lining should be between 40 and 60 ppm if it is 180 ppm in pot bath. This is well within the observed range for many pot types and is especially so for those pots which have been meticulously cross-sectioned with the samples kept dry. Note that since BeF_2 and Na_2BeF_4 are quite soluble in water, spent pot liner that has come in contact with moisture will not yield accurate results for Be content.

Turning once again to our example smelter the pot life is 2000 days and spent pot liner generation is 33 kg/mt Al produced. If the average concentration of Be in the spent pot liner is 60 ppm then 413 kg/yr of Be leaves this way, 34% of the total Be_{IN}.

Losses to spent pot liner, sales of excess bath and metal flow can bring our accounting for net Be losses up to >50% of the total Be_{IN}. The remaining Be must be tracked down by looking at bath that exits with metal, bath losses at various concentrations, bath dust losses, carbon dust losses, pot fume losses and secondary alumina losses. A mass balance on Be may not close completely, but this may often be due to the lack of good information on exit streams. A thorough mass balance is best done during steady state conditions and should close such that net outputs are within 15% of net inputs.

Discussion – Disruption to Steady State

A valuable way to gain insight into the mechanisms of beryllium accumulation and decay is to gather and analyze data when changes occur that disrupt steady state conditions. Such case studies may be based upon the introduction of a new source of Be to the pot room bath or the elimination of a source of Be with a change in raw materials.

Increases in Be concentration are relatively simple to model when a new source is added so long as the concentration of Be or beryllium mineral in the source has been accurately determined. An initial spike in beryllium concentration in the liquid bath is expected, followed by some relatively rapid decay as Be migrates into the frozen forms of bath throughout the smelter. This period is followed by a slow, asymptotic period of decline.

An example is offered in figure #4. This smelting location had a background level of beryllium in pot room bath was relatively low, at 3.5 ppm. When a single shipment of a different alumina source was received it provided the opportunity to follow the event and confirm the prediction for the increase in Be concentration. Eventually this information and additional data led to the development of a simple algorithm to that is used for making predictions for Be concentrations in similar cases.

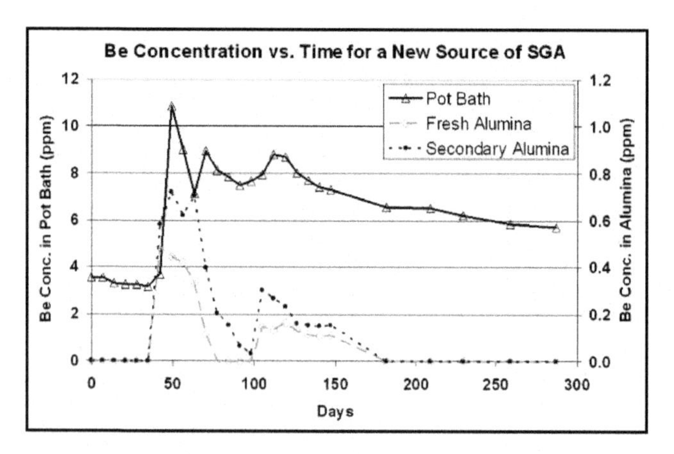

Figure 4 – Single Shipment Beryllium Accumulation and Decay

The algorithm is: Be_{PEAK} (ppm) = $Be_{INITIAL}$ (ppm) + [Be_{IN} (kg)/Total Liquid Bath in Smelter (kg) *1,000,000/100 * % of Be_{IN} consumed in 25 days]

In this case Be_{IN} represents the total mass of beryllium in the alumina shipment. If it is consumed over more than a twenty-five day period then the input mass is limited only to the amount consumed during the first twenty-five days.

Another way to examine the dynamics of beryllium concentration in pot room bath is to examine what happens when a major source of Be contamination is suddenly removed. Figure #5 includes the decay trace that is shown in figure #4 after the shipment was consumed along with examples from six other locations that had initial concentrations of Be in bath ranging from 10 to 200 ppm.

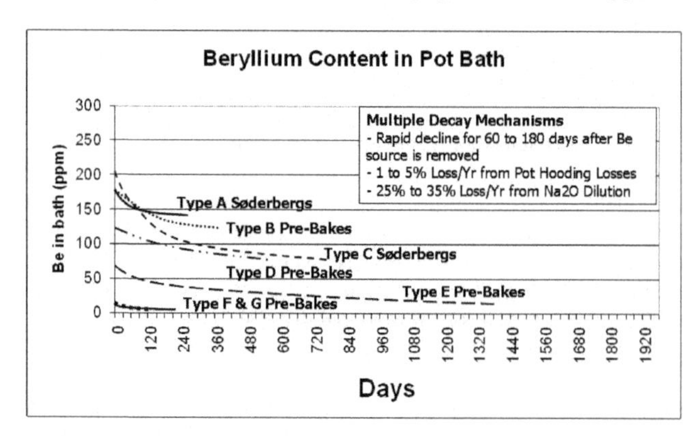

Figure 5 – Comparisons of Be Decay Rates in Tapped Bath

The examples in figure #5 all represent Be in tapped bath from both pre-bake and Søderberg pot types. Included are smelters that have used an individual shipment of alumina containing Be in concentrations over 1 ppm, smelters that have used alumina containing Be between 3 and 4 ppm for many years, and a smelter that had never used alumina containing beryllium, but that had used crushed tapped bath from a location that contained Be at concentrations greater than 200 ppm.

Although the circumstances are varied some generalities emerge when a major source of beryllium contamination is suddenly removed. There is a period of rapid decline of two to six months followed by a period of prolonged steady decline. The initial rapid rate of decline characterizes a shift from the Be input rate being a primary factor for determining the steady-state Be concentration in pot bath to Be dilution rates being the primary determining factor for residual concentrations in pot bath.

Note that there appear to be two main factors that determine the rate of change of Be concentration. The first can be called the "bath factor", or the ratio of tons of bath to tons of alumina consumed. The second is the rate of sodium input to the smelter.

These two factors may be compared to a glass bowl full of colored water that has a clear dilution stream entering and causing the bowl to overflow to a drain. The "bath factor" is analogous to the size of the bowl and the sodium input is analogous to the flow rate of the dilution stream.

A smelter with a relatively small amount of bath per ton of alumina consumed will have a relatively rapid rate of decay in beryllium content. A small bowl will be diluted more quickly than a large bowl when both have equal overflow rates.

The initial rapid rate of decrease in Be concentration lasts two to six months while the system adjusts to the new balance between inputs and outputs. The ratio of tons of bath to tons of alumina consumed, or the size of the bowl, plays a large role in modeling this period of decay in beryllium concentration.

Smelters with a relatively high Na_2O content in alumina will have a more rapid rate of steady decline in Be concentration. This is like having a dilution stream with a relatively high flow rate. The point is illustrated in figure #6 where the average Na_2O content is compared to the prolonged decay rate of the curves shown in figure #5.

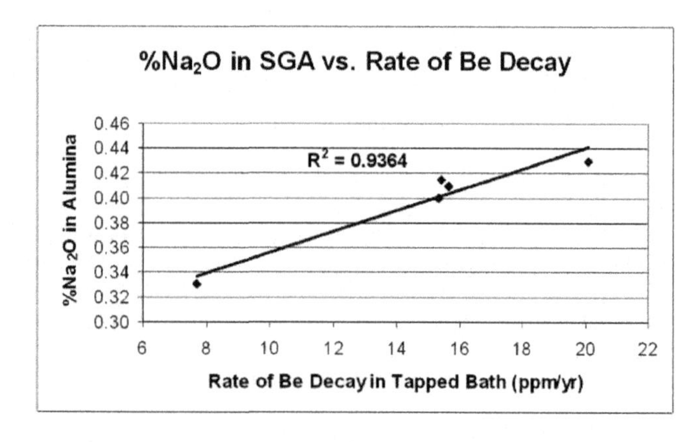

Figure 6 – Relation of $\%Na_2O$ to Be Decay Rate

There are multiple sources of Na that contribute to the rate of dilution. Sodium input in alumina is typically the largest factor by far. The consumption rates of sodium carbonate or soda ash, cryolite and imported bath also play a role in dilution.

When these factors are combined a rough rule of thumb emerges. Six months after quitting a source of Be input the rate of decay of beryllium concentration in pot room bath will be 35% +/-5% per year. In other words if the Be concentration begins the year at 100 ppm it will drop to 65 +/-5 ppm at the end of the first year, 42 ppm at the end of the second year, 27 ppm at the end of the third year and so on. A more general rule might be that *one-third of the concentration of beryllium will be lost each year relative to the starting point for that year*. Of course this varies from case to case, but having such a general guideline is often useful.

A third factor, the ratio of bath in liquid form to bath in solid form is also suspected as being of importance to the initial rate of change of beryllium concentrations in tapped bath illustrated in figure # 5. This has to do with the concentrated levels of Be that can be found in crust bath and the concentrating effect that it may have on Be in liquid bath as it is consumed.

Consider one last time our hypothetical smelting example in which the liquid bath has an average Be concentration of 180 ppm. One ton of crust bath from this smelter has an average of 140 ppm Be and 33% alumina content. When this material is consumed to the pots the 0.67 tons of liquid bath that is generated will have the equivalent of 209 ppm of Be. A plant that has large reserves of crust bath is likely to decay at a slower initial rate than a plant with little inventory of crust, and basement, bath.

Other

Although it is open to further study high concentrations of beryllium have not been found in materials other than pot bath. Carbon dust may be an exception to this based upon limited evidence. It has been well documented in the literature that other contaminants such as Fe concentrate on carbon dust that floats on liquid bath. Beryllium may do likewise. This is only theory at this point and subject to further study.

The Be concentration of particulate material in the inlet ducts to dry scrubbers indicates that the Be concentration in bath determines the concentration in this material and thus in reacted or secondary alumina.

One other factor that is open to further investigation is treatment of beryllium laden bath with water. The beryllium found in SGA is essentially insoluble in water. However, in individual bath samples beryllium content may range from being nearly insoluble to as much as 95% soluble in water. Most samples are 50% to 90% water soluble. This property may open some avenues to beryllium extraction by giving pot bath a "bath" if the reader will excuse the expression. Of course this also begets the question of what to do with the beryllium fluoride once it is in wash water.

Conclusions

There are generally two common sources for beryllium accumulation in pot bath. The most common is Smelting Grade Alumina. The other primary source is via purchase of pot bath from a smelter that has used alumina containing beryllium. In either example Be in alumina is a common root.

Beryllium is highly attracted to fluoride and has a low partition coefficient. Thus Be concentrates in pot bath in the range of 40X to 90X that of the concentration in incoming alumina. The most common concentration is 60X that of alumina, but this can vary based upon the rate of net sodium input and losses from a smelter.

Beryllium also tends to concentrate in what may be isotherms in both the pot crust and the spent pot lining. The melting point and vapor pressure of beryllium fluoride in combination with the temperature of molten pot bath are suspected as the driving forces of this phenomenon.

A small portion of beryllium is lost to the product metal. It is not uncommon for a smelter with 150 ppm of Be in pot bath to have less than 1 ppm of Be in metal. The primary sinks are with bath losses to cathodes and spent pot lining, bath transferred with metal that goes to dross, bath sales, spent anode cleaning residue, carbon dust skimmings and all other forms of bath losses. Following the losses of bath from a smelter is the key to understanding the losses of beryllium.

Changes in Be concentration in liquid bath can be readily modeled using data on a new raw material stream that contains Be.

Once beryllium is present it pot bath the primary mechanism to reduce its concentration once the input source has been removed is dilution and loss to the cathode which one might consider a form of dilution as the bath must be replenished. In general smelters bring in enough sodium in various forms to replace 33% and 50% of their total bath inventory each year. The amount of sodium entering and leaving a smelter is an important factor in determining how quickly beryllium concentration in pot bath can be reduced.

The other main factor in the determination of how quickly the beryllium concentration will change is the size of the "bowl" or the relative amount of bath that is kept in circuit per ton of alumina that is consumed. Those plants with greater economy of pot bath inventory will see more rapid increases or decreased in Be content when sources of Be are added or removed than those smelters that have relatively large reserves of pot bath kept in circuit. The inventory of various forms of solid bath relative to the total inventory of liquid bath may play a role in the dilution rate as well.

Six months following the termination of a raw material containing beryllium the concentration of Be in bath will decline by about one-third of the initial value of any given year.

Finally, there are factors about beryllium in pot room bath that require further study including mechanisms of how it concentrates in crust bath or on carbon dust and if its relatively high solubility in water may become an effective means at controlling the risks that may be associated with worker exposure.

References

1. Eyer, S. L., Nunes, M. D., Dobbs, C. L., Russo, A. V., Burke, K., "The Analysis of Beryllium in Bayer Solids and Liquids" ,

Proceedings of the 7th International Alumina Quality Workshop, 2005, pp 254 – 257

2. United Nations Environment Programme, the International Labour Organisation and the World Health Organization, International Programme on Chemical Safety, "Environmental Health Criteria 106 Beryllium", Section 2.2, Table 3 Physical and chemical properties of beryllium and selected beryllium compounds
http://www.inchem.org/documents/ehc/ehc/ehc106.htm

Acknowledgements

Luca Bellan, Luciano Nalon, Moreno Niero, Roberto Nalesso, Lee Blayden, Mike Palazzolo and Programma Ambiente for assistance with sampling.

Greg Kraft and Giuliano Biasutto for sponsorship

Miland Chaubal, Sherwin Alumina Company, "Beryllium in Aluminum" Aluminum Association Be Task Group, August 2002

Merino, Dr. Margarita R. (Ph.D. – Florida State University) – for her encouragement, dedication and support

Light Metals 2006 *Edited by Travis J. Galloway* ***TMS (The Minerals, Metals & Materials Society), 2006***

AUTOMATED CRYSTAL OPTICAL TECHNIQUE FOR QUANTITATIVE DETERMINATION OF PHASE AND PARTICLE SIZE COMPOSITION OF ALUMINA

Margarita V. Tsvetkova[1], Lidiya A. Popova[1], Alexander G. Suss[1], Vladimir G. Teslya[1], Andrey V. Panov[1]

[1]Russian National Aluminium & Magnesium Institute (RUSAL-VAMI), 86, Sredny Pr., St. Petersburg, 199106, Russia

Key words: alumina, aluminium hydroxide, phase and particle size composition, technique for determination

Abstract

For many years specialists of VAMI have been involved in developing of automated techniques and new equipment arrangement for determination of phase and particle size composition of products in alumina manufacture. Crystal optical analysis being a complicated technique requiring professional background however is not free from subjective estimation as far as the quantitative analysis is concerned. State-of-the art computer diagnostics and processing allow overcoming these problems.

Now at several alumina refineries a technique for automated quantitative determination of alumina phase composition using *Image Analysis System* is tested. This technique is a combination of latest achievements in the instrumentation art, software and many-years practical experience.

The following automated techniques were developed with the help of *Image Analysis System*:

1. determination of alpha-modification content in of all grades alumina irrespectively of method of production of alumina and way of its formation;
2. determination of quantitative phase composition of smelter grade alumina.
3. Automated determination of particle size distribution of aluminium hydroxide containing fine particles.

The developed methods solve the problem of phase composition monitoring in the course of phase transformation from hydrargillite to alpha-alumina in the calcination process and control of particle size distribution of aluminium hydroxide in full fraction range.

Introduction

Every crystal substance has the defined set of optical constants. For diagnosis of alumina products the light refraction can be considered as the determinative optical constant. The light refraction – is the passage of light rays from one media to other (solid substance – liquid) with light refraction on the border of those media. The value N represents the light refractive index of one media relative to other. For every two solid and liquid media this value is strongly constant.

Figure 1 shows the image of waves refraction under their oblique downfall on the media boundaries.

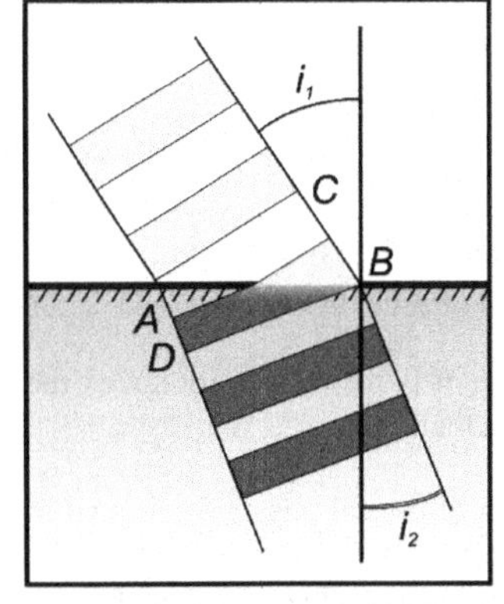

Figure 1. Diagram of light refraction

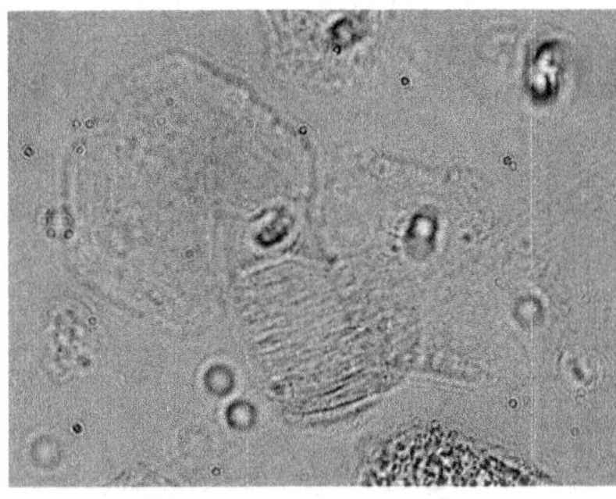

Fig. 2 a. Congruence of two media refractive indices. In immersion and in passing light, magn. 625ˣ

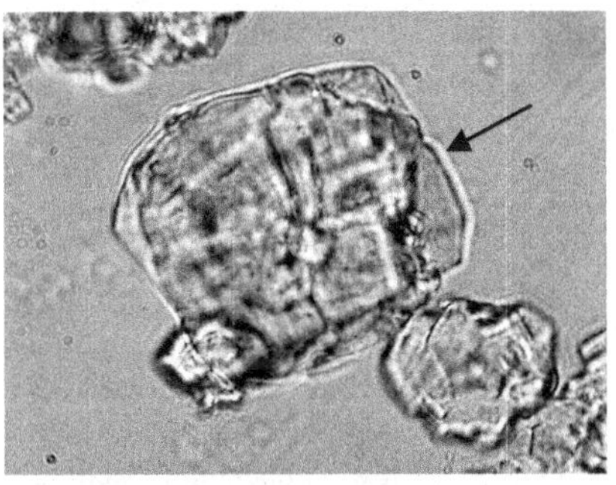

Fig. 2 b. Light line on the border of two media (solid – liquid) under unequal refraction conditions. In immersion, in passing light, magn. 625ˣ

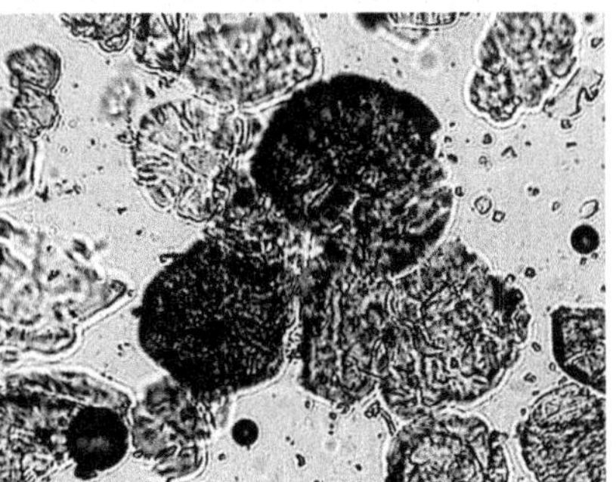

Fig. 2 c. The contrast of particles color with different refractive index with regards of immersion media. In immersion, in passing light, magn. 250 ˣ

Fig. 2. View at different refractive indices of media and crystals

From the triangles ABC and ABD can be seen that

$$\frac{CB}{AD} = \frac{\sin i_1}{\sin i_2} = \frac{\lambda_1}{\lambda_2} = n_{12},$$

where n_{12} – media refractive index relative to other media;
λ_1 и λ_2 – wavelength in the media,

notably the refraction law Descartes - Sinelius is a consequence of above hypotheses on the fact than the length of light wave is variable when the light passes from one media to other.

When the light passes from the vacuum to the media the refractive index (N) is presented as follows:

$$N = \frac{c}{v} = \frac{\lambda_0}{\lambda}$$

where N – media refractive index;
c and λ_0 –wave speed and length in the vacuum;
v and λ –wave speed and length in the media.

As far the light speed in the vacuum is the maximum speed the refractive index N is always higher than 1.

On the practice the light refraction of crystal phases is defined in the immersion specimen by the selection of the immersion media with known refractive index to obtain the image of two defined media and immersion media refractive indices congruence.

Under equality of those media refraction indices the crystal phase particles are practically merged with liquid phase and only in this case the solid and liquid phases have the equal refraction indices and brightness values (fig. 2 a). In case of difference in light refraction for solid and liquid substances on the border of two media the light line appears (Becke line) – see fig. 2 b.

The particles will be highlighted on the background of the liquid phase obtaining the relief and color (optical effect). The intensity of the light line and grain contrast can indicate on the difference of the refraction parameters and their brightness with regard of particular immersion media. With very high difference in the refraction parameters of two media the particles in the passing light of polarization microscope have the color from brown to non- transparent black (fig. 2 c).

1. The methodological scope for determination of quantitative phase composition

It is common knowledge that the aluminum hydroxide under influence of high temperatures during the thermal treatment goes through modification of physical and chemical properties and as effect the alteration of optical constants. The calcination of gibbsite up to 1200 – 1300 °C leads to the triple rearrangement of lattice up to the maximum compression and total removal of crystal water. The rearrangement degree of initial product depends on the temperature and duration of the heat ambience. The chain of Al_2O_3 phase transformations in the furnace is formed as further:

The solid phase transformations of gibbsite under thermal decomposition:

Gibbsite	T=200°C →	Non transparent aggregation (transformation into boehmite)	300-400°C →	Boehmite and amorphous alumina	500°C →	Amorphous alumina + γ- Al_2O_3 (active)	600-800°C →
$N_{aver.}$ = 1.58				N = 1.62 –1.64		N = 1.63-1.65	

γ- Al_2O_3 + traces of amorphous alumina	900-1100°C →	(δ, æ, θ, etc.) Al_2O_3 High temperature alumina modifications and starting transition into α- Al_2O_3	1100-1200°C →	α- Al_2O_3 (starting of formation)	1200-1300°C →	α- Al_2O_3 and corundum
$N_{aver.}$ = 1.64-1.70		N = 1.70-1.74		N = 1.74		N = 1.740-1.756

The aluminum hydroxide calcination covers the removal of external and chemically bound humidity with further alumina recrystallization. During dehydration the polyphase crystallization product is formed with different homogeneity degree, presented by aggregates and monocrystals, forming pseudo morphosis on the initial gibbsite with practically identical morphological signs. The low temperature "active" products of dehydration with cryptocrystalline structure are X-ray amorphous, as far as the crystallization is not over.

It is known that the fluidized bed furnaces are producing alumina that is more completely calcined and more homogenous by phase content, with minimum tails, represented by the products of initial dehydration and boemite. It is conditioned by the fact that the fluid bed allows under instant heating to produce constant temperature field for the whole volume of the layer with high range control of material holding time.

The final stage of gibbsite solid phase transformation is the α-alumina and it can be considered as the most stable modification of aluminum oxide. α- Al_2O_3 has the densest lattice and the highest refraction index, corresponding to N – 1.740-1.756 (corundum). The actual crystal optics methods for determination of α- modifications in the alumina the iodic methylene is used with refractive index of 1.740. In iodic methylene the particles of α- alumina are practically merged with immersion media (fig. 3) and all other dehydration products will have much low refractive indices and much low brightness values (fig. 4).

In this case we can indicate the direct correlation between the values of refraction and brightness characteristics of solid phase with relation to the liquid phase – iodic methylene.

Each intermediate stage in the process of Al_2O_3 generation is accompanied by the interval of refraction variations, reflecting the specific crystallization step in the process of aluminum hydroxide dehydration.

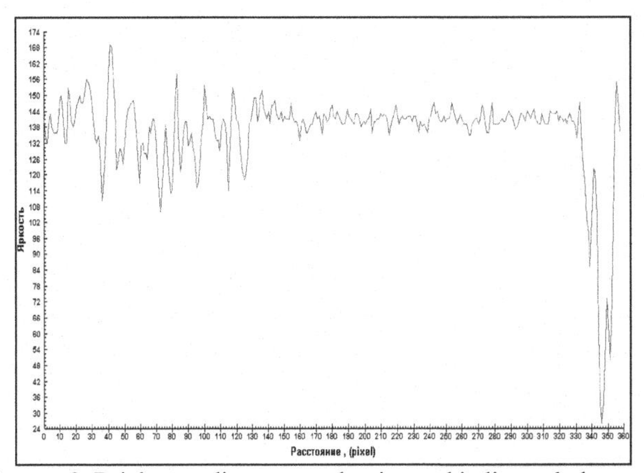

Fiugure 3. Brightness diagram α- alumina and iodic methylene

For elaboration of the method to obtain the most complete characteristic of the alumina phase content the following stages of the dehydration products separation have been adopted (using optical constant – refractive index):

Stage 1 - N from 1.585 to 1.630 –primary dehydration product of $Al(OH)_3$ and amorphous alumina;

Stage 2 - N = 1.640 – boehmite;

Stage 3 - N from 1.640 to 1.670 – gamma low temperature alumina;

Stage 4 - N from 1,670 to 1.700 – gamma high temperature alumina;

Stage 5 - N from 1.700 to 1.740 – high temperature Al_2O_3 modifications (δ, æ, θ etc.) and α- Al_2O_3 – starting of formation;

Stage 6 - N from 1.740 to 1.756 - α- Al_2O_3 different stage of crystallization (up to corundum).

Phase identification at each stage in the Al_2O_3 phase transformation chain is realized by the selection of the immersion liquid with the appropriate to this phase refractive index.

Image Analysis System (thereinafter IAS) allows to see and allocate on the brightness histogram the brightness thresholds of different media (solid and liquid). The brightness section should not present the variations of the brightness in the homogenous media. (brightness threshold – null). In practice there is almost no homogenous media. The brightness section inevitably will have minimum deviations of the brightness values.

IAS soft- and hardware (fig. 5-6) can be represented as follows:

- IAS allows to obtain the image of the analyzed object using the color TV camera and polarization microscope for the digital processing;
- IAS software (including measurement of different optical parameters and linear sizing) allows having reliable identification of the crystal phases using optical constants, and through statistic processing to compute the number of the above phases. It is critical for diagnostic and phase estimation for the calcined products with incomplete crystallization.

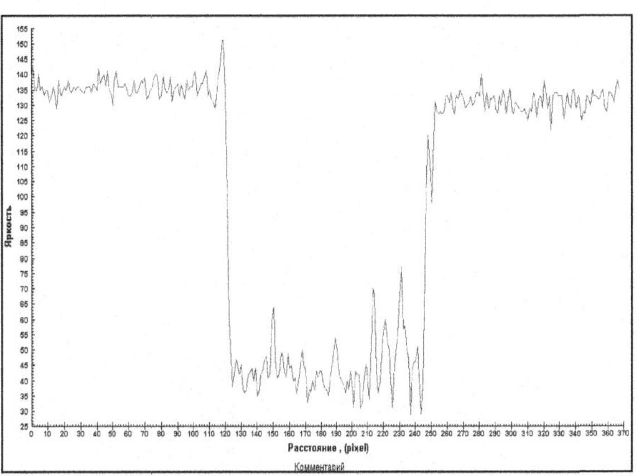

Figure 4. Brightness diagram of low refraction particle in iodic methylene

The basis of the method is the identification of the analyzed phase using the refractive indices with further highlighting of its brightness thresholds on the histogram with automated computing of phase content values.

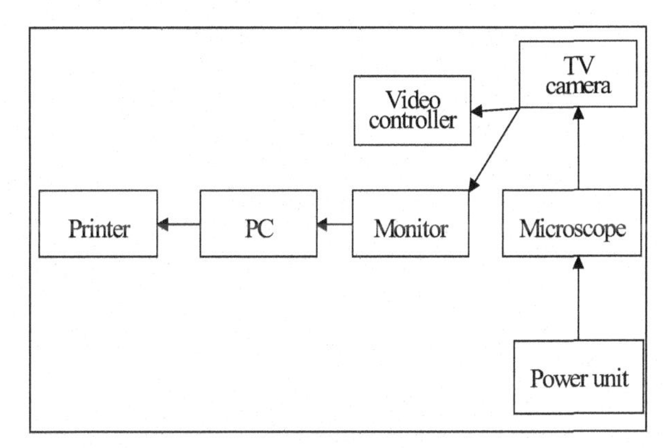

Figure 5. Block diagram of *Image Analysis System* hardware

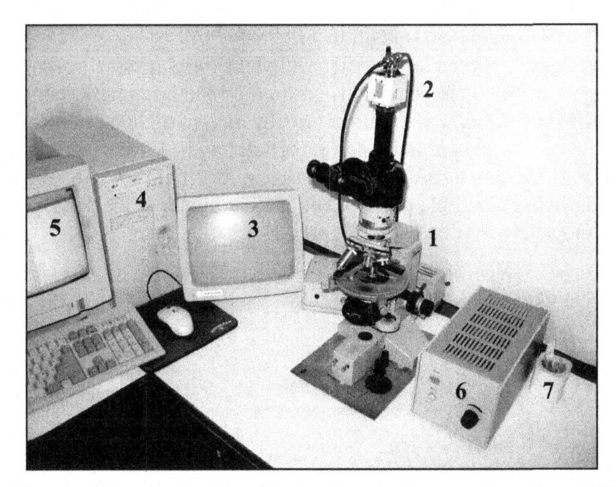

Figure 6. *Image Analysis System* (1- microscope, 2 – colour TV camera, 3 – colour video controller, 4 – PC, 5 – PC monitor, 6 – power unit for microscope, 7 - set of immersion oils)

The use of color TV camera in for IAS allows to bring closer the color image in microscope and on PC monitor. The color picture of solid phase and immersion media with similar refraction parameters has the identical brightness and same blue coloration.

The increased intensity of the particles blue color indicates on the certain increase of the refraction index (up to corundum) relative to immersion media.

The particles with the refraction index slightly below immersion will have slight pinky- yellow coloration. Under high refraction indices difference of particles and immersion media the particles will differ from immersion by relief and color with transformation to the deep brown up to black (fig. 7)

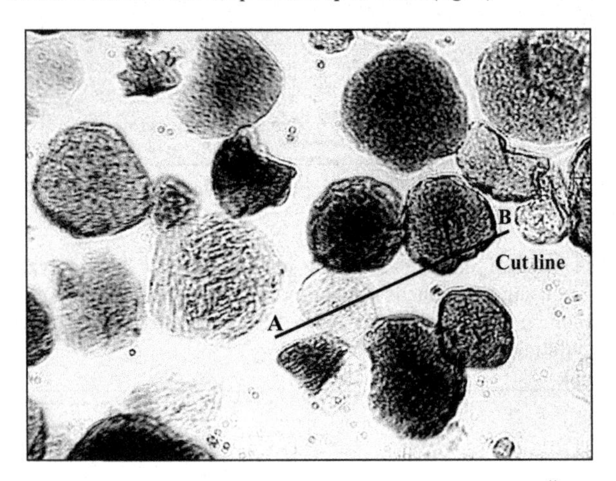

Figure 7. Alumina speciemen color view (magn. $250^{\times}$)

The method of alpha modifications determination in the alumina provides the distinction in the iodic methylene the particles of α-Al_2O_3 on the background of lower temperature calcination products with further computing.

The alpha-modification is a conventional phase with refraction index within 1.740 up to 1.756. In view of the refraction parameters adjacency for particles and liquid media their brightness thresholds on the histogram will always be very close and within the highest brightness values (top of histogram) The brightness thresholds for all other phases with much low refraction index and much low brightness will be located below the alpha alumina and iodic methylene brightness threshold. Figure 8 shows the thresholds position on the brightness section (A-B on the fig 7) of an alpha modification particle on the iodic methylene background and a particle with lower refractive index. Deviation of the brightness in the field of the alpha modification particle threshold brightness indicates the degree of homogeneity of this particle.

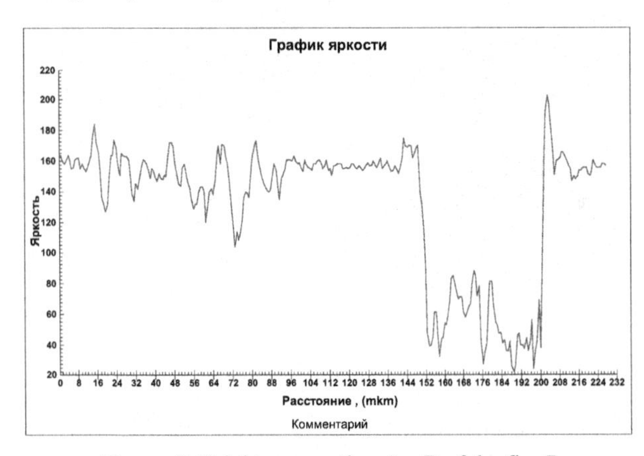

Figure 8. Brightness section A – B of the fig. 7

Table 1 shows the determination of alpha modification for alumina sample from Nikolaev alumina plant (NAP), obtained by different methods at NAP and VAMI.

Table 1. Determination of alpha modification for NAP samples using X- ray analysis in 2 independent laboratories and *Image Analysis*

No.	No. of sample	α-Al_2O_3, % NAP	α-Al_2O_3, X- ray	α-Al_2O_3, % IA
1	6	4.9	7.0	5.0
2	7	20.0	18.0	20.0
3	8	7.9	3.0	8.0
4	9	16.4	15.0	17.0
5	10	1.7	2.0	2.0
6	4	52.0	46.0	50.0
7	56	27.0	26.0	28.0
8	549	2.5	4.0	5.0

The quantitive determination of ♋-Al_2O_3 in the alumina using XRD ray method is based on the fact that the intensity of ♋-Al_2O_3 diffraction line is proportional to the amount of ♋-Al_2O_3 in the sample. The concentration is calculated using the formula:

$$C♋ = \frac{I^{♋}\ \text{sample}}{I^{♋}\ \text{standard}} \times \text{Cstandard} \qquad (1)$$

where: C♋- concentration of ♋-Al_2O_3 in the sample, mass %;
$I^{♋}$sample –intensity of ♋-Al_2O_3 analytical peak on the alumina sample X-ray diagram;
$I^{♋}$standard–intensity of the same peak on the standard sample X-ray diagram;
Cstandard – concentration of ♋ - Al_2O_3 in standard. alumina sample, mass %.

The analytical line is the ♋-Al_2O_3 line with interplanar spacing d=1,74 Å, as this line is totally free from the overlap of γ - Al_2O_3 lines (fig. 9)

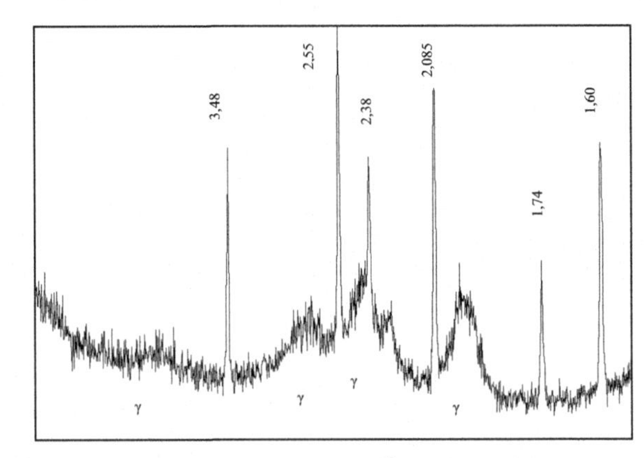

Figure 9. X-ray diagram ♋- and γ - Al_2O_3

The previously produced and studied specimens form VAMI and other laboratories were used as standard samples for the studies. The standard samples were presented by the set of 2 specimens for X-ray analysis and for determination of the ♋-Al_2O_3 mass content in smelting grade alumina by X-ray quantity analysis.

Presently the sufficient number of statistical and method accuracy data are accumulated based on the alumina from different refineries with the help of *Image Analysis System*. As independent method of alpha modification determination the XRD analysis is used, using VAMI methodology. The developed methods are based on the standardized crystal – optical principals "Crystal – optical determination of α - Al_2O_3 in the smelting grade alumina" (GOST 25733-89 Alumina. Crystal –optical determination of alpha alumina), certified at the CIS refineries.

The IAS method of quantitative phases determination in the alumina is applied to obtain the whole quantitative characteristic of the phase content for smelting grade alumina and is aimed on the phase identification in the chain of phase transformation of Al_2O_3 using optical phases constants with further computing of its content. The method is based on the principal of correlation between the refractive indices of the analyzed phase and immersion media, corresponding to each step in the chain of phase transformation. The study starts from determination of alpha modification content. At this stage the target alpha phase has the refraction parameters and brightness thresholds to the maximum approaching the iodic methylene. The phases with much low refraction indexes will be located on the brightness diagram below the brightness thresholds of alpha alumina. In the next stages phase identification and calculation of phase content are done in other immersion media. Every stage study is to allocate the brightness threshold as the starting point for calculation of content of analyzed phase in its immersion media.

The method allows more precise monitoring the alumina phase content to estimate the degree of calcination and crystallization for calcined products. The content of α- Al_2O_3 and LOI, being the factors of the calcination degree together with phase – quantitative parameter of the calciner products allow to obtain the most accurate estimation of alumina and to conduct input and output product control at the alumina refineries.

This technique is a combination of latest achievements in the instrumentation art, software and many-years practical experience of VAMI specialists.

The automated phase control and microstructure analysis methods are possible using *Image Analysis System*. As far as the alumina industry is using the products of poly- phase composition with complex microstructure their diagnostic is rather complicated. The generating of software for such kind of products required high level programming expert.

2. Method for determination of fine fractions content

Image Analysis System allows improving and automating the crystal optical method for determination of powder products grain size. Developed software provides determination of aluminum oxide and hydroxide practically for all fractional range. Using the results of statistic and stereologic parameters the grain size distribution by preassigned fractions is provided with graphic response.

The most complicated methodology is connected with determination of grain size, when the product has variable grain size with particles bigger than 125 μ together with significant quantity of fine fraction less than 10 μ. Without knowledge of micro structural features of such complex product as aluminum hydroxide in is impossible to make objective appraisal of its grain size.

The detailed analysis of aluminum hydroxide microstructure has been conducted under high magnification (1150 и $1800^{\times}$). Aggregations in all samples are rather dense, plate shaped. In the growth process the plate-like monocrystals are structured using the most developed surface forming stepwise, round or oval shape (fig. 10).

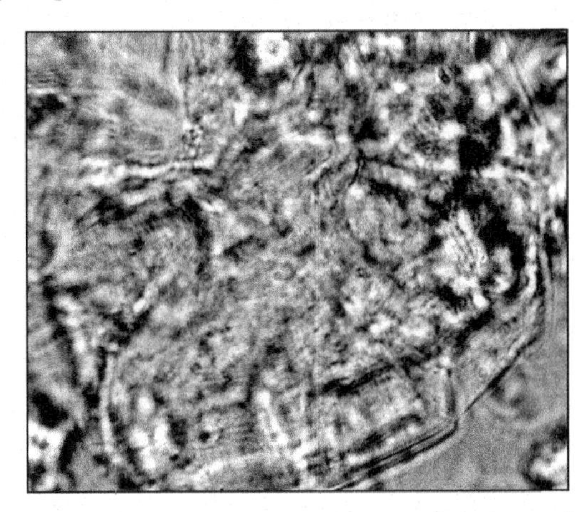

Figure 10. Layer shape of Al $(OH)_3$, sample in immersion liquid, magnification $1150^{\times}$

The monocrystals in such aluminum hydroxide are presented in form of rectangles, plates and fine hexagons, the largest can be 100 μ and higher. The macro laminar crystals are the result of large particles (aggregates) cleavage breaking (001), specific for gibbsite. On the edges of the majority of the plates the traces of the layers growth are seen in form of continuous layers growth (fig. 10).

To conduct the grain size analysis of powder products using laser instruments the anticoagulant is usually applied to prevent the coagulation (such as sodium pyrophosphate). The application of sodium pyrophosphate as immersion media in this methodology conducted to the positive results and allowed to separate and compute the particles for the whole range of grain size. Fig. 11 demonstrates the grain size picture for the industrial samples of aluminum hydroxide. The 2% sodium pyrophosphate has been used as immersion media.

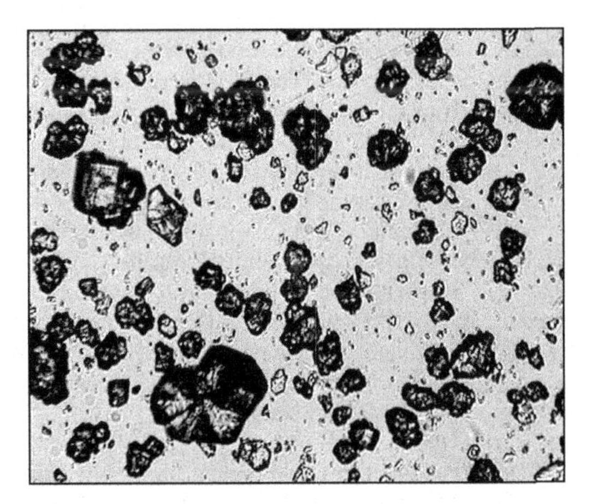

Figure 11. $Al(OH)_3$ grain size, sample 1 in anticoagulant. Magn. $230^{\times}$

It is quite possible to use different immersion media with anticoagulant properties. However for the selection of the media it should be considered that regardless of the powder product gain size measurement method in any liquid media where mechanical impacts occurs, the marginal results alteration are unavoidable due to the distraction of polycrystal particles or their coarsening. As for the fine fraction it should be noted that its content will be a function of the amount of released structurally bounded fines. The distorted results will be direct function of the aggregation strength.

The selection of the immersion media is only one of sample preparation conditions to conduct grain size analysis. All other conditions should be observed to obtain the reliable result: homogenizing of the devided sample and final weighted (1 g) specimen; exact execution of immersion specimen preparation. In correctly prepared specimen the particles should be located in one layer with minimum contact and uniform distribution in the liquid phase (anticoagulant). Together with sampling preparation procedure the physical properties of analyzed matter should be considered, namely: fragility of the alumina hydroxide aggregates and variety of its grain size. The existence of even small number of particles with the size more than 200 μ can complicate the analysis method.

Image Analysis System software solves the problem of the grain size determination. This program is based on the computing of the image of immersion specimen, installed in the sampler of polarization microscope. The high-resolution TV camera transmits the image to the PC and in parallel to Video Controller. The advantage of this method is the possibility to have the visualization of the object with simultaneous software computing. For maximum magnification of the computer and Video Controller images (microscope) the program has the the manual correction feature to correct the particles contour using the virtual cut of stuck particles or drawing of open circuits. This method automates process of grain size determination with gathering of necessary statistic for stable outcome. The software package includes the set of essential parameters to conduct the grain size analyze: diameter of the equivalent circle, length and width of the particles, number of particles per square/volume unit, maximum/minimum particle size, tabulation of the particle distribution by the random classes. The factor of particles shape is considered.

Results reliability is achieved by the computing of the images, representing the whole grain size range of analyzed sample. The *Image Analysis System* software covers the measurement of the particles size for fixed picture. The fixed picture means the shot of TV camera from the microscope and transmitted to PC for computing. Under moderate grain size distribution in the analyzed sample (not higher than 50 μ) the appropriate magnification can be selected when all the particles will be focused with sharp contours, required for analysis. The grain size composition with such granulometry can be considered as particular case. Practically we need to analyze the products with considerable grain size spreading. This problem can be overcome by the computing of the images, taken under different magnification to produce the clear contours of coarse and very fine fractions in the pictures for computing.

The developed method allows combining the results using the "cross-link" method through combination of data, generated during two magnifications.

Conclusions

1. Using *Image Analysis System* the following methods have been developed:

- ❖ automated determination of alpha modification in specialty grade and smelting grade aluminas & hydrates of all kind regardless the fabrication method and aluminum oxide generation method;
- ❖ automated determination of phase quantity analysis in smelter grade alumina;
- ❖ automated determination of grain size for aluminum hydroxide with higher fines.

2. Elaborated methods allow monitoring the alumina phase content within the range of phase transformation from hydrargillite to alpha modification under calcination process and aluminum hydroxide grain size for whole fractional range. Therefore the suggested methods are the instrument for fine-tuning of calciner.

3. As for aluminum hydroxide the possibility to monitor the distribution of ultra fines allows to manage the production of sandy alumina.

ALUMINA & BAUXITE

Plant Design, Operation and Maintenance

SESSION CHAIR
Jean-Pierre Riffaud
Alumina Partners of Jamaica
Spur Tree, Jamaica

Light Metals 2006 *Edited by Travis J. Galloway* ***TMS (The Minerals, Metals & Materials Society), 2006***

OPERATING COST – ISSUES AND OPPORTUNITIES

TWS Services and Advice, Peter-Hans ter Weer
TWS Services and Advice, Imkerweg 5, 1272 EB Huizen, The Netherlands

Keywords: Operating Cost, Opex, Economics

ABSTRACT

The operating cost of a bauxite and alumina project is often regarded as a burden to have to deal with or at best as an unavoidable sometimes even bureaucratic management requirement. However in a sense the operating cost and its underlying rationale provides tools to assess the "health" of a project and they may offer improvement opportunities.

This paper provides an insight in facets and issues related to the operating cost of a bauxite and alumina project and it explores ways for improvement.

INTRODUCTION

Economic evaluations of a project require generating (discounted) cash flows over its lifetime and applying several criteria to ascertain if it meets threshold levels.

Generally the primary economic evaluation criterion of a project is its Net Present Value (NPV), which is calculated as follows:

$$NPV(i)=\sum a_j*(1+i/100)^{-j} \qquad (1), \text{ with}$$

i = discount rate, %;
$\sum$ = summation from j=0 to n (project duration, no. of years);
a_j = annual cash flow (after tax) at time (yr) j.

Key inputs to annual cash flows are:

- On the revenue side: the amount of product generated and its sales price.
- On the cost side: capital cost, operating cost and tax payable.

In other words operating cost is an important element in the economics of a project, including bauxite and alumina projects.

Exchangeability of Opex & Capex

Both the project total operating cost ("opex") and initial capital cost ("capex") represent negative cash flows. Therefore a form of exchangeability exists between the two with respect to their effect on NPV and thus project economics.

To illustrate this effect, the following is assumed:

1.5 Mt/yr Alumina capacity expansion project
Alumina Price: 210 $/tA
Evaluation period: construction time + 25 yrs
Construction time: 3 yrs (capex spread equally)
Project construction starting next year

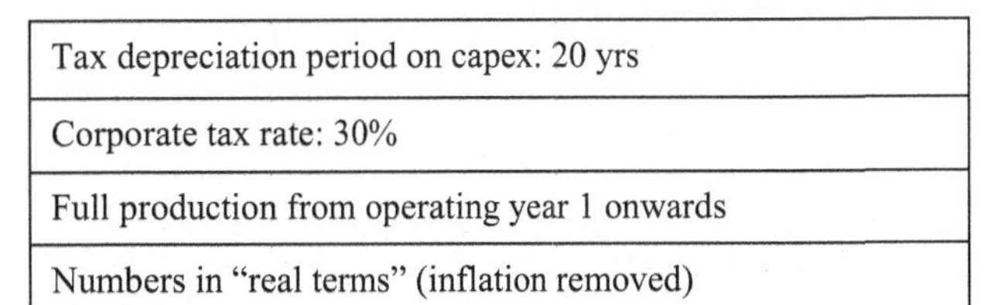

Tax depreciation period on capex: 20 yrs
Corporate tax rate: 30%
Full production from operating year 1 onwards
Numbers in "real terms" (inflation removed)

Table 1 – Main Assumptions

Assuming that project NPV (10%) should remain unchanged the following correlation can be found for a brownfield type project:

$$Capex = -6.5*Opex + 905 \qquad (2), \text{ with}$$

Capex in \$/Annual tA expansion capacity, and
Opex in \$/tA.

Correlation (2) means that an increase in opex of 10 \$/tA for a brownfield project would require a drop in capex of 65 \$/AnntA if NPV (10%) was to remain unchanged, as illustrated in Figure 1.

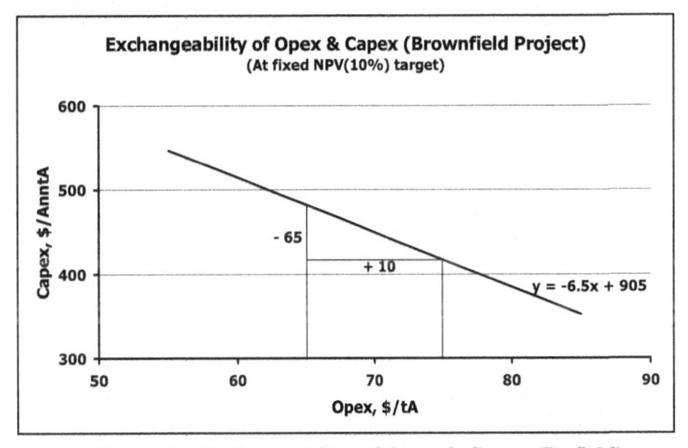

Figure 1 – Exchangeability of Opex & Capex (Br field)

The slope of the line in Figure 1 (-6.5) applies to a fixed NPV (10%) target. At other discount percentages this slope would be different (steeper at lower discount percentages).

A similar approach for a greenfield project, assuming a 10 year tax holiday (reasonable for this type of project), with otherwise the same assumptions as Table 1, results in the following correlation for a fixed project NPV (10%):

$$Capex = -7.7*Opex + 1567 \qquad (3).$$

In other words an increase in opex of 10 \$/tA for a greenfield project with a 10 year tax holiday would require a drop in capex of 77 \$/AnntA if NPV(10%) was to remain unchanged, as illustrated in Figure 2.

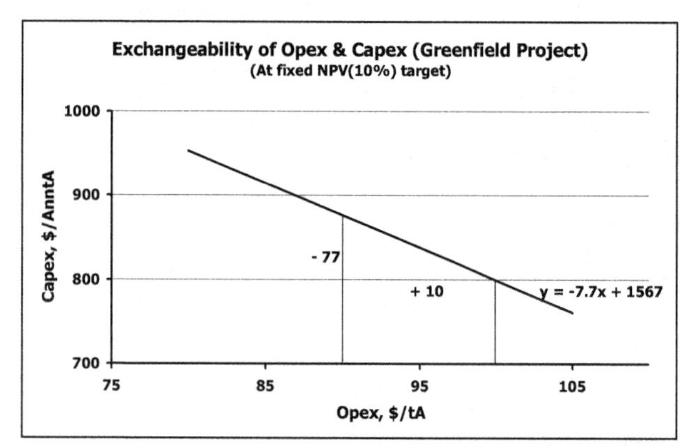

Figure 2 – Exchangeability of Opex & Capex (Gr field)

The following conclusion may be drawn from correlations (2) and (3): in order to maintain NPV(10%), an increase in opex for a greenfield project with a 10 yr tax holiday would require a significantly larger decrease of capex than a brownfield project would.

The difference between correlations (2) and (3) (reflected in the different slopes) is the consequence of the 10 year tax holiday assumed for the greenfield project.

Another way of putting the conclusion above is that under the assumptions mentioned, a change in opex has a more profound effect on the NPV of a greenfield project than on the NPV of a brownfield project.

This may be illustrated as follows: a decrease of 10 $/tA in opex for the above mentioned example of a 1.5 Mt/yr alumina capacity expansion project would result in an NPV (10%) increase of:

- 92 M$ for a greenfield project (10 yr tax holiday);
- 72 M$ for a brownfield project.

Robustness of Low Operating Cost

Projects with low operating cost under otherwise the same conditions, are inherently robust with respect to adverse economic conditions such as a drop in the alumina price.

This is illustrated in Table 2 for a greenfield project based on Table 1, with a 10 year tax holiday:

Low Opex Robustness (greenfield)		Low Opex Project	High Opex Project
BASE CASE	Aa Price, $/tA	210	210
	Opex, $/tA	90	105
	Capex, $/AnntA	800	800
	NPV(10%)	146	7
	IRR	11.8	10.1
Price Sensitivity	*Aa Price -10 $/tA*	200	200
	NPV(10%)	54	-85
	IRR	10.7	8.9

Table 2 – Low Opex Robustness (greenfield project)

As can be seen from Table 2, the economics of the low opex greenfield project remain attractive whereas the economics of the high opex project become unattractive.

Table 3 illustrates the same for a brownfield project, comparing a typical low opex, high capex option with a high opex, low capex option having similar base case IRR values.

Low Opex Robustness (brownfield)		Low Opex Project	High Opex Project
BASE CASE	Aa Price, $/tA	210	210
	Opex, $/tA	60	100
	Capex, $/AnntA	550	400
	NPV(10%)	469	348
	IRR	17.1	17.2
Price Sensitivity	*Aa Price -30 $/tA*	180	180
	NPV(10%)	255	133
	IRR	14.0	13.0

Table 3 – Low Opex Robustness (brownfield project)

Again the low opex option proves to be more robust with respect to alumina price fluctuations.

Opex, NPV and IRR

In the paper "Greenfield Dilemma – Innovation Challenges" [1] correlations were discussed between Opex and NPV (linear correlation) and Opex and Internal Rate of Return (IRR) (polynomial of the second order).

Conclusions

Conclusions are that the operating cost of a bauxite/alumina project has a significant effect on project economics (more so on those of a greenfield project than those of a brownfield project), and low opex projects are inherently robust.

OPEX BASICS

Conventionally project operating cost may be split in different ways, e.g. "controllable" and "uncontrollable", fixed and variable. This serves a purpose for an existing bauxite/alumina project. When a greenfield projects is being considered however it makes good economic sense, as outlined above, to focus on identifying projects with inherently low opex. Let us therefore consider the basics of operating cost.

The opex of a bauxite/alumina refinery project is not a stand-alone item but is an integral characteristic of the context of the project because it is influenced by:

I. Resource quality;

II. Country infrastructure;

III. Logistics of raw materials import and alumina export;

IV. Alumina production capacity;

V. Technology.

(Note that commodity prices – e.g. caustic soda, coal, fuel oil, lime – are not included because FOB prices are basically the same for all projects)

Simply put, opex reflects the efforts required to deal with the "imperfections" of the bauxite resource when converting it to smelter grade alumina with the selected technology, at the desired quality, in a responsible way with respect to safety and environment.

[1] Refer "Greenfield Dilemma - Innovation Challenges", paper by P.J.C. ter Weer, Light Metals 2005, San Francisco, pp 17-22

I. Resource Quality

Resource quality includes:

- **Location** and accessibility of the resource with respect to the port from which the alumina is exported (distance, height, mountain and river crossings).

  This is reflected in a transportation cost (either of the raw materials and alumina to the refinery site at the resource or of the bauxite to the refinery at the exporting port). Depending on resource particulars this transportation cost may range from 3-8 \$/tA.

- **Deposit characteristics** (uniform versus pockets, overburden thickness, bauxite horizon thickness, beneficiation requirement). These affect cost of mining, crushing, storage, beneficiation, rehabilitation, etc.

  This aspect is reflected in the overall \$/tBx cost CIF refinery (bauxite transportation may either be included – e.g. if the mine is close to the refinery, or be covered by a separate transportation cost referred to above), and may range from 6-40 \$/tA.

- **Bauxite** refinery feed **characteristics** (hardness, available alumina – level and % gibbsite/boehmite, reactive silica, impurities).

  These aspects are mainly reflected in caustic soda cost (partly) and energy cost (partly), and may range from 8-50 \$/tA.

Note that if an alumina refinery project does not participate on an equity basis in a bauxite mine, the first two aspects mentioned above do not apply. In that case a market price will have to be paid for the bauxite, which will be more than the sum of the costs mentioned above for these two elements. For reasons of economics it is unlikely that new greenfield alumina refinery projects would be based on export bauxite.

II. Country Infrastructure

In the context of this paper, country infrastructure includes:

- **A "hardware" aspect.** This covers items such as the presence and state of repair of a port (via which raw materials are imported and alumina exported), rail road (if the resource is a significant distance from the port), a town site (incl. hospital), roads, and water availability.

  This aspect is reflected in infrastructure operating and maintenance costs.

- **A "software" aspect.** This refers to elements such as royalty (on bauxite or alumina), levies, duty on raw materials, and the effect of legislation (e.g. with respect to environmental requirements).

  Taxes (and tax holidays) are very important, but more so from an overall economics perspective than from the point of view of the opex.

Total infrastructure related costs may range from 0-11 \$/tA.

III. Logistics

Accessibility of and logistics to the port via which raw materials are imported (e.g. caustic soda, coal, fuel oil, lime) and alumina (or bauxite) is exported.

This item is affected by location of the port (e.g. industrial area), ship size (water depth) and the proximity to frequently used sea lanes, and it is reflected in the cost of raw materials CIF the importing port and/or in freight charges for export alumina (or bauxite). In other words it is affected by refinery location in case the refinery is built in the same country as the bauxite resource.

Total logistics costs cover a wide range from 2-35 \$/tA.

IV. Alumina production capacity.

Refinery production capacity has an impact on opex because it "dilutes" fixed operating and maintenance costs and may also result in efficiency improvements. It also has a positive "economy of scale" effect on capital cost (outside the scope of this paper).

A larger production capacity therefore generally results in (significantly) more attractive overall project economics.

Resource size is an important aspect in this context. In order to capture the opex (and capex) advantage of a large production facility, a bauxite resource should be able to support an alumina refinery for its lifetime (typically 50^+ years).

However, a project's initial and final production capacities are also the result of a company's view on alumina demand and supply developments and access to capital, and to a lesser extent the selected process technology.

V. Technology

In terms of operating cost, "technology" relates to plant design and layout, process and equipment technologies and operating and maintenance philosophies. These aspects are reflected in:

- **Energy and raw materials** costs (caustic soda, lime, etc) – albeit partly in some cases. These are affected by technology choices such as co-generation of steam and power (when possible), choice of energy carrier (coal/fuel oil/gas), residue circuit thickener/washer, precipitation, residue disposal, heat recovery, return of disposal run-off water etc.
- **Other variable** costs (crystal growth modifier, filter cloth, acids, etc);
- **Maintenance materials** and contract services costs, affected by the complexity of the refining plant, choice of layout, etc;
- **Labour** and other fixed costs: as maintenance materials affected by technology and layout choices.

The costs related to capacity and technology may range from 55-105 \$/tA.

Summary

Table 4 summarises the above with typical ranges of these five elements.

Element	Range (typ) ($/tA)
I. Resource Quality	25 – 95
II. Infrastructure	0 – 11
III. Logistics	2 – 35
IV&V. Capacity and Technology	50 – 105
Total Opex incl. sustaining capex	**95 – 230**

Table 4 – Typical ranges of opex elements

In other words, if aspects of the infrastructure of the country of the bauxite resource and the location of the exporting port are also considered as part of resource quality in its widest sense, the typical ranges of "Resource Quality" and Capacity and Technology costs (incl. sustaining capex) are as shown in Table 5.

Element	Range (typ) ($/tA)
"Resource Quality"	35 – 130
Capacity and Technology	50 – 105

Table 5 – Typical opex ranges of Resource Quality & Capacity and Technology

Table 5 shows that the cost range of "resource quality" is even wider than that of "capacity and technology". Taking the capacity effect out (which may range from 2-15 $/tA) would narrow the latter range may be to 60 – 100 $/tA.

In other words "bauxite (resource) quality" in its widest sense has a more profound effect on operating cost than technology. Or to put it differently, for a greenfield project it makes good economic sense to primarily focus on identifying the "right" bauxite resource.

This may be done by ranking bauxite resources on the basis of a set of selection criteria taking the above elements into account.

Bauxite Resource Selection Criteria

A set of bauxite resource selection criteria should focus on main criteria and provide target threshold values. Such a set is not meant to be applied rigidly. In other words a bauxite resource not meeting one (or possible more) of the threshold values should not necessarily be discarded. The overall result of the ranking of available bauxite resources should be considered.

Criteria relating to the strategic importance of a resource to a company, which could result in a different outcome of a resource ranking exercise, have not been included because those fall outside the scope of this paper.

Table 6 presents a set of bauxite resource selection criteria addressing the resource quality elements I-III and the alumina production capacity element IV discussed above. These criteria address both opex as well as capex related aspects as the two are linked.

This set of selection criteria may be used as ranking tool for bauxite resource evaluation purposes.

Criterion	Target
1. Distance Resource to Alumina Export Port	200 km max
2. Average Bauxite Horizon Thickness	5 m min
3. Total Material Handled	4 t/tA max
4. Bauxite Beneficiation	Not required
5. Alumina in Boehmite	3 % max
6. Ratio Extractable Organic Carbon / Available Al2O3	0.008 max
7. Residue to Disposal Factor	1.2 t/tA max
8. Total Caustic Soda Consumption (100% NaOH basis)	70 kg/tA max
9. Country Infrastructure	Mostly in place
10. Resource Contained Alumina	200 Mt* min

* M=million

Table 6 – Bauxite Resource Selection Criteria

As mentioned above these criteria and their targets should not be looked at in isolation but as part of an overall analysis of a bauxite resource.

The rationale behind these criteria is as follows:

1. **Distance bauxite resource to alumina export port maximum 200 km.** This criterion addresses the resource location aspect of element I above. If the transportation distance increases much above 200 km, raw materials and alumina or bauxite transportation costs increase prohibitively. If bauxite slurry pumping as planned by CVRD for the Paragominas mine in Brasil proves to be successful the maximum distance may increase (costs of bauxite de-watering and additional evaporation should be included).
2. **Average bauxite horizon thickness minimum 5 m.** This criterion addresses the deposit characteristics aspect of element I above and has economics and environmental angles: if the bauxite horizon thickness is much below 5m, the mining opex increases significantly and the acreage of land mined (and rehabilitated afterwards) per tonne alumina becomes environmentally prohibitive.
3. **Total material handled maximum 4 t/tA.** This criterion also addresses the deposit characteristics aspect of element I above and includes the bauxite proper (affected by available alumina and alumina recovery in the refining plant), the overburden (affected by the overburden/bauxite ratio) and tailings from the beneficiation plant if applicable (affected by beneficiation recovery – see below).
4. **Bauxite beneficiation not required.** This criterion also addresses the deposit characteristics aspect of element I above. Bauxite beneficiation has several drawbacks: additional installations are required involving extra capex, extra opex and environmental issues (water usage, tailings disposal). In addition if the beneficiation recovery is low (may be as low as 25%) these aspects quickly become

prohibitive. When the overall economics of a bauxite/alumina project improve, applying bauxite beneficiation may be appropriate.

5. **Alumina in boehmite maximum 3%.** This criterion addresses the bauxite refinery feed characteristics aspect of element I above. There are several aspects to this point, the most important one being that below this maximum the large majority of the available alumina is gibbsitic requiring a low digestion temperature. This has two advantages, i.e. 1. Required temperature and pressure for steam to digestion are at a relatively low level, providing the opportunity to first use high pressure boiler steam for co-generation of steam and power, which has a positive impact on overall energy cost and thus opex, and
2. Capex of digestion and power & steam generation equipment is lower than would be required for high digestion temperatures.

   Another aspect is that below this maximum boehmite in bauxite feed level, processing technology and operating conditions can be chosen such that boehmite does not dissolve in digestion and should not give rise to boehmite reversion in residue decanters and washers (which would otherwise negatively affect efficiencies and operational conditions). Higher boehmite levels would require high temperature digestion with consequential higher opex (energy and maintenance costs) and capex (high pressure refinery and powerhouse equipment).

6. **Ratio of extractable organic carbon to available alumina maximum 0.008.** This criterion also addresses the bauxite refinery feed characteristics aspect of element I above. Below this level impurity removal may either take place sufficiently by natural balances (e.g. with residue) or by relatively simple removal methods (e.g. concentrating plant spent liquor in a salting out evaporator). At higher ratios extensive facilities may be required (oxalate removal, organics removal) at significant opex and capex. In addition precipitation yield may be affected negatively impacting on overall energy efficiency and capex.

7. **Residue to disposal factor maximum 1.2 t/tA.** This criterion also addresses the bauxite refinery feed characteristics aspect of element I above and impacts on residue handling and disposal facilities, i.e. has both economics and environmental angles. It is affected by the available alumina and reactive silica in bauxite refinery feed and alumina recovery in the refining plant.

8. **Total caustic soda consumption maximum 70 kg (100% NaOH basis)/tA.** This criterion also addresses the bauxite refinery feed characteristics aspect of element I above because caustic soda is one of the main operating cost components, mainly influenced by the ratio of reactive silica to available alumina in bauxite refinery feed, physical soda losses with bauxite residue (affected both by available alumina and alumina recovery in the refining plant and by the technology chosen in the residue wash circuit) and soda losses with removed impurities (oxalate, organics).

9. **Country infrastructure mostly in place.** This criterion addresses elements II and III above, i.e. refers both to physical installations and a country's legal framework. Its effect on opex may range from quite low (e.g. with limited royalty requirements and no physical infrastructure to run) to significant (with bauxite levies and extensive infrastructure opex). Its effect on capex may also be significant.

   Some of these elements may be negotiable and therefore more difficult to quantify at an early stage of a potential project. In addition a country's government and/or other third parties may be willing to assume responsibility for (some of) the physical infrastructure requirements.

10. **Resource contained alumina minimum 180 Mt.** This criterion addresses element IV above. In order to capture the opex (and capex) advantage of a large production facility, a bauxite resource should be able to support a typical "mature" alumina refinery production capacity of 4 Mt/yr. Alumina refineries are long lifetime facilities (50^+ years). These two combined form the rationale for 200 Mt contained alumina in resource. At a bauxite factor of typically 3 tBx (in situ)/tA, this would convert to a bauxite resource of about 600 Mt.

Conclusion

Conclusion is that for greenfield bauxite/alumina projects the focus should primarily be on identifying the "right" bauxite resource using a limited set of appropriate selection criteria.

OPEX OF EXISTING PLANTS

For existing plants and brownfield expansion projects it may be more appropriate to use a more conventional approach to opex. Let us therefore consider the opex build-up.

Opex Build-up

Project opex may be broken down in the following main components: variable costs, fixed costs and sustaining capital.

Variable Costs

In $/year these costs vary with plant production, at least within certain plant production rates (typically $\pm$ 10-15%).

Variable costs comprise bauxite, energy, caustic soda, other raw materials (e.g. lime, flocculants) and process & operating supplies (e.g. crystal growth modifier, grinding media, and filter cloth). They cover a wide range from typically 60-160 $/tA.

Transportation and handling costs of raw materials and product are sometimes accounted for in the raw materials costs (i.e. raw materials costs CIF plant site), but sometimes presented by a separate variable cost item "materials transportation and handling cost".

A significant part of the variable costs are directly or indirectly the result of "bauxite quality" in its widest sense (see above). They are therefore difficult to improve once the bauxite quality is fixed and technology choices have been made for the various processing steps of the alumina refinery.

In other words structurally improving the variable cost in an existing alumina refinery processing a given bauxite type requires an adaptation to or complete change of technology of one of the processing steps (e.g. deep thickening technology in stead of conventional thickeners, seed recycle in stead of solids retention in precipitation, etc).

This also means that the scope to improve variable costs significantly in an existing refinery without spending significant capex is limited. A good opportunity is sometimes presented when a brownfield capacity expansion project of a refinery is considered. In some cases it is possible to apply new technology of the brownfield project also to existing units, thus more than proportionally improving economics.

Fixed Costs

In $/year these costs do not vary with plant production, at least within certain plant production rates (typically ± 100,000 t/yr). They typically range from 20-70 $/tA.

Fixed costs comprise labour, plant maintenance materials, contract services, infrastructure operating and maintenance costs, overheads and other fixed costs.

Plant design and layout affect fixed costs to some extent through simplicity of design, distance to get to facilities, central control room, etc. At the same time plant management controls fixed costs by their choice of operations management systems, number of shifts, maintenance procedures, outsourcing etc.

As mentioned earlier, another way of improving fixed costs (in $/tA) is "dilution" through increased plant production. In fact the "phenomenon" of a long term decrease of plant opex is to a large extent the result of plant production "creep" (the small but steady increase of plant production on a long term basis).

Sustaining Capex

Sustaining capital expenditure comprises items required to maintain the project at target production level, the product at target quality and HSE on target. Although it may be treated differently for tax reasons from than variable and fixed operating costs, it represents an annual cost to a project and should be included in the operating cost.

It ranges typically from 4-9 $/tA and includes:

- Mine sustaining capex;
- Plant (including powerhouse) sustaining capex, and
- Infrastructure sustaining capex.

Sustaining capex is a consequence of equipment installed, i.e. of design and technology choices and thus also difficult to structurally improve without spending significant capex. In addition the scope for improvement is relatively limited.

CONCLUSIONS

The conclusions from the above may be summarised as follows:

- The operating cost of a bauxite/alumina project has a significant effect on project economics (more so on those of a greenfield project than those of a brownfield project);
- Low opex projects have inherently robust economics;
- Greenfield bauxite/alumina projects should focus primarily on identifying the "right" bauxite resource using a limited set of selection criteria;
- The scope to improve variable cost significantly in an existing refinery without spending significant capex is limited;
- Plant management controls to some extent fixed costs. "Dilution" of fixed costs by production increase is a tool to lower opex.

Light Metals 2006 *Edited by Travis J. Galloway* ***TMS (The Minerals, Metals & Materials Society), 2006***

DIGITAL FIELDBUS IMPLEMENTATION FOR MINERAL PROCESSING

Manoj Pandya

Alcan Engineering Pty Ltd, Level 10, 119 Charlotte Street, Brisbane 4000, Australia

Keywords: Digital, Fieldbus, CHAZOP, Interoperability, Planning, Pre-assembled Modules

Abstract

A technological change is emerging in the process control and monitoring environment. The enormous economic benefits of the application of digital communication for instrumentation equipment are influencing decisions to abandon the conventional point-to-point technology, making the shift to digital technology imminent.

To date the large-scale use of digital technology has been limited to the petro-chemical sector; however, Alcan Inc has taken steps to keep abreast of this technological change as demonstrated by the introduction of FOUNDATION™ fieldbus and Profibus DP as part of the AUD2bn expansion of Alcan Gove's alumina refinery.

This paper discusses the implementation of digital communication technology for process control equipment from concept to commissioning. The primary focus of the paper is a technology plan developed by Alcan Engineering Pty Limited (ALCENG). The topics covered include migration to new technology and associated risk management, engineering design process and commissioning, as well as the benefits gained and "tips and traps" for a successful project implementation.

Introduction

Fieldbus is a robust, digital, two-way, multi-drop communication link between sophisticated industrial measurement and control devices. It allows multiple industrial devices to connect to a single pair of wires resulting in significant reductions of cabling, intrinsic safety barriers, I/O cabinets, marshalling panels and field junction boxes. The combination of these physical entity reductions, the time-savings produced from using fieldbus function blocks, and the additional process and device "health" information results in lower costs of engineering, installation, commissioning, documentation and ownership of instrumentation and control systems (I&C).

ALCENG was commissioned by Alcan Gove (formerly NABALCO) to develop a technology upgrade plan as part of its continuous improvement process. Following preliminary investigations, ALCENG realized that the current legacy distributed control system (DCS) installed at Alcan Gove needed to be upgraded or replaced, thus creating an opportune time to consider a plant-wide technology upgrade. The plan also identified that continuing to use traditional 4-20mA I&C devices and wiring practices would place Alcan Gove at a technical disadvantage in the not too distant future.

The investigation emphasized that staying in the "comfort zone" simply wasn't an option and that there was an eminent need to invest resources in learning more about fieldbus technologies. Alcan needed to understand the organizations backing the various fieldbus technology platforms, and the offerings and plans of the control system manufacturers supporting the various fieldbus communication protocols.

Technology Migration

In 1999, when ALCENG began formally assessing the current and future I&C requirements for the Alcan Gove facility, fieldbus technology was still quite new, multiple fieldbus technologies were being developed and promoted, and Standards development groups were struggling to sort out the best, long-term solution(s) for technology implementation.

At about the same time, I&C manufacturers were merging, acquiring, divesting, retooling, refocusing, obsolescing and generally leaving users scratching their heads trying to figure out which organizations and manufacturers, and more importantly, which control systems would survive in the long-term.

Before ALCENG engineers began their investigative analysis, they identified the risks associated with migrating to fieldbus technology and developed strategies to mitigate each risk. The mitigation strategies targeted the following key areas:

- Alcan Inc's fundamental criteria of placing Environment, Health and Safety (EHS) "first" and not compromising EHS at any cost.
- Continued utilization of and integration with business management systems at Alcan Gove.
- Evaluating and ensuring that fieldbus devices will provide the same level of interoperability as traditional point-to-point 4-20mA technology.
- Retaining fieldbus technology expertise to assist in developing operationally secure segment designs.
- Ensuring that I&C equipment selected will be capable of supporting multiple fieldbus technologies, their connectivity and integration with current I&C systems.

ALCENG management also expected the investigation to confirm fieldbus proponents' claims about cost savings – specifically cost savings applicable to Alcan Gove in the following areas:

- **EHS** – Reduced frequency and severity of EHS incidents as a direct result of improved diagnostics, more information and increased operator awareness.
- **Engineering** – Reduction in the number of drawings, cable schedules, cabinets, footprint and configuration efforts.
- **Material and labor** – Increased use of multi-variable devices resulting in fewer process penetrations, devices,

junction boxes, marshalling cabinets, terminations, control system I/O cards and engineering activities.

- **Installation and commissioning** – Reduced times to install and commission I&C systems.
- **Incremental expansion** – Increased flexibility and simplification when adding I&C devices at a later date as necessary to improve and/or optimize the process.
- **Maintenance** – Reduced maintenance and related documentation resulting from embedded fieldbus device diagnostics, alarms and alerts.

The investigative analysis focused on evaluation of different bus systems and interoperability issues. Visits were arranged to existing fieldbus installations in Australia and Singapore to learn about the experiences of early fieldbus adopters. At the end of the investigation, ALCENG engineers were convinced that while the fieldbus technology was still evolving, it was:

- Technically superior to traditional point-to-point solutions;
- Likely to garner continued enhancements for years to come; and
- Providing early adopters the opportunities to gain a competitive advantage.

The ALCENG investigations eventually revealed that the fieldbus technologies best suited to Alcan Gove operations were FOUNDATION fieldbus for process-related measurements and control, Profibus DP for intelligent discrete devices, such as "smart" motor control centers (MCC) and variable frequency drives (VFD), and AS-Interface (AS-I, www.as-interface.com) for simple discrete indications and functions.

While these technologies provide robust solutions, have an extensive installed base and an impressive list of manufacturers offering compatible products (even in 1999), it was comforting to learn that no individual company, nation or regulatory body "owns" Fieldbus Foundation (www.fieldbus.org) or Profibus Nutzerorganisation e.V. (www.profibus.com).

Each is a not-for-profit organization supported by end-users and instrumentation equipment manufacturers. While each organization exists to advance its respective fieldbus technology, they work in a cooperative manner. For example, both technologies use files created using Device Description Language (DDL) to communicate among the respective fieldbus networks.

In 2003 the two organizations formed a cooperative project to enhance DDL (thus creating EDDL) to facilitate the increasing sophistication of such devices as control valves, radar level gauges and variable frequency drives (VFD). The project has now been completed and device manufacturers recently began releasing sophisticated devices that plug-and-play on their respective fieldbus.

A noteworthy aspect of DDL, and now EDDL, is that the files created using DDL permit instrument number one to work right alongside instrument number 15 million[1] – never once requiring an upgrade, revision or patch of any kind.

Technology Implementation

Satisfied that migrating to fieldbus technology would better position Alcan Gove competitively, ALCENG developed a plan to introduce and familiarize Alcan Gove operations with fieldbus technologies.

Not wanting to commit the entire facility all at once, the plan identified the following tasks:

- Secure executive management support;
- Ensure adequate ownership and resources were thoroughly prepared to embrace the technology. This included arranging for appropriate hardware and software training;
- Retain the services of a fieldbus expert/consultant;
- Evaluate candidate control systems suitable to replace Alcan Gove's installed legacy control systems as well as to provide an interface for the digital bus devices for the expanded plant; and
- Develop a multiple-phase fieldbus technology implementation plan that mitigated risk by the time Alcan Gove was ready to begin its 3rd stage expansion project, tentatively scheduled for 2004.

Phase 1 was to be a small pilot project consisting of ~20 monitoring devices, carefully selected for fieldbus implementation and not critical to plant operation upon loss of device or fieldbus segment.

Assuming phase 1 results were successful, phase 2 was to be part of a planned project in the gas conversion area. This would serve as a medium-size project (~2,000 field devices).

ALCENG engineers recognized that it would be easy to begin installing the phase one project immediately, thus introducing fieldbus technology to the plant very quickly. However, the goal of the pilot project wasn't just to get fieldbus technology installed as quickly as possible; it was to develop a complete fieldbus framework of practices and procedures for designing, engineering, installing, commissioning, operating and maintaining fieldbus technology.

One of ALCENG's original investigation goals was to learn just how well manufacturers of Alcan Gove's current I&C systems would be able to support connectivity to multiple fieldbus technologies. In this regard, it had become apparent that the current system was in "legacy system" status, meaning that while parts were available, the system manufacturer was not intending to invest in new enhancements.

It is nearly impossible to investigate the various fieldbus technologies without being exposed to a variety of different manufacturers' control systems. Thus the investigation of one leads to a very good understanding of the others and provides valuable insight into how well manufacturers have embraced and embedded various fieldbus technologies into their control systems architecture.

[1] As of May 2004, an estimated 15 million devices were installed using files created using DDL. Source: May 2004 issue of InTech Protocol supplement, p.6, "Demystifying a protocol – Fieldbus device descriptions provide interoperability, by Stephen Mitschke."

Once the decision to use FOUNDATION fieldbus, Profibus DP and AS-I, had been made, the list of control systems that might be used to replace the installed legacy control systems had already been narrowed.

ALCENG engineers noted that it was interesting how European and North American manufacturers of I&C equipment had established a niche product presence in Australia, not always offering the same complement of products and/or services available to other parts of the world.

Following an extensive technical evaluation, ALCENG chose to replace the installed legacy control system with Emerson Process Management's DeltaV™ digital automation system (www.easydeltav.com). Key factors influencing this decision included:

- Major fieldbus installation experience;
- System and device support in Australia;
- Diversity of approved fieldbus equipment;
- Breadth and depth of installed devices, systems and support available in Australia;
- Availability and completeness of fieldbus related standards and guidelines that Alcan Gove could utilize;
- Commercial considerations.

With the fieldbus technologies and control system manufacture selection process completed, ALCENG's next step was to hire a fieldbus expert/consultant. Eventually an Emerson consultant was chosen, one who was familiar with the three fieldbus technologies of choice as well as the inter-workings of the DeltaV system.

Phase one

Using traditional CHAZOP (Control Hazards and Operability) study techniques, ALCENG engineers identified 20 monitoring points that could be retrofitted to fieldbus technology without serious ramifications to plant operations in the event of a device or segment failure.

Because interoperability had been identified as a key technology migration risk, ALCENG carefully selected devices from multiple instrument manufacturers in order to gain a first-hand experience of the interoperability issues.

An added benefit of selecting devices from multiple instrument manufacturers was the opportunity to:

- Gain experience with installation, set-up, configuration and troubleshooting of multiple devices from multiple manufacturers.
- Review and use device specific documentation from a variety of manufacturers.

Phase two

We all have experienced that even the very best of plans are subject to change: the multiple-phase fieldbus implementation plan was no exception.

The gas conversion project earmarked as the platform for the phase two of the fieldbus implementation was placed on hold. However, the 3rd stage expansion project that had been identified as the final phase of fieldbus implementation had received approval to proceed.

This had ALCENG engineers facing a major question: "Had the phase one fieldbus project, combined with the plant visits and interaction with I&C manufacturers provided ALCENG engineers and Alcan Gove management with sufficient confidence in the technology to commit to using it on the large scale project?"

Phase three

The phase one had produced positive results in the areas of:

- Development of standards for documentation, wiring methods, equipment, etc.;
- Development and use of CHAZOP to ensure the robustness of the installed fieldbus;
- Effective training for engineering and maintenance personnel;
- Efficient, quick and easy commissioning of devices and segments; and
- Verification that significant cost savings could be realized on future projects.

When everything was considered, ALCENG engineers and management were convinced that the phase one project had established sufficient confidence in the fieldbus technology. It had also reinforced the fact that it was an appropriate path for Alcan Gove to continue pursuing. The decision was made: "Utilize fieldbus technology on the green-field plant of the AUD2bn 3rd stage expansion project (G3), while retaining the conventional 4-20mA system for the brown-field retrofit." Also the choice of bus technology was further narrowed to utilize FOUNDATION fieldbus and Profibus DP.

The I&C scope for the G3 expansion project consisted of ~5,000 field devices, VFD and motors, which equated to ~15,000 traditional I/O points.

While ALCENG had been using a conventional CHAZOP process for quite some time, the introduction of fieldbus technology introduced some important new considerations. Working in conjunction with Emerson fieldbus consultants, ALCENG engineers modified the CHAZOP process to accommodate fieldbus technologies and the DeltaV digital automation system.

In a traditional HAZOP study, piping and instrumentation diagrams (P&ID) are the main documents being studied. In the CHAZOP for Alcan Gove's 3rd stage expansion project, fieldbus segment design along with the control system functionality and operator interface became the focus of the studies.

While viewing a segment design, CHAZOP participants were presented with a "flash card" developed to address a specific aspect of the segment design. For instance,

- "What would be the consequences if the segment failed at a particular connection point?" or
- "Does this segment have sufficient critical measurement design segregation?"

Segregation design policies often insist that critical, congruent measurements not share a common point of failure. For example, the temperature and pressure measurements inside a vessel may be congruent, meaning the pressure increases as the temperature rises and vice-versa. Good engineering practices would insist that these two measurements do not share a common point of failure, such as sharing the same segment or be connected to the same I/O interface (H-1) card. The CHAZOP flash card review helped identify where these congruent conditions exist and helped ensure such measurements are appropriately segregated.

Through the use of the flashcard CHAZOP approach, ALCENG engineers were able to thoroughly review such things as:

- Implementation of correct segregation of segments, devices, I/O cards, cables, etc.;
- Interface for "smart" MCC equipment, as well as motor stop/start and protection designs; and
- Documentation completeness of design, installation, testing and maintenance.

One of the unique approaches adopted for the G3 expansion project was the concept of modular construction, also referred to as Pre-assembled Modules (PAM).

PAMs are fully fitted out off-site to form a complete plant module including electrical and instrumentation equipment. Upon shipment from the assembly area, each PAM is fully pre-commissioned and ready for installation at site. (See Figure 1)

Figure 1 Fully fitted-out PAM

More than 300 PAM assemblies, ranging in size from 200 to 1,200 tonnes are being built and delivered to the Gove site. This is producing such benefits as:

- Improved safety by reducing the number construction personnel and activities required on-site;
- Improved project schedule through the simultaneous use of multiple unit assembly locations;
- Improved efficiency and quality by implementing a "controlled factory" construction environment;
- Reduced costs by assembling where people and equipment are already located;
- Reduced social impact by avoiding the mobilization of a large construction work force to the Gove site;
- Reduced environmental impact by minimizing on-site personnel and activities; and
- Reduced on-site material lay-down and staging requirements.

Though it was not initially part of the fieldbus investigation or identified as a cost saving factor, the use of fieldbus technology proved to be an excellent fit with the PAM concept. It delivered benefits that were beyond initial expectations and that had not been included during the investigation and approval process.

PAMs could be fully assembled including field instruments, control valves, conduit, wiring, junction boxes, terminations, etc. The completeness of the assembly facilitates an exhaustive factory acceptance test (FAT), and makes on-site connection of the PAM to the DeltaV system much easier.

To minimize on-site work and achieve a trouble-free plant commissioning, ALCENG developed:

- A device testing procedure that commences when a device is originally removed from its box and extends until the device is shipped to the site as part of the assembled PAM;
- A collection of field test reports designed to aid stage one and stage two commissioning activities;
- Detailed segment health check procedures that facilitate a more efficient commissioning of the PAM on-site as well as establishing a future "segment health" reference signature;
- Documentation for sign-off just prior to the shipment of a PAM; and
- Sign-off and ownership transfer process for each PAM.

A portable DeltaV system was used to facilitate the testing of I&C equipment contained within each PAM prior to its shipment to the Gove site.

Because DeltaV was designed by Emerson to be fieldbus aware, its use made PAM testing highly efficient. For example, when field devices are connected to a fieldbus segment and powered up, each device completes its on-board diagnostics, establishes communication with the segment, and automatically reports its presence and identity to the DeltaV system.

Once all devices have initialized, a comparison between the devices that actually reported and the segment's design documentation is made. Any discrepancies are easily spotted and corrective action is taken. Once all devices are verified to be electrically present, the next phase of testing can begin.

ALCENG also utilized sophisticated 3D models for the design of each PAM. The software used to produce the 3D models supports establishing rules that ensure even the smallest detail of engineering consistency is maintained throughout the project, including such items as cable segregation. This is helping to ensure fieldbus segment design requirements are being consistently applied.

This attention to detail resulted in the elimination of the risk of electrical noise on fieldbus segments during detail design, and allowing ALCENG to engineer reliability into the installed solution.

ALCENG is one of the first companies to implement the automated generation of segment connection drawings and other engineering deliverables in a true fieldbus project environment.

It has been ALCENG's aggressive use of technology, automated engineering processes, attention to detail and extensive planning exercises that are providing better management of preliminary and detailed design changes while, at the same time, eliminating the risk of fieldbus segment performance issues.

Business Case

The days of installing technology for the sake of technology have long since disappeared. In today's highly competitive business environment, every capital investment requires analysis and a justification that the expenditure will produce an acceptable cost benefit.

Acceptable cost advantages are achieved through quantified improvements in product throughput and quality, as well as reduced variability, installation, maintenance and operating costs, when upgrades to process instrumentation equipments is being considered.

The initial economic value analysis for migrating/upgrading the Alcan Gove's alumina refinery to fieldbus technology included both hard (quantified tangibles) and soft (intangible) entities.

Hard entities include actual costs associated with engineering, purchasing, installation and commissioning; while soft entities include estimated costs associated with improved operations resulting from additional process and device diagnostics, and reduced efforts to gather, assemble and maintain accurate, up-to-date calibration and maintenance records.

Additionally, ALCENG was interested in establishing cost reductions for both the green-field (new) and the overall project, which included both the green-field and brown-field retrofit.

The estimates produced indicate that by using digital fieldbus technology, ALCENG could expect to produce hard cost savings greater than 60% for the green-field part of the project, which was implemented using the fieldbus technology; however, the overall cost savings for the project would be around 40% when both green-field and brown-field aspects of the work are considered.

Despite the slightly higher cost of fieldbus products (less than 20%), overall savings were determined to come from a significant number of engineering, construction, implementation, maintenance and operational areas. From among these areas, ALCENG engineers determined the greatest hard savings would be achieved in the areas of engineering drawings, material and labor, and commissioning.

The material and labor savings alone, which amounted to >40% from an overall project perspective, were largely delivered due to the reduction in the following:

- Cabling requirement;
- Cable terminations;
- I/O cards and system cabinets;
- DCS interface hardware; and
- Field junction box assemblies.

However, there are other areas expected to produce savings. When considered, these would result in even higher saving projections, but were not included in developing the overall savings calculations. These include:

- Reduction in other engineering deliverables, e.g. cable schedule, system configurations, etc.
- Cable tray reductions;
- Reduced real estate;
- On-site versus off-site cost of delivery.

Although the G3 project is still in the construction phase, soft benefits already worthy of mention include:

- The first fully tested control system cabinet was designed, assembled and delivered to site six months after project approval, a feat ALCENG engineers are convinced could not have been achieved using traditional 4-20mA technology.
- The first fully fitted-out PAM, including 98% of instrumentation work, was completed and shipped to site eight months after receipt of project approval (See Figure 2). The fieldbus installation enabled the reduction of on-site work to just twelve trunk cables wired back to the DCS and eight 24V power supply cables to feed the local distribution boards.

Figure 2 PAM being delivered to Gove

The value proposition of soft entities, such as improved operations resulting from additional process and device diagnostics, can be difficult to quantify. However, according to a report by independent industry experts at ARC Advisory Group (www.arcweb.com), operations and maintenance are the very areas where fieldbus technology provides its greatest benefits. In other words, the introduction of fieldbus technology becomes the enabler to achieving a heightened level of production effectiveness.

Fieldbus technology expands the quantity of process and device data that is available. At the same time it simplifies the gathering of that data for plant asset management systems designed to watch over devices, tracking their "health," and even detecting problems before they adversely affect operations and/or product quality – a capability that is known as predictive or proactive maintenance.

The effective use of plant asset management systems results in increased process availability, thus preventing unnecessary process costs that often are incurred as a result of unexpected process excursions and/or shutdowns.

Fieldbus technology also enables easier, quicker and more accurate instrument calibrations to be performed. In fact, digital technology and advanced diagnostics have allowed some instrument manufacturers to extend the recommended period between instrument calibrations to as much as six years, thus freeing maintenance personnel for other activities.

Fieldbus technology, when combined with plant asset management systems, also enables developing and maintaining more accurate records of device calibrations and maintenance activities. Where traditional calibration and maintenance record-keeping relies on handwritten entries, calibration and maintenance activities using fieldbus technology and plant asset management systems are automatically recorded and saved.

Developing a sound business case for migrating/upgrading to fieldbus technology requires identifying benefits beyond reduced wiring and installation costs. It requires identifying and quantifying costs associated with:

- Preventable process excursions and shutdowns
- Control valve turnaround maintenance
- Instrument calibration and maintenance
- Calibration and maintenance documentation.

Tips and Traps

Sharing of lessons learned doesn't occur nearly as often as everyone would like, thus opportunities such as those provided by The Minerals, Metals and Materials Society are extremely valuable.

ALCENG believes that the factors that played a significant role in the success of its migration to a digital fieldbus include:

- **Planning** – The amount and level of detailed planning that went into every aspect of this migration journey is absolutely vital and is producing significant dividends from concept through commissioning.
- **Standards and documentation** – The simultaneous development of fieldbus specific standards and associated documentation templates is helping to ensure new documentation has a similar look and feel to the existing I&C documentation – something maintenance personnel appreciate.
- **Tools and processes** – Taking the time to develop structured robust engineering tools, processes and rules for such activities as segment design and segregation, automated drawing development and management, CHAZOP and commissioning is producing an installed solution that is consistent, easier to engineer, test and maintain, and will be less costly to own and operate.

In 1999 ALCENG was concerned about interoperability issues resulting from installing different manufacturers' products on the same fieldbus segment. However, it has now become more apparent that Fieldbus Foundation and Profibus Nutzerorganisation e.V.'s cooperative use of DDL and EDDL, combined with the robust independent testing provided by the two fieldbus organizations, provides users the freedom to choose the best field devices for each application with little, if any, concern for interoperability issues.

ALCENG engineers also believe avoiding the following "traps" is playing a significant role in the overall success of its on-going fieldbus migration initiative.

- **Technology upgrade considerations** – For any large operating facility, there is a very close integration between field instrument equipment, MCC, plant control systems, emergency shutdown, burner management, telemetry, and other resource planning systems. The migration to newer technology in any of these entities must not occur in isolation. The technology upgrade path must ensure that a sound and supportable integration solution has been applied; one that is the best overall choice while protecting the investment into current systems.
- **Retain expert help** – It is very easy to falsely believe that we have the know-how to implement a robust fieldbus installation through books, journals and attending an occasional two-day course or a seminar. Also everyone has a good understanding of "office" type networks. However, office networks and fieldbus networks are not the same. Ensuring each and every fieldbus segment is designed and implemented correctly as well as operationally robust requires heeding the advice of experts.
- **Crawl before you walk** – Fieldbus has been around for nearly two decades, thus there is a tendency to jump in with both feet. Unfortunately, no two operating facilities or installations are exactly alike. New adopters can save themselves a lot of anguish by starting with a small installation and progressing from there. What they learn about fieldbus specific to their company, plant and application will prove invaluable in the long run.

Conclusion

Since ALCENG initially began investigating the feasibility of migrating to fieldbus technologies, the installed base, capabilities and number of manufacturers supporting FOUNDATION fieldbus and Profibus DP have continued to increase, especially in Asia and Australia.

When all these aspects are viewed as a whole, it confirms ALCENG's wisdom of deciding to move out of the "comfort zone" and begin the migration to fieldbus technologies at the Alcan Gove refinery – one that will help Alcan Inc. obtain and maintain a competitive advantage well into the future.

MAXIMISING BAUXITE GRINDING MILL CAPACITY TO SUSTAIN PLANT PRODUCTION

Pierre G. Cousineau[1], Jean Larocque[1], Colin Thorpe[2]
[1]Alcan Bauxite and Alumina, Alcan Inc., Vaudreuil Works, 1955 Mellon boulevard, Jonquière, QC, G7S 4L2, Canada
[2]Alcan Bauxite and Alumina, Alcan Inc., Alcan Engineering, 119 Charlotte Street, Brisbane, Qld, 4000, Australia

Keywords: Bauxite, Grinding, Production

Abstract

In 2001, a new bauxite wet grinding circuit was commissioned in Alcan's Vaudreuil alumina plant. It replaced the original 60-year-old dry grinding circuit. The former circuit was labor intensive and restricted the type of bauxite that could be processed. The new installation, consisting of two rod and ball mills, enabled the plant to improve alumina recovery, process ores with higher moistures and reduce bauxite supply costs.

In March 2004, important premature equipment damage was noticed on both bauxite mills. In order to avoid significant production losses, it was necessary to upgrade each mill to handle full plant needs, as the repairs on each mill were estimated to last about 6 weeks.

A task force from Vaudreuil in Canada and Alcan Engineering in Australia was formed with the aim of identifying quick and efficient solutions. Alcan Engineering's milling model was used to evaluate the following options:

- Closed circuit grinding (screening product slurry)
- Wet overflow (removal of discharge grate)
- Removal of expected oversize prior to digestion
- Finer pre-crushing of ore
- Increasing rod diameter for better impact
- Steel charge equilibrium
- Maximizing mill power by sustaining charge S.G.

Conclusions and recommendations coming out of the study and plant results are presented.

Introduction

A new bauxite wet grinding circuit was commissioned at the Vaudreuil alumina plant in 2001 replacing the original 60-year-old dry grinding installations (one in each alumina plant). The new installation, consisting of two combination rod and ball mills, each designed to operate at 190t/h, enabled the plant to process ores with higher moisture content and to improve process efficiency. Consequently Vaudreuil could realize significant savings in bauxite supply costs. The grinding circuit was designed to operate with two mills.

Each mill has two compartments, the first for rods and the second for balls. Each mill has an overall length of 14.6m, a diameter of 4.6m and is powered by a 2250kW motor.

The Problem

In March 2004 after approximately three years in operation, significant premature equipment damage was noticed on both mills:

- 1.5" diameter liner bolts were falling to the plant floor, creating a major safety hazard.
- Caustic process liquor was leaking from liner bolt holes onto the floor.

These observations prompted a more thorough inspection. A liner was removed and the equipment was subjected to an ultrasonic inspection. This specialized examination revealed:

- Microscopic cracks in the shells;
- Premature inside shell wear;
- Bolt hole elongation;
- Missing rubber liner on rod side of the mill;
- Broken liner bolts; and
- Retaining wall had moved from perpendicular and retaining bolts were cracked or broken (87% and 65% of the retaining bolts of the South and North mills respectively) (see Figure 1).

Figure 1. Displaced retaining wall bolt on South mill showing 8mm gap. The bolt washer should be flush to the shell surface.

Immediate action was required. The structural integrity of the two mills was compromised. As a safety measure, access to the mill area was restricted. The scheduled addition of mill grinding media (balls and rods) was suspended in order to decrease stress on the mill shells.

The situation was critical. Both mills exhibited identical structural failures. It was impossible to estimate how long mill operations could continue. Each mill would have to be taken out of service for an extended period of time for repairs. The manufacturer had never seen this type of failure. No methodology existed for repairing the damaged mills. Catastrophic mill failure was a major possibility. Vaudreuil was faced with a potential major production loss.

The Challenge

The challenge was to restore the integrity of the mills, increase equipment reliability and avoid production losses while protecting employee safety and the environment. This challenge was to be met within the following context:

- At the end of April 2004, production was below the plan.
- Repairs were expected to take about 45 days per mill, or 90 days for both. The scope of the work was unknown. The manufacturer had never undertaken to repair this type of failure.
- Nobody knew how long the remaining lone operating mill would last.
- The vacation period for operating personnel was eminent, July and August.
- The bauxite supply was set for the next six months. This left few options to change the supply mix by choosing softer bauxite types to aid a single mill in operation.
- A strong confident relationship between the task force, plant operators and the repair team was essential.
- The plant requires about 380t/h of bauxite at normal flow. The proven maximum capacity of one mill was 300t/h. At this capacity a single mill could not meet the plant requirement for bauxite.
- The plant had never operated with one mill for an extended time. In March 2004, a 3.5 day shut-down of the South mill caused many operational difficulties with approximately 650 tonnes of lost production;
- Equipment blockages;
- Abnormally high rates of scale basket cleaning; and
- Increased risks to personnel and equipment.

Vaudreuil management assembled a task force team consisting of Vaudreuil, Alcan's port facility near the refinery (Port-Alfred) and Alcan Engineering (ALCENG) staff. Team members possessed a sound technical and practical understanding of the principles of grinding and of plant operation. They needed to work closely together and collaborate with the repair and operation teams.

The task force mandate was to:

- Satisfactorily control the risks to the personnel and to the equipment;
- Keep production losses at a minimum; and
- Ensure that repairs would permit sustainable plant operation.

Options

The team was set up to quickly explore the possible alternative, recommend a path forward and follow-up throughout the repairs. ALCENG developed a milling model to study milling configurations. This model was quickly adopted and used to quantify the impacts of identified options.

Closed circuit grinding

Closed circuit grinding recycles the oversized product back to the feed. Detailed engineering was carried out by Alceng at their Brisbane office, in close cooperation with Vaudreuil staff. After lengthy discussions, this option was discarded because:

- Approximately 100t/d of oversize slurry containing caustic liquor would have to be returned to the mill feed. Environmental risks and safety were deemed unacceptable due to space constraints and the quantity of material being handled.
- The time line to implement the option was considered to lengthy considering the risks to the integrity of the mills.

Wet overflow

The removing the wet overflow grate was consider as a method to increase mill throughput. This would have allowed the mills to be pushed by an additional 50 t/h. However, it was ruled out due to:

- The uncertainly in controlling coarse particle discharging from the ball section of the mill; and
- The extent on modification to the existing mill by removing the grate discharge.

Removal of expected oversize prior to digestion

The removal of the oversize material prior to digestion was quickly examined and rejected because the risks were deemed to high due to the hot caustic (approaching atmospheric boiling).

Finer pre-crushing of ore

The Vaudreuil plant process a blend of bauxites with very different grinding properties. The ALCENG simulations indicated that if the size of the CBG ore could be reduced to less than 37 mm, mill capacity would increase by more than 30t/h. The bauxite supplier (CBG) was visited but it was not practical for CBG to crush the bauxite due to the high throughput rates required, 3,000t/h. The only alternative was to pre-crush at the receiving end of the shipment either at Port-Alfred or Vaudreuil. An important issue was the probability of dust generation, especially at Port-Alfred which is surrounded by an urban area. The most practical solution was to lease a mobile crusher used by road-building contractors. In a matter of two days, three contractors were visited and a successful test was carried out with 30t of bauxite in a local quarry. Dust generation was minimal.

Increasing rod diameter for better impact

One of the key drivers in modeling the mill was to evaluate the combined mills as two stage of milling. This meant reverse calculating the desired ball mill feed size to obtain an acceptable grind for the process. This assessment determined the rod section needed to maximize the impact energy at maximum rod charge.

During the assessment a charge equilibrium study was carried out to maximize the rod bulk density. Whilst this was achievable by regular charging, the rod bulk density did improve. The increase in charge was insufficient. Even greater impact forces were needed to increase the mill capacity. The CBG ore work index was the key driver and after assessing the work index and the CBG feed size it was determine that the 90mm rod had insufficient energy to reduce the particle to the a desired ball mill feed. The decision to increase the largest rod diameter from 90mm to 110 mm gave a 60% increase in impact forces allowing addition 10 t/h in mill feed.

Steel charge equilibrium

The mill charge was initially reduced to avoid further stressing of the mill shells. The ball and rod charges to the mills were optimized. The ALCENG simulation identified a further bauxite feed increase of 20 t/h.

Maximizing mill power by sustaining charge S.G.

Before the mill failure, this parameter had not been controlled adequately. The implementation of a simple, inexpensive control loop enabled us to increase throughput by approximately 10t/h by maintaining the slurry density constant.

Adopted Solution

There was no easy solution. 80t/h of extra grinding capacity was required: a 25% increase in capacity. By early July, the team knew that no single solution would suffice. Four potential plant modifications and operating methods were identified as offering the most potential for achieving the production objectives.

Pre-crushing

Normally, bauxites from various sources are blended at the Port before being transported via railroad to Vaudreuil. Port-Alfred was selected for logistical and throughput rate considerations for pre-crushing operations. Port-Alfred management agreed to allow pre-crushing provided that strict dust and noise emissions were met.

With the cooperation from Port-Alfred, preparatory work quickly started. This included:

- Moving approximately 25,000t of bauxite to clear sufficient space for installation of the crusher and dust control equipment inside the shed;
- Adjusting the logistics of bauxite handling in the sheds; and
- Training the operators and updating the standard practices.

Optimization of mill operations

Prior to the mill failures, the mill operations only drew 1570 kW of power. After three years of operation at this level, the plant personnel were not comfortable at increasing the mill charges to the point were someone stated "... you break the mill in half."

The use of the ALCENG milling simulation gave the plant personnel confidence in increasing the grinding media charge and the rod diameter (with the consent of the manufacturer). This was done in steps and validated each time.

Throughput increases

Throughput increases were decided on a daily basis. Each increase meant an additional risk for personnel running downstream operations (potentially more coarse bauxite).

The risks were managed with:

- Daily monitoring of critical indices;
- Immediate discussion of results with plant staff and back-tracking when potential difficulties threatened;
- Development and implementation of simple and practical sampling and analysis devices to monitor mill performance; and
- Execution of test work on a continuous basis to ensure that the plant was operating under safe conditions

Results

The required bauxite throughput of 380 t/h was exceeded up to 410t/h. The mill power consumption started at a comfortable 1570 kW and finished at 2100 kW or 92% of motor amps. Considering the initial fears by the plant personnel, this far exceeded the expectations. Close monitoring of the process, gradual increase in grinding media and the increase in bar diameter contributed to the increase in capacity of the mills. The operations quickly gained confidence in the early days. The ALCENG model results were confirmed at each step of the ramp-up in bauxite charge.

The bauxite slurry size distribution had shifted to a coarser curve at the increased throughputs compared to the two mill operation. The down stream processes were able to cope with the coarser material. At times, the bauxite charge was reduced to respond to the unacceptable situations in the down stream processes.

The plant management were expecting to loose production. The task force team was set out to minimize these loses. At the end of the repairs, only minimal production loses were attributed to the one mill operations. The plant however produced ABOVE the production plan for each of the three months of the repairs and even approached production records (see Figure 2).

Conclusions

The task force approach to critical problems is used by Alcan and has proven to be an effective means of solving difficult problems. The success lies in choosing people with a mix of skills and experience from operators all the way to consultants. Each makes valuable contributions to safely and quickly resolve the problem. Teamwork is essential. The team not only prevented production losses, but made an important contribution to produce above the production plan.

The use of tools like Alcan's milling model proved to be valuable in quickly assessing the benefits, feasibility and viability of the various options identified. The milling model helped in raising the confidence of the operation as it was validated at each increment in throughput.

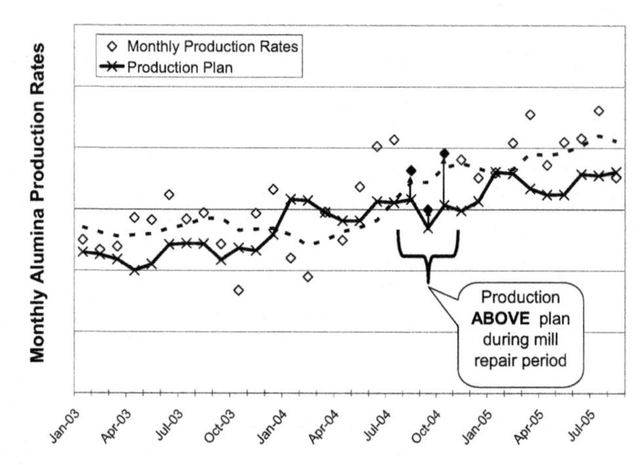

Figure 2: Actual and planned monthly alumina production rates for Vaudreuil prior, during and after the mill repairs.

One mill operation is possible for Vaudreuil if the CBG bauxite is crushed to minus 25 mm. Long term operation with one mill will save power, reduce emergency mill repairs with one mill on stand-by and result in higher equipment availability.

Recommendations

The task force approach works well for complex and critical problem and should be used as an effective problem solving tool.

One mill operation at Vaudreuil should be evaluated on an economic basis. Technically it does work.

Acknowledgement

The authors acknowledge the contribution of the task force team members and the many individuals who together turned a difficult situation into a success: Vaudreuil plant operations, the Vaudreuil project, Vaudreuil Technology, Port-Alfred operations and Alcan Engineering personnel.

Special thanks to the spouses of some of the team members who were away from home for extended periods.

Upgrade of Existing Circulating Fluidized Bed Calciners at CVG Bauxilum without Compromizing Product Quality

Vladimir Hartmann[1], Lugo Guzman [1], Olivier Hennequin[2], Andrew Carruthers[3], Michael Missalla[4], Hans-Werner Schmidt[4]
[1]CVG Bauxilum C.A. Ciudad Guayana / Venezuela
[2]ALCAN Bauxite and Alumina, Alcan Engineering Gardanne / France
[3]ALCAN Bauxite and Alumina, Montreal / Canada
[4]Outokumpu Technology GmbH, Oberursel / Germany

(Lurgi Metallurgie GmbH was acquired by Outokumpu in 2001 and changed its name in 2004 to Outokumpu Technology GmbH)

Keywords: Alumina Calciner, Upgrade, Product Quality, Particle Breakage

Abstract

To date 50 Circulating Fluidized Bed (CFB) Calciners have been installed in the alumina industry worldwide or are under construction. They represent an alumina production capacity of 28 million tons per year (tpy).

Recently, the capacities of the three initial CVG Bauxilum Calciners in Puerto Ordaz were increased to 2004 tons per day (tpd) each from a nominal capacity of 1500 tpd. The upgrade concept achieves a capacity increase without compromizing the product quality especially regarding particle breakage. This paper describes the concept for the capacity increase and the modifications. Performance figures of the existing calciners before and after the upgrade modification as well as the relevant product quality data are presented.

The particle breakage achieved with the upgraded calciners was reduced while the capacity was increased by approximately 30 %. All other product quality data such as loss on ignition (LOI) or specific surface area (BET) remained unchanged. The alumina discharge temperature was significantly lowered. Furthermore, the specific energy consumption was reduced. Particle emission with the waste gas of the calciners was reduced by adding a third field to the existing Electrostatic Precipitator (ESP).

The close co-operation between Owner, Contractor and Calciner Designer during the upgrade project resulted in a major success: the project schedule was met without compromizing the high safety standards. All performance targets were achieved.

Introduction

At the CVG Bauxilum Alumina Refinery three Lurgi CFB Calciners were installed when the refinery was built as a greenfield plant in 1983. The total design capacity was 1.0 million tpy of alumina. Originally, the nominal capacity of each calciner was 1250 tpd. In 1986 the three calciners were expanded to 1500 tpd each by installing one additional blower for each calciner.

In 1990 the CVG Bauxilum Alumina Refinery was expanded to 2.0 million tpy. The capacity of the calcination facility was extended by adding a new calciner with a nominal capacity of 1700 tpd of alumina.

Both designs, the three older calciners from 1983 with 1250 tpd each and the new calciner which was installed in 1990, represent different design milestones during the Lurgi Calciner evolution: starting with the first CFB calciner introduced to the alumina industry already in 1970, the capacity of the CFB calciners was steadily increased [1].

The evolution concept was based on operating results which confirmed that the existing calciners had significant unused capacity. These results formed the basis for the design of the next generation of calciners with higher throughputs. Besides the steady capacity increase, major design changes were made mainly in the plant arrangement with the target to achieve a competitive calciner with low capital cost. Apart from that target the demand for product quality as well as steady and stable operating performance was always the final and overruling aim for the design of all CFB calciners.

Since 1970, 50 calciners have been built, seven of them are currently under construction and will be commissioned during the next 12 months. They represent a production capacity of 28 million tpy of alumina which is more than 40% of the world capacity for smelter grade alumina. Over the years 17 existing calciners were upgraded to higher capacities.

The latest and most successful upgrade was realized with the three original CVG Bauxilum Calciners in Puerto Ordaz, Venezuela, built in 1983. Those calciners were so-called "flat arrangement" CFB calciners and were designed to operate under positive pressure. Earlier calciners installed in other alumina refineries had a "tower arrangement" and were operated with negative pressure [2].

CVG Bauxilum together with Alcan Bauxite and Alumina with its Engineering Group in Gardanne / France selected Outokumpu Technology to upgrade the initial calciners to a nominal capacity of 2004 tpd each. The more recent fourth calciner was modernized at the same time.

The objectives of the upgrade project were:

- increase existing calcination capacity to achieve a production of 2.15 million tpy
- reduce the specific energy consumption

- decrease alumina discharge temperature below those experienced at 1500 tpd
- improve operating conditions
- improve maintenance
- maintain existing product quality
- meet new obligations regarding environmental and emission standards
- maintain high safety standards.

Concept for Upgrade

The energy for the conversion of aluminium-trihydrate to alumina is generated by direct fuel combustion in the CFB furnace. A low specific energy consumption is reached due to the optimized counter current flow principle of hydrate / alumina and off gas / combustion air realized in the CFB process. Thus the air does not serve only for combustion but also for transportation purposes in the process [3]. Designing plants from scratch means to optimize investment costs versus required product quality and consumption figures. Usually, particle breakage which is a key parameter in calcination plants increases with throughput and increased gas velocities. This parameter would require increased dimensions while from a cost point of view dimensions should be minimized. In an existing calciner not all dimensions can be easily and thus economically viably enlarged. This explains the difficulty of upgrading an existing calciner within a reasonable budget and time period.

In general, an upgrade of the capacity for an existing calciner can be achieved by two methods:

- reduction of specific energy consumption (Method 1)
- increased air flow to the calciner to allow more fuel combustion (Method 2).

The latter measure for an existing calciner always results in an increase of the velocities in the system if no further modifications are made.

The reduction of the specific energy consumption is the more effective method because more hydrate can be introduced to the system with the same fuel and air flow. Although this method appears to be more elegant, it has its limit because the scope of reduction of specific energy consumption in modern calcination units is relatively small. The chosen concept was therefore a combination of both measures.

A detailed investigation of the calciner was carried out to determine bottlenecks and necessary modifications to achieve the targeted capacity if the calciner was effectively used.

Figure 1 shows the basic flow diagram of the existing calciner before the modification. Gas flows are depicted with thin lines while the hydrate / alumina flow is drawn in bold line. The pre-dried hydrate from venturi stage 1 is separated in a two field ESP and by an airlift transported to the 2nd venturi preheating stage. In venturi 2 solid and transportation air are directly introduced, the transportation air is mixed with the waste gas and its oxygen is not used for the combustion.

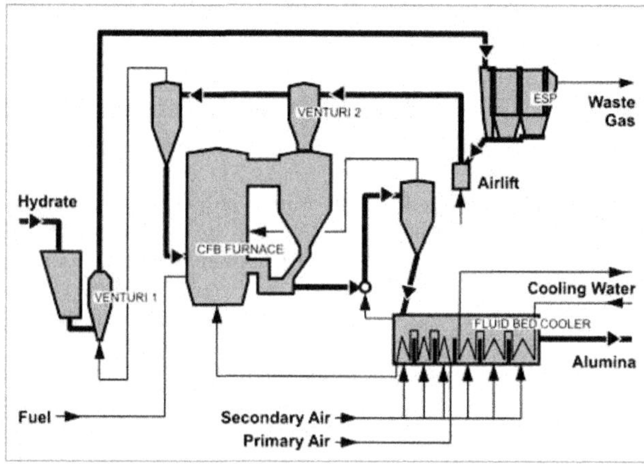

Figure 1: Process Flow Diagram before modification

The final calcination takes place in the CFB furnace [4]. Energy is supplied by direct fuel injection [5]. The combustion air is delivered by air blowers and preheated as primary and secondary air. The fluid bed cooler recovers the enthalpy from the hot alumina by preheating the combustion air flows.

The fluid bed cooler consists of two sections, an air-cooled and a water-cooled section. The air-cooled section preheats the primary air indirectly whilst the secondary air is used as fluidizing air for the cooling chambers. The water-cooled section provides the final cooling of the alumina in fluidizing chambers with water-cooled heat-exchanger bundles.

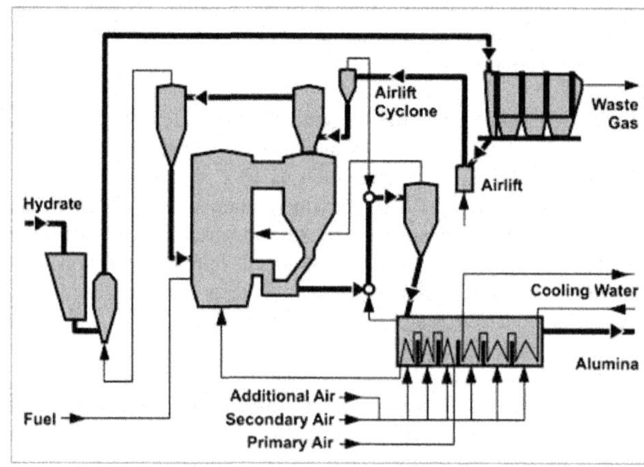

Figure 2: Process flow diagram after modification

Figure 2 shows the process flow diagram of the calciner after the upgrade modification. The significant change is the use of the transportation air of the airlift as combustion air. The concept provides a cyclone which separates the pre-dried hydrate from the transportation air. The transportation air is then routed to the secondary air duct where it is introduced to the furnace and serves as combustion air. Venturi 2 was modified with respect to the new inlet point of the pre-dried hydrate and to allow for an improved mixing of hot off-gases from the CFB furnace with the pre-dried hydrate. This modification principle results in a higher thermal efficiency of the process and decreases the actual gas volume flow to the ESP. A new airlift blower had to be installed capable of overcoming the higher pressure drop due to the new cyclone and the targeted higher nominal capacity of 2004 tpd. The existing airlift blower was relocated and serves as an additional combustion air blower.

The investigation before the modification furthermore indicated that the calciners were operated with a high excess air ratio, which could be significantly reduced. This allowed an additional increase in production capacity. The flexibility gained by these modification measures was then used to feed more combustion air and fuel to the furnace to achieve the targeted nominal capacity of 2004 tpd for each of the three calciners.

Furthermore, the heat-exchanger area in the fluid bed cooler had to be increased to guarantee the final alumina product temperature. The limit in the cooling-water outlet temperature as well as the available space had to be considered for the new design.

Because of the increased normal gas volume flow (referred to normal temperature and pressure, NTP) in the calcination system, it became also necessary to investigate gas velocities in all critical areas of the calciner to ensure that product quality was maintained, in particular regarding particle breakage. The experience with other calciners and investigations with different hydrates have shown that velocities in specific areas have to be limited to avoid particle breakage [6].

With the upgrade of the calciner the normal gas volume flow of the off-gas in the ESP was also increased by about 30 %. As the waste gas temperature was lowered by the other upgrade modifications, the velocity in the ESP was actually increased by only 6.5 %. To meet the new required dust emission standard of 50 mg / m³ (NTP, wet basis) at the stack outlet of the calciner, it was necessary to implement an additional electrostatic precipitation field to the existing ESP.

The installation of the additional third electrostatic precipitation field and the limited height under the ESP resulted in the implementation of a conveying system with low height to collect the pre-dried hydrate from the three ESP hoppers. Therefore chain conveyors were selected to replace the earlier screw conveyors.

Reference Values for Particle Breakage before Upgrade

Particle breakage is defined as the difference in the alumina mass fraction < 45 μm in the product and the hydrate mass fraction < 45 μm fed to the calciner [7].

Improved hydrate quality was a major objective of the CVG Bauxilum Alumina refinery upgrade. Since hydrate quality has a large impact on particle breakage, it was necessary to establish a reference platform so that a valid comparison could be made between the existing and the modified calciners.

Normally, particle breakage increases when the throughput of a calciner is increased. Assuming that hydrate of same quality is being fed, the expectation for the project was that the particle breakage would increase with capacity by a certain value. To identify a change in hydrate quality before the start of the upgrade project, a calciner comparison test was carried out. It was agreed, that this comparison test had to be done just before and after the implementation of the modification to analyze the impact on product quality, especially on particle breakage.

For this purpose calciner 1 (to be modified) and Calciner 101 (reference) were compared. It was agreed that if the particle breakage of the reference Calciner 101 would increase significantly during the performance test period, hydrate quality should be adapted accordingly.

Both calciners were operated at their actual maximum capacity over a period of 7 days. The production rate of the calciners 1 and 101 was measured. Tolerances of the measurement were taken into account.

The results of the comparison test can be seen in Table 1

	Calciner 1 before modification	Calciner 101 (reference)
Average production in tpd	1564	1566
Average hydrate fraction < 45 μm in %	3.52	3.45
Average alumina fraction < 45 μm in %	9.81	12.32
Average particle breakage in %	6.29	8.87

Table 1: Comparison test between Calciner 1 (before modification) and Calciner 101 (reference)

The comparison test shows relative large figures for particle breakage for both calciners but also a remarkable difference of the breakage figures between the two units.

Today's standard for smelter grade alumina is a mass fraction < 45 μm of less than 10 %. Depending on the market situation higher values in this fraction could lead to substantial discount when selling on the world market.

The particle breakage will rise with higher capacities assuming that no changes are made to the calciner and to the hydrate quality. Consequently, the alumina fraction < 45 μm will rise. On the basis of having the figures depicted in Table 1 it was foreseeable that the alumina fraction < 45 μm would rise above the maximum of 10 % in the absence of calciner design modifications and hydrate quality improvement.

Therefore the Calciner Designer, Outokumpu Technology, together with the main contractor, ALCAN Bauxite and Alumina, and the owner, CVG Bauxilum, identified all critical areas in the unmodified calciner with respect to particle breakage. Deviations were systematically identified and rectified. Main emphasis was the reduction of gas velocities in key areas to achieve a reduction of particle breakage at the increased nominal capacity.

Project Execution

It was a major challenge for the upgrade project that all modification work had to be executed during the normal production of the refinery. For this purpose a time schedule was developed which minimized the shutdown time of one single calciner while the other three calciners continued operating.

The overall schedule for the whole project had a time frame of 12 months on site. Modification of one calciner was carried out in three months including re-commissioning whereas the downtime was limited to max. 6 weeks. The plant shutdown was combined with the scheduled overhaul of the calciner. The schedule also allowed repair works of unforeseen damages which could only be detected after the respective calciner was shut down. The last in the succession of the modification, Calciner 101, achieved a downtime of less than 4 weeks.

Infrared inspections were made during normal operation of the calciner in order to be prepared for repair work particularly in refractory lined areas.

Detailed check lists including target dimensions for the respective areas and their acceptable tolerances were helpful in evaluating potential influence on particle breakage. The target was to reduce velocities in these critical areas in an economic way [8]. Also high differential velocities between gas and solids had to be avoided.

In the cooler area the heat-exchanger bundles in the water-cooled sections were replaced by bundles with a larger surface area. Also the alumina outlets of the fluid bed coolers were modified to fit the increased capacities.

The 3rd ESP fields were preassembled in the Bauxilum refinery next to the calciners. This included the insulation and the installation of the collecting plates as well as the charging electrodes. After the 3rd ESP field had been lifted into place the new stack was attached. The stack was also prefabricated including the insulation, the necessary ladder and the platform for the emission measurement. The ducting between the last fields and the stacks, and the old stacks themselves, were not reused. This preassembly method allowed a significant reduction of the downtime for each calciner and a reduction of the overall project schedule.

Figure 3: Bauxilum calciners with dismantled ESP outlet and stack at Calciner 101. The calciner on the far right side is Calciner 1, followed by Calciner 2. The left calciner is Calciner 102

Figures 3 and 4 demonstrate the erection of the 3rd field to the existing ESP with 2 fields. As can be seen in Figure 4 the 3rd ESP field was installed in a modular form.

Further to the implementation of a new third field the existing screw conveyors at the ESP were replaced by chain conveyors. The dust collection points of the existing fields of the ESP, the new 3rd ESP field as well as hydrate inlet at the air slide had to be adapted to match the new chain conveyors.

Figure 4: Lifting of 3rd field as prefabricated unit

To reach the high safety standards which are a permanent issue of the calciner designer, a HAZOP was carried out and new interlocks have been proposed to the owner and the main contractor. They were implemented with the upgrade. As an example of this high safety level applied to the upgraded plants the interlock at the airlift cyclone is explained: hot secondary air might enter the airlift cyclone and flow back to the airlift causing significant damage to the vessel and exposing people to risk. Therefore, an interlock has been implemented which protects the plant in this case of the reversed air flow via the airlift cyclone.

Further interlocks were implemented to ease operation. For example the interlock at the fluid bed cooler: when the alumina discharge at the fluid bed cooler is closed due to short maintenance works on the alumina belt conveyors, the hydrate feed to the plant is not reduced. Thus the level in the fluid bed cooler rises. To prevent the rise in the fluid bed cooler above a certain level, the alumina outlet of the CFB is closed. After opening the outlet of the fluid bed cooler the level is normalized and the discharge from the CFB starts again.

The following list highlights the modifications of each calciner:

- installation of airlift cyclone
- installation of one new air blower for service purpose
- installation of new airlift blower
- use of existing airlift blower for additional combustion air
- replacement of water-cooling bundles in fluid bed cooler
- modification of fluid bed cooler discharge
- installation of a third field in ESP
- replacement of the screw conveyors at the ESP by chain conveyors
- modification of critical areas to reduce gas velocities
- modification of hydrate inlet in venturi 2 to improve gas and solid mixing
- implementation of new interlocks.

Commissioning and Results

Following the above described schedule the individual calciners were re-commissioned after the modification during the year 2004.

Commissioning was successful. Changes had to be made to the chain conveyors which were initially not properly laid out for such an application. When the re-designed equipment was in operation, it was observed on the long term that significantly less maintenance has to be provided compared to the originally installed screw conveyors.

Table 2 shows the results of the seven days plant performance test before and after modification. Calciner 1 has been chosen since it was the first calciner for the upgrade and Calciner 101 because it initially showed exceptionally high particle breakage.

	Calciner 1		Calciner 101	
	before modification	after modification	before modification	after modification
Average Production in tpd	1564	2029	1566	2059
Particle breakage in %	6.29	5.15	8.87	5.39
LOI in %	0.75	0.72	0.78	0.71
BET in m^2/g	72	74	74	74

Table 2: Plant Performance of Calciner 1 and Calciner 101 before and after upgrade modification

As depicted in Table 2 the target capacities of 2004 tpd were excelled. The average particle breakages in the calciners were reduced significantly. In particular the exceptional high particle breakage of Calciner 101 before the modification was brought with the modification to the same level as the two other upgraded calciners 1 and 2. Other product quality data like LOI and specific surface area (BET) remained unchanged after the modification.

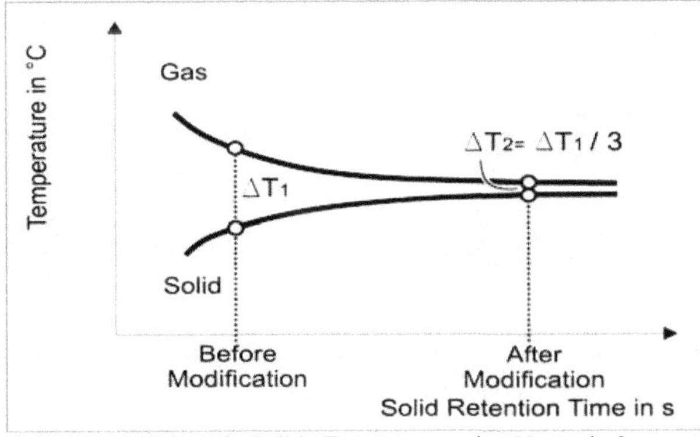

Figure 5: Gas and Solid Temperature in Venturi 2 versus retention time of solids before and after solid inlet modification

Figure 5 demonstrates the temperature difference between gas and solids before and after modification of the hydrate inlet of venturi 2. The increased solids retention time has reduced the temperature difference between gas and solids significantly. This measure also had a positive impact on the specific energy consumption because it resulted in a lower waste gas temperature.

With the example of Calciner 1 the contributions of the two general methods to increase the production of a calciner can be evaluated:

- improve specific energy consumption (Method 1)
- increase air flow to the calciner to allow more fuel combustion (Method 2).

	Calciner 1 before modification	Calciner 1 after modification
Average production in tpd	1564	2029
Average hourly natural gas consumption in m^3_{NTP} /h	6074	7000
Specific gas consumption in m^3_{NTP} / t (Al_2O_3)	93.20	82.78

Table 3: Gas consumption of Calciner 1 before and after the modification

In Table 3 the gas consumption of Calciner 1 before and after the modification is shown. To achieve the upgraded capacity the specific gas consumption was decreased by about 11 %. The gas flow rate increased by about 15 %.

Calculating the alumina production increase gained due to the improved specific energy consumption (Method 1) shows:

$$\left(\frac{1}{82.78\frac{m^3_{NaturalGas,NTP}}{t_{Al_2O_3}}}-\frac{1}{93.20\frac{m^3_{NaturalGas,NTP}}{t_{Al_2O_3}}}\right)$$

$$*6074\frac{m^3_{NaturalGas,NTP}}{h}\cdot 24\frac{h}{d}=197tpd$$

Assuming that the plant would have been upgraded without reduction of the specific energy consumption the rise in production (Method 2) can be calculated as follows:

$$\frac{(7000-6074)\frac{m^3_{NaturalGas,NTP}}{h}}{82.78\frac{m^3_{NaturalGas,NTP}}{t_{Al_2O_3}}}\cdot 24\frac{h}{d}=268tpd$$

The production rise of 465 tpd for Calciner 1 can now be related to the two methods.

	Method 1	Method 2	total
Increase in tpd	197	268	465
Related to total rise of 465 tpd in %	42	58	100

Table 4: Production rise related to the two general methods to increase the capacity of a calciner, calculated from data achieved with Calciner 1

The above data shows that the upgrade approach chosen was a balance of the two general methods for upgrading a calciner.

During the whole project and in particular during the re-commissioning periods the actual operating data were continuously compared against the theoretical values. This procedure was extremely helpful in achieving the final aim of the upgrade project and quickly identified the areas which needed adjustments.

Furthermore, it should be pointed out that the close co-operation between the owner, CVG Bauxilum, the main contractor, ALCAN Bauxite and Alumina with its Engineering Team Aluminium Pechiney from Gardanne in France, and the Calciner Designer, Outokumpu Technology, was a key element for the success of the whole project. This excellent co-operation generated a team spirit between owner, contractor and designer in which project schedule and project objectives were achieved without compromizing the high safety standards [9]. The Calciner Designer contributed to this area with its experience of calciners in other alumina refineries. Owner and main contractor contributed with the organization of the construction and paid special attention to risk analysis and the close follow up during the construction.

The capacity of the calciners was increased by approx. 30%, so that the total capacity of 2.15 million tpy of the refinery was easily achieved. At the same time the particle breakage was reduced by approx. 20 % on average for each upgraded calciner.

The outlet temperature of the fluid bed coolers which had been a critical bottleneck before the upgrade project was significantly reduced.

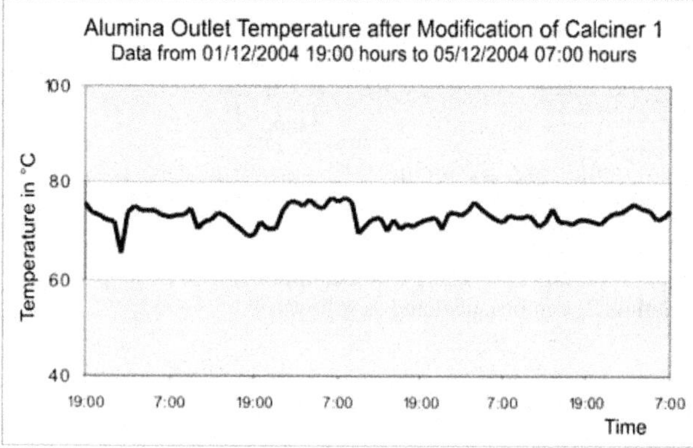

Figure 6: Cooler discharge temperature after calciner modification versus time

Figure 6 demonstrates the cooler discharge temperature over a representative time period of 84 hours after the upgrade modification at a capacity of about 2000 tpd: the outlet temperature is continuously below 80°C even with short time disturbances.

Summary and Conclusion

An upgrade of existing calciners with a capacity increase of 30% was executed during normal industrial operation of the alumina refinery. At the same time it was possible to improve the specific energy consumption and the product quality.

In particular particle breakage was reduced because modifications were made in critical areas regarding the gas and solid velocities. The specific dust emission was lowered by enlarging the existing ESPs with a 3rd precipitation field.

The major results are:

- increase of capacity by 30%
- no change of product quality regarding LOI and BET
- reduction of specific energy consumption by 11 %
- reduction of particle breakage by approx. 20 %
- reduction of dust emission to 50 $mg/m^3_{NTP\ (wet\ basis)}$
- improvement of maintenance
- improvement of process safety.

The described methods can be applied to other existing calciners as well. The project reports high economic profitability with a return of investment of 30 %.

References

[1] J. Yetmen, H.W. Schmidt, "Large Units for Alumina Calcination Plants with improved Technology" Int. Conference on New development in Metallurgical Process Technology Proceedings VDEM, pp 369-375 (1999)

[2] E. Guhl, R. Arpe, "Nearly 30 years of experience with Lurgi Calciners and influence concerning particle breakage", TMS Light Metals (2002)

[3] A. Squires, "Origins of the Fast Fluid Bed, Advances in Chemical Engineering", Vol. 20, Fast Fluidisation (Ed. Kwank, M.), Academic Press (1994), pp 4-35

[4] L. Reh, H.W. Schmidt, "Application of Circulating Fluid Bed Calciners in large size Alumina Plants", TMS Light Metals (1973)

[5] H.W. Schmidt, "Combustion in a Circulating Fluid Bed" Proceedings of 3rd International Conference in Fluidized Bed Combustion, Environment Protection Technology Series, EPA-650/2-73 pp 11-14 (1973)

[6] A. Saatci, H.W. Schmidt, W. Stockhausen, M. Ströder, P. Sturm, "Attrition Behaviour of Laboratory calcined Alumina from various Hydrates and its influence on SG Alumina Quality and Calcination Design", TMS Light Metals, pp 81-86 (2004)

[7] H.W. Schmidt, H. Beisswenger, F. Kämpf, "Flexibility of Fluid Bed Calciner Process in View of Changing Demands in Alumina Market", Journal of Metals, vol 32, No2 (1980)

[8] V. Martinent-Catalot, J.-M. Tamerant, S. Favet-Cossant, L. Ferres, "A New Method for smelter grade Alumina (SGA) Characterization", TMS Light Metals, pp 87-92 (2004)

[9] H.W. Schmidt, M. Ströder, G. Singh, M. Sant'ana, J. Ribeiro, J. Piestrzeniewicz, M. Cable, "Improved Health and Safety Conditions and Increased Availability in Large Alumina Calcining Units", TMS Light Metals (2005)

Crystal Growth Modifying Reagents; Nucleation Control Additives or Agglomeration Aids?

James A Counter[1]
[1] Nalco Australia, 2 Anderson St., Botany, NSW, 2019, Australia

Keywords: precipitation, crystallisation, nucleation, agglomeration, precipitation additive

Abstract

Crystal growth modifiers (CGM) are commonly used in the Bayer process to increase the average gibbsite particle size and improve the particle size distribution during crystallisation from supersaturated sodium aluminate solutions. With respect to the mechanism of action there has been much debate as to whether this result could be due to the additives aiding agglomeration or reducing the number of secondary nuclei.

Presented here is work completed on both synthetic and real Bayer plant liquors, over various experimental conditions in order to study the mechanism of action of CGM additives. Crystallisation kinetics, secondary electron microscopy (SEM) and atomic force microscopy (AFM) data are collated to examine the impact that these surface active additives have on the precipitation process.

Introduction

Crystal growth modifiers were introduced to the Bayer industry over 20 years ago [1] and have gained general acceptance as additional control tools within the crystallisation section of the Bayer process. These products have exhibited widespread utility under high-agglomeration and all-growth precipitation strategies. The effect of addition of CGM has lead many plants to a reduction in ultra fines, tougher particles, liquor productivity increases via either reduced fill temperatures or increased seed charges and alumina trihydrate coarsening along with many other benefits using doses ranging from 5-50 ppm.

Three major crystallisation mechanisms, nucleation, growth and agglomeration of crystals are in action for producing gibbsite as the preferred phase. The yield of gibbsite is largely determined by nucleation and growth of particles, while agglomeration is responsible for producing a coarse crystal of commercial interest. The effect of optimising the Bayer plant precipitation parameters in each of these crystallisation scenarios is the most important part of the entire industrial process. The relationship between the operating parameters and gibbsite yield has been detailed in a number of publications over many years [2-6]. Optimum conditions for maximum yields do not always coincide with optimum product size specifications, demanded by the smelters. CGM additives have allowed Bayer plants around the world to operate under optimum conditions without product size compromise.

Although there have been a number of recent publications postulating the CGM mechanism of action, [7-9] there still exists a degree of uncertainty as to the impact of these programs and their influence under various agglomeration and secondary nucleation strategies practiced within the industry. The main thrust of the present studies is to examine the influence of a CGM on each of the precipitation scenarios in order to better understand the mechanism of action, leading to further development of new and improved products.

This paper presents combined work of the impact of CGM on both synthetic and real Bayer plant liquors under specific crystallisation scenarios over a temperature range experienced in industry. Results from a number of experimental techniques are collated to answer the questions posed by industry and propose a general mechanism of action for these additives.

Experimental

Synthetic Bayer Liquor Preparation

Pure synthetic, supersaturated sodium aluminate solutions were used in this investigation. They were prepared from analytical grade and high purity grade reagents: aluminium trihydroxide, $Al(OH)_3$, (C31 grade, 0.01% SiO_2, 0.004% Fe_2O_3, 0.15% Ba_2O; ALCOA, Arkansas USA); sodium hydroxide, (99.0% pure, 0.01% Si, Merck, Australia); Milli-Q water (surface tension 72.8 μNm^{-1} at 20°C, specific conductivity < 0.5 μscm^{-1} and pH = 5.6).

The synthetic liquors used in the crystallisation measurements were prepared in a 2.5 dm^3 well sealed stainless steel digester immersed in an oil bath at 150°C. A known mass of NaOH was dissolved in 1.3 dm^3 of Milli-Q water in the digester with an agitation rate of 250 rpm. This was followed by slow addition of a known mass of the aluminium trihydroxide. After complete dissolution of 3 – 4 hours, the solution was transferred to a 2.0 dm^3 volumetric flask and made up to the mark with Milli-Q water.

Bayer Plant Liquor Preparation

The Bayer plant liquor experiments were conducted using reconstituted plant spent liquor. Spent liquor is the term used in the Bayer process to describe the liquor after the final classification stage which returns back to digestion. A desired weight of spent liquor was measured into a stainless steel beaker and the volume was reduced by evaporation to approximately 30%. To this a set weight of $Al(OH)_3$ solid was added and the mixture stirred until it was dissolved. This solution was removed from the hot-plate and placed on a weighing balance and de-ionised water added until a desired weight was attained. The pregnant liquor was filtered to remove any insoluble material. Final liquor composition was such that A/C = 0.66 ± 0.05 where:

A (aluminium hydroxide) = 150 ± 10 g/L of Al_2O_3
C (total caustic) = 230 ± 10 g/L of Na_2CO_3
S (soda, total alkali) = 260 ± 10 g/L of Na_2CO_3

Seed Preparation and Characterization for Agglomeration

Seed used for the agglomeration experiments was the well characterized Alcoa C31, having BET surface area of 0.381 m^2/g and d(0,1) = 6.3 μm, d(0,5) = 37.1 μm and d(0,9) = 92.3 μm, determined by a Malvern Mastersizer.

Seed Preparation and Characterization for 2° Nucleation

Broken seed fragments, formed via secondary nucleation, provide the main source of new crystals in the Bayer process [10,11]. In this type of secondary nucleation process, fluid shearing action, crystal-crystal and crystal-vessel wall/impeller blade collisions are dominantly responsible for removing nuclei from the growing seed crystal surfaces into solution. It has been stated that particles > 20 μm in diameter are less likely to agglomerate as they do not have sufficient surface energy to do so [2]. Therefore in order to focus on the effect of the CGM additive on secondary nucleation alone, especially in the early stages of crystallisation it was necessary to use a large screened seed, high supersaturation, low seed charge and enough agitation to allow the seed to remain in the bulk solution, without settling.

Uniform coarse, gibbsite seed crystals were prepared from high purity $Al(OH)_3$, (C31) by wet screening using both 55 and 90 μm sieves to give crystals in the size range 30 – 105 μm. These crystals were then treated with ultra-sonication, followed by decantation to remove the fine particles (<10 μm). The resulting product (washed and dried at room temperature in a desiccator) consisted of agglomerated, pseudo hexagonal-shaped crystals, in the size range 30 - 100 μm with reasonably smooth surfaces (Figure 1) and a corresponding BET surface area of 0.1445 m^2g^{-1}.

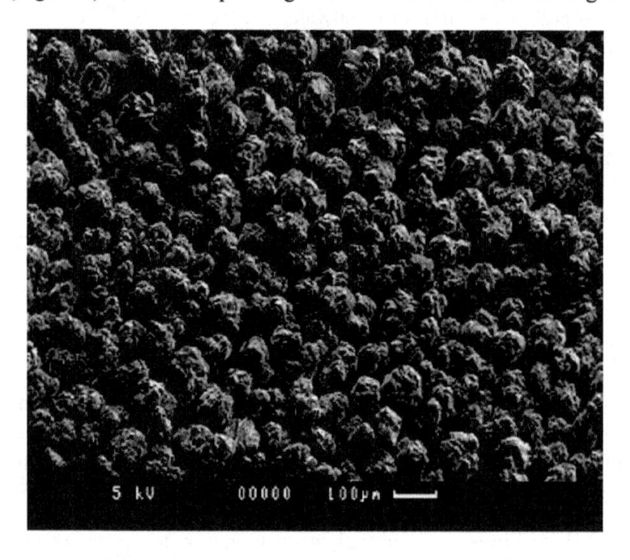

Figure 1. SEM photomicrograph of gibbsite seed crystals for 2° nucleation experiments.

Batch Crystalliser

A baffled, well sealed, 2.5 dm^3, 316 stainless steel vessel was used for the crystallisation experiments. A central, 4 blade, 45° – pitch turbine impeller driven by a 70W, multi-speed motor provided a constant agitation speed at 400 rpm as well as fully developed axial flow, moderate shear and a high degree of suspension uniformity within the crystallizer. A thermocouple sensor and a conductivity probe were fitted through the lid of the crystallizer. The entire vessel was submerged in a 15 dm^3, thermostatically-controlled oil bath, maintaining a constant temperature to within ± 0.05°C.

The crystal growth modifier used in this study was the commercially available Nalco product, 7837, unless otherwise stated.

Models for the Kinetic Study

Quantification of the crystallisation kinetics data produced from the crystallisation experiments was performed using the semi-empirical power law model for secondary nucleation and growth of gibbsite as shown below:

$$-\frac{1}{S}\frac{d\sigma}{dt} = k(\sigma)^n \quad (1)$$

where σ is the Al(III) relative supersaturation which is defined as $(C - C_e)/C_e$, where C and C_e are the instantaneous Al(III) concentration and Al(III) equilibrium solubility, respectively; S is the total particle surface area; k is the nucleation rate constant and n is the reaction order. To quantify the influence of temperature and estimate the activation energy from the rate constant k, the Arrhenius expression was used.

$$k = k_o \exp\left(-\frac{E_a}{RT}\right) \quad (2)$$

where k_o is the pre-exponential factor, E_a is the activation energy, R is the gas constant and T is the absolute temperature.

Atomic Force Microscope (AFM) Measurements

Interaction force measurements between two gibbsite particles (sphere and prism) were measured using a Nanoscope III Atomic Force Microscope (Veeco). A gibbsite spherical particle (hydrated alumina sphere [12]) was glued onto a triangular silicon nitride cantilever and an aluminium disk with Epikote heat sensitive resin. Particle interaction forces were measured using a fluid cell filled with sodium aluminate liquor at 25°C. The cantilever spring constant, 0.1 Nm^{-1} was determined using the Cleveland method [13].

Results and Discussion

Agglomeration Experiments

The majority of plant test work over the last decade has been completed by simulating agglomeration conditions of the particular Bayer site. Fine seed charge into pregnant liquor at high temperature is the standard method for CGM evaluation. From the work completed here over a temperature range (58-78°C) typical to Bayer plants, coarsening of the fine C31 seed was evident after 18 hours. The results given in Figure 2 below and taken from the 68°C test, indicate the increase in the d(0,1) values with respect to the CGM dose. It is evident that the increase in particle size and the subsequent decrease in available seed surface area results in an overall decrease in yield in the isothermal batch experiment. Therefore in continuous Bayer plant conditions, process changes are made to increase yield up to a higher level without size compromise.

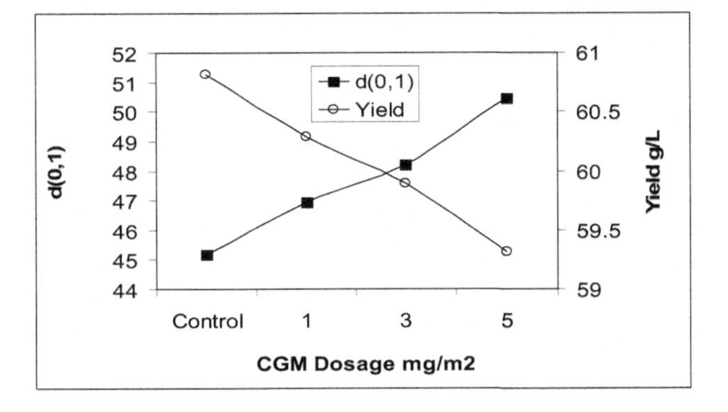

Figure 2. d(0,1) and yield results for 68°C agglomeration test.

SEM images given below show the difference between the control and CGM dosed test at 68°C. The control, given in Figure 3 below exhibits large amounts of fine particles, yet to be fully incorporated into the crystal structure of the parent seed. It is clearly evident that over the 18 hour experiment the CGM additive has reduced the number of fine particles and achieved a more robust, block-like hexagonal morphology, under these conditions.

Figure 3. SEM images of control agglomeration test for 68°C.

Figure 4. SEM images of CGM treated agglomeration test for 68°C.

Data from the kinetic analysis of this system is given in Table I. The reaction order at each temperature was found to be n = 2, corresponding to values reported extensively in literature for systems under comparable conditions. Activation energies for the control and CGM additive tests were estimated from Arrhenius plots. Values fall comfortably in the 53 -83 range reported in the literature for pure sodium aluminate solutions [3, 6, 14].

Table I. Kinetic parameters for seeded gibbsite agglomeration experiments at 58, 68 and 78°C.

Terms	T(°C)	Control	7831 Treated	7837 Treated
k_o ($m^{-2}h^{-1}$)	58,68,78	1.10×10^5	32.6×10^5	7.78×10^5
Reaction Order (n)	58	2.0 ± 0.2	2.0 ± 0.2	2.0 ± 0.2
	68	2.0 ± 0.2	2.0 ± 0.2	2.0 ± 0.2
	78	2.0 ± 0.2	2.0 ± 0.2	2.0 ± 0.2
Ea ($kJmol^{-1}$)		61.2	70.3	66.6

Higher activation energy means greater sensitivity to temperature in the presence of the additive. The increase in the pre-exponential factor of k_o indicates the additive aids in increasing the successful collision rate of fine seed particles during the agglomeration stage. In this case the product 7831 has a higher k_o possibly indicating a greater successful collision rate than 7837.

Consider the information thus far in terms of the mechanism of agglomeration put forward by Low [15] and given in Figure 5 below. The action of the CGM additive would be aiding the cementation process during agglomeration and in terms of attrition the additive appears to successfully reduce the breakdown of particles, forming a fully cemented agglomerate, which is evident in the morphology of the SEM images given in Figure 4.

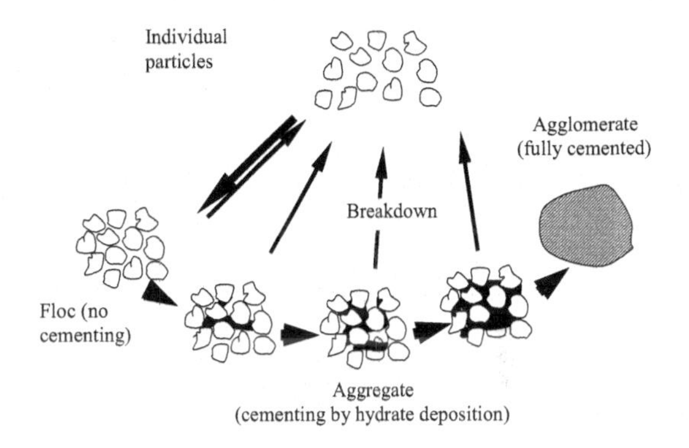

Figure 5. Mechanism of agglomeration, Low [15].

These tests have only considered one of the possible crystallisation mechanisms present under Bayer conditions. It was necessary to perform experiments under nucleating conditions in order to establish a more conclusive mechanism of the action of the additive.

Secondary Nucleation Experiments

To investigate the influence of the additive on secondary nucleation, isothermal batch crystallisation experiments were carried out over 24 hours. The tests were conducted in pure synthetic liquors, under constant conditions of NaOH = 4 M, initial $\sigma_{Al} = 1.5$, seed surface area = 7.23 m^2dm^{-3} at temperatures of 60, 65 and 70°C. Results are discussed in terms of new crystal formation, kinetics and morphology.

New Crystal Formation

The observation of new fine particles (secondary nuclei) in the size range of 0.1 – 30 μm during all of the crystallisation tests over the first hour are summarised in Table II. There is a marked difference between the control and treated samples for the 60 and 65°C runs, however no discernable difference could be detected over the first hour for the higher 70°C temperature run.

Table II. Number of new particles generated for all crystallisation experiments under constant conditions over the first hour.

T (°C)	New Particles < 30 μm ($10^{10}h^{-1}$ per 25 mL)	
	Control	With CGM
60	3.1	1.1
65	10.1	3.7
70	17.5	18.4

The induction times for the appearance of secondary nuclei were measured by conductivity. At a constant initial $\sigma = 1.5$ and seed charge of 7.23 m^2dm^{-3} the induction time was found to decrease with increasing temperature for both the control and treated experiments. However, the induction time was significantly shorter in the control experiment than in the treated solutions at 60 and 65°C. At 70°C there was no detectable difference in the induction time measurement for the two scenarios. This observation would imply that the CGM additive could be affecting solution speciation and/or crystal surface processes during the secondary nucleation. It is plausible that the CGM additive initially deactivated the seed surface to a high extent resulting in a lengthened time for generating sufficient active sites for fast growth. The negative effect of additive on the processes was negated at the higher temperature.

The net, total crystal surface area with time, is shown in Figure 6 and appears to contradict the number of new particles generated given in Table II. The control tests displayed a lower total surface area than that of CGM treated tests at the same temperature, which would tend to indicate the possible impact of agglomeration during the crystallisation. However, it is proposed that initially a larger number of secondary nuclei formed in the treated solutions and these nuclei immediately were either adsorbed back on the seed surface or rapidly agglomerated with each other to form aggregates with a high porosity. This resulted in a low number and a high net surface area of the crystals.

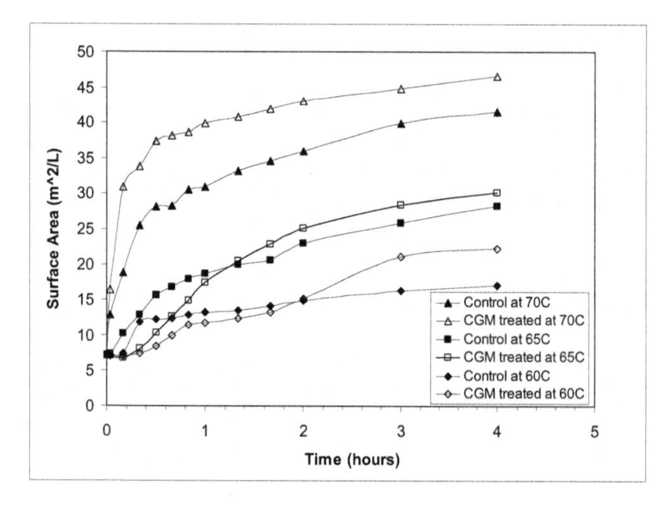

Figure 6. The variation of total available surface area for the crystallising slurry as a function of temperature for seeded, control and CGM treated tests over time.

SEM analysis strongly supported the above observations. The images of the product crystals revealed the existence of significant differences in the development and growth of the new surface layer and microstructures at the seed crystal surfaces, between the control and CGM treated tests. After 4 hours the SEM images for the control experiments exhibited expected temperature dependences. At 60°C only 2° nucleation was observed, at 65°C a mixture of 2° nucleation and separate aggregates and at 70°C the majority of particles formed were separate aggregates. The primary crystals were pseudo-hexagonal plates (Figure 7b) which agglomerated and grew to secondary crystals with a size range of 10 – 30 μm. Very few, if any secondary nuclei were found to attach and grow on the parent seed surface (Figure 7a). In general for the control solutions it was found that as the isothermal batch temperature was increased the aggregation of the small nuclei, broken off the parent seed surface, increased. These aggregates are evident in the Figure 7a, with sizes of up to ~30 μm

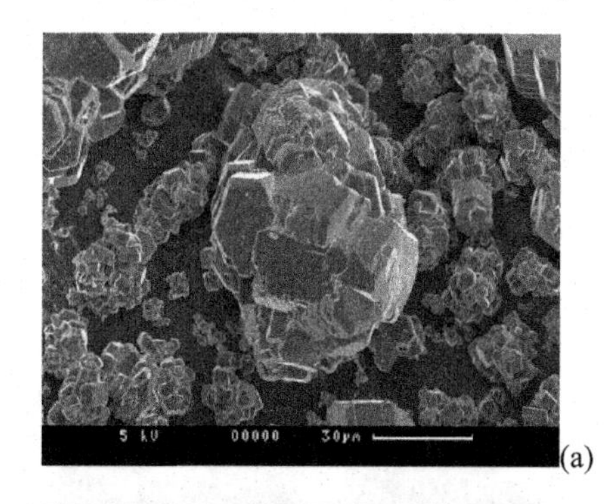

(a)

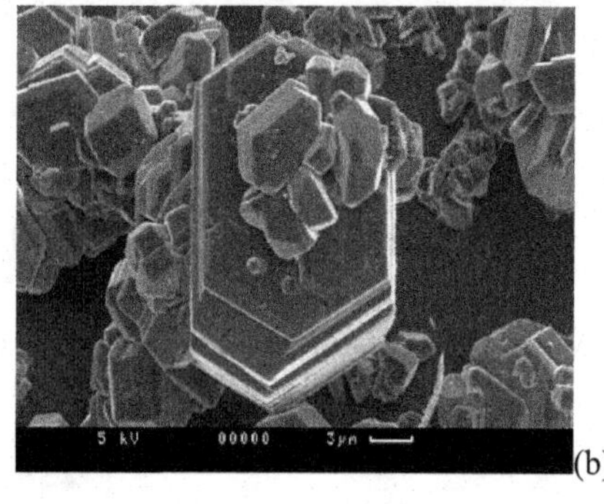

(b)

Figure 7. SEM image for the control test after 4 hours at 65°C. Low (a) and high (b) magnification.

In contrast, the SEM images for the CGM dosed experiments showed that at 60°C only 2° nucleation on the surface was observed, at 65°C a mixture of 2° nucleation and aggregates, but not separate from the parent seed, they had re-adsorbed and at 70°C the majority of particles were secondary nuclei or aggregates on the seed surface. The primary crystals predominately exhibited pseudo-hexagonal plates with some elongated hexagonal shaped crystals produced. The parent seed surfaces were partly or fully covered by fine particles (Figure 8b). These particles exhibited stick or needle-like morphology along with the pseudo-hexagonal plates growing on the surface of seeds. The size of these primary crystals were in the range of 0.5 – 10 μm, implying quick attachment as soon as the nuclei formed.

In the presence of CGM a majority of the fine particles adhered and grew on the seed surface, the degree to which the nuclei remained on the parent seed surface increased with increasing isothermal batch temperature. The remaining fraction of fine particles formed aggregates in the bulk solution, with some re-attaching to the parent seed surface. This is clear evidence for the observation of the higher net surface area and correlates well with the lower number of new crystals in the treated systems compared with the control.

(a)

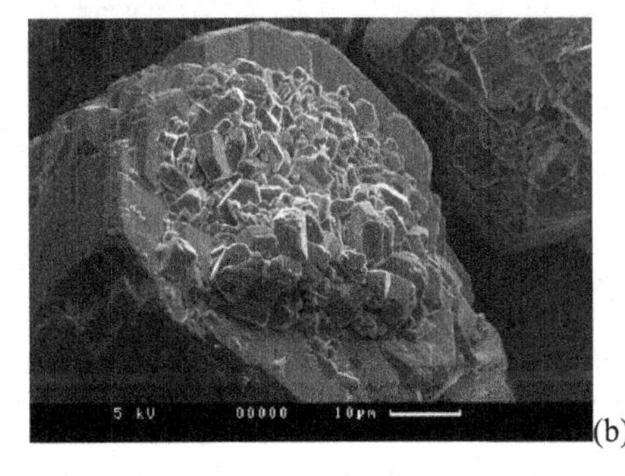
(b)

Figure 8. SEM image for the CGM dosed test after 4 hours at 65°C. Low (a) and high (b) magnification.

The conclusion from this evidence is that the involvement of the CGM additive in the seeded gibbsite crystallisation for all temperatures not only enhanced the parent seed growth mechanism, through accelerated secondary nucleation, with less attrition and hence particle surface roughening but also enhanced the secondary nuclei agglomeration behaviour.

Kinetics of Secondary Nucleation

As the growth rate of gibbsite is relatively slow when compared with other inorganic crystals [3,10,16-18] it is feasible to assume the initial kinetics to be dominated by secondary nucleation at high σ (> 1.15). Therefore, the birth rate of 3-dimensional nuclei, reflected in a dramatic crystal population increase, is considered to far exceed the rate of growth (2-dimensional nucleation process) of both parent seed crystals and the nuclei [16,17].

The crystallisation kinetics data produced from the seeded crystallisation at a high initial $\sigma = 1.5$ for which secondary nucleation occurred were analysed. Plots of $\ln[(-1/S)(d\sigma/dt)]$ versus $\ln\sigma$ for both the control and CGM treated solutions over 4 hours at each of the specific temperatures yielded linear relationships. The secondary nucleation dominated-reaction order (n) was derived from the slopes of these plots and reproducibly found to be equal to 4. The results are summarised in Table III. Similar values have been reported for secondary nucleation of gibbsite from sodium aluminate solutions under comparable conditions [14,19].

Table III. Kinetic parameters for seeded gibbsite secondary nucleation experiments at 60, 65 and 70°C.

Terms	T (°C)	Control	CGM Treated
k_o ($m^{-2}h^{-1}$)	60, 65, 70	3.81×10^{22}	4.50×10^{34}
React Order (n)	60	4.2 ± 0.2	4.1 ± 0.2
	65	3.9 ± 0.2	4.0 ± 0.2
	70	4.0 ± 0.2	4.1 ± 0.2
E_a ($kJmol^{-1}$)		162 ± 30	239 ± 40

Activation energies for gibbsite secondary nucleation for the control and CGM additive tests were estimated from Arrhenius plots. E_a values of 162 ± 30 and 239 ± 40 $kJmol^{-1}$ for secondary nucleation were calculated for the control and additive tests, respectively. These values are higher than the 132 ± 15 $kJmol^{-1}$ reported for secondary nucleation from synthetic sodium aluminate solutions under similar conditions to the control [19], but are much closer to the value of 160 $kJmol^{-1}$, reported in earlier work [14]. The increase in E_a in the presence of the additive suggests that more temperature-induced, chemical perturbations occur in these treated solutions. It is proposed that the additive is not only involved in the clustering of aluminate ions in supersaturated sodium aluminate solution but also the interactions between the clusters and surfaces of seed crystal, which are responsible for the development of microscopic growth units. As was the case for the agglomeration work, the pre-exponential factor of k_o was found to be much higher for the CGM treated gibbsite secondary nucleation than for the control tests. It is therefore obvious that the successful collision frequency between the nuclei forming, Al(III) containing species was enhanced by the presence of additive.

Interaction Forces

The direct interaction forces between two gibbsite particles under Bayer conditions were measured in-situ and the effect of a crystal growth modifier determined, using the atomic force microscope (AFM). Synthetic liquor was prepared containing 4 M NaOH and $\sigma = 1.5$ at 25°C. The fluid cell was allowed to equilibrate for 20 minutes prior to collection of data.

Initially the interaction force between the gibbsite particles was first measured in pure sodium aluminate liquor. As illustrated by the force curves (repeat approaches) in Figure 9, there is only a small repulsion on approach between the particles, which would be expected as the electrostatic repulsive forces are screened as a result of the high ionic strength of the sodium aluminate liquor. On separation of the surfaces, a very weak adhesive force is observed. In addition, there appears to be some hysteresis in the approach and retract curves. This may be a result of the viscous nature of the solution introducing some hydrodynamic forces into the system.

The CGM was mixed with the synthetic liquor for 10 minutes prior to addition into the fluid cell. It is clearly evident the difference the CGM additive has on the interaction forces by examining Figure 10. On approach, as for the case in the control experiment, there was very little interaction. However, upon separation of the gibbsite surfaces an adhesion is observed that varies with position on the surface. This is a common observation in systems where a chemical additive is used that does not form an even monolayer of coverage.

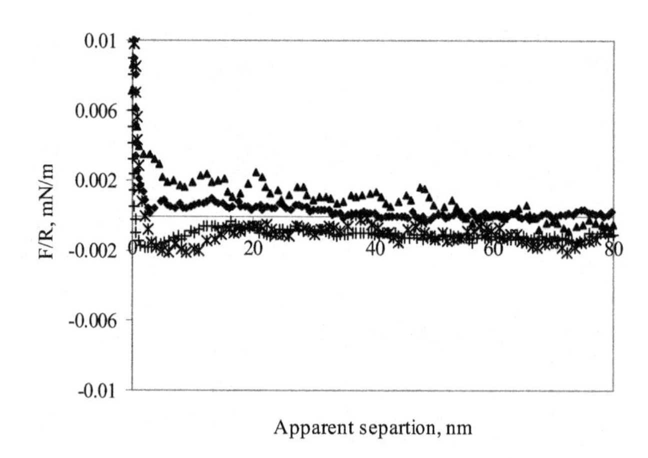

Figure 9. Interaction forces on approach (filled symbols) and separation (crosses) between gibbsite particles in-situ at 25°C.

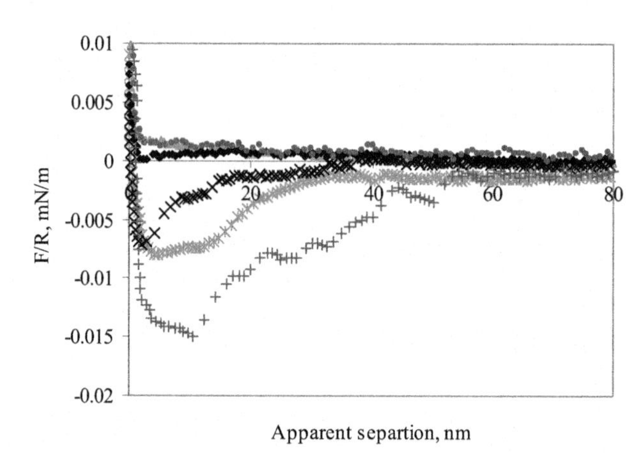

Figure 10. Interaction forces measured on approach (filled symbols) and separation (crosses) of gibbsite particles in CGM treated liquor at 25°C.

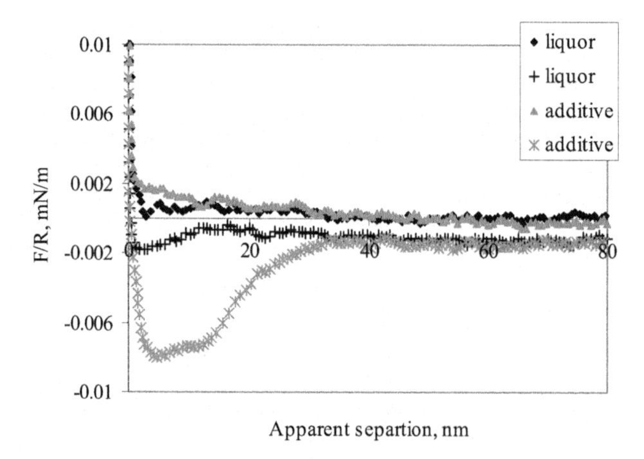

Figure 11. Comparison of average interaction force curves between gibbsite particles in the control (liquor) and dosed (additive) tests at 25°C.

A comparison of the interaction between gibbsite particles in the control and CGM dosed experiments is illustrated in Figure 11. On approach there is little difference between the interaction force curves. The repulsive force is very small and would allow the particles to contact each other under normal processing conditions. On separation however, there is a significant increase in the adhesion observed between the gibbsite particles in the presence of the CGM additive. Under process conditions, the gibbsite particles would therefore be likely to agglomerate in the presence of a CGM.

Conclusion

In a process where the chemical inclusion of the particle onto the surface of the seed is the rate limiting step and where particles are colliding and continually breaking off parent seed crystals, the addition of CGM may just allow the particle sufficient time to chemically bond with the seed surface and be incorporated into the structure. Evidence has been presented here which shows the increase in the successful collision frequency of not only Al(III) containing species but also small fractured gibbsite particles and fine seed crystals in the presence of the additive.

Two gibbsite particles will agglomerate if they have sufficient surface energy to do so and also have sufficient time to allow cementation to take place [2]. It is postulated that the CGM modified seed surfaces possessed a high surface energy interacting with the nuclei strongly with little or no repulsive energy barrier. Therefore to conclude, addition of a CGM additive induces strong adhesive interaction forces between the gibbsite particles and nuclei, and under specific crystallization process conditions, this would markedly enhance both agglomeration and coarsening of the particles and also allow for secondary nucleation control.

References

1. W.J. Roe, D.O. Owen and J.A. Jankowski, "Crystal Growth Modification: Practical and Theoretical Considerations for the Bayer Process", *First International Alumina Quality Workshop*, Gladstone (Australia), (1988).
2. C Misra, "The Precipitation of Bayer Aluminium Trihydroxide", (Ph.D. Dissertation, University of Queensland, 1970).
3. C. Misra and E.T, White, "Kinetics of Crystallisation of Aluminium Trihydroxide from Seeded Caustic Aluminate Solutions." *American Institute of Engineers Symposium Series 438*, 67(110), (1971), 53-65.
4. M Kanehara, "Formulation of Alumina Hydrate Precipitation Rate in Bayer Process for Plant Design and Operation", *Light Metals*, (1971), 87-105.
5. J Anjier and H Roberson, "Precipitation Technology", *Light Metals*, (1985), 367-375.
6. E.T. White, and S.H. Bateman, "Effect of Caustic Concentration on the Growth Rate of $Al(OH)_3$ Particles." *Light Metals*, (1988), 157-162.
7. Y Xie et al., "Study on the Application and Mechanism of Cationic Surfactant on the Precipitation of Sodium Aluminate Liquor", *Light Metals*, (2001), 135-137.
8. Y Xie et al., "Research on the Application and Mechanism of Crystal Growth Modifier on the Precipitation Process in Sodium Aluminate Liquors, *Light Metals*, (2002), 157-160.

9. Y Xie et al., "Research on the Mechanism and Optimum Adding Method of Additives in Seed Precipitation", *Light Metals*, (2003), 87-91.
10. P.I.W. Loh, H.M. Ang and E.A. Kirke, "Secondary Nucleation of Alumina Trihydrate in a Batch Crystallizer", *Australia's Bicentennial International Conference for the Process Industries*, Sydney, (1988), 304-309.
11. A.R. Hind, S.K. Bhargava and S.C. Grocott, "The Surface Chemistry of Bayer Process Solids: A Review", *Colloids and Surfaces*, 146, (1999), 359-374.
12. J. Addai-Mensah et al., "Particle Interactions and Surface Forces During Gibbsite Precipitation From Bayer Liquors", *6th International Alumina Quality Workshop*, Bunbury (Australia), (1999).
13. J.P. Cleveland et al., *Rev. Sci. Instru.*, 64, (1993), 403-405.
14. A.V. Nesterov and V.G. Teslya, "Mathematical Model for Crystallisation of Aluminium Hydroxide from Aluminate Solutions", *Zhurnal Prikladnoi Khimii*, 62, (1989), 1999-2004.
15. G.C. Low, "Agglomeration Effects in Aluminium Trihydroxide Precipitation", (Ph.D. Dissertation, University of Queensland, 1975).
16. A Halfon and S Kaliaguine, "Alumina Trihydrate Crystallisation, Part 1 Secondary Nucleation and Growth Rate Kinetics", *Canadian J. Chem. Eng.*, 54, (1976), 160-167.
17. A Halfon and S Kaliaguine, "Alumina Trihydrate Crystallisation, Part 2 A Model of Agglomeration", *Canadian J. Chem. Eng.*, 54, (1976), 168-172.
18. W.R. King, "Some Studies in Alumina Trihydrate Precipitation Kinetics", *Light Metals*, 2, (1973), 551-563.
19. J Li et al., "Secondary Nucleation of Gibbsite Crystals from Synthetic Bayer Liquors: Effect of Alkali Metal Ions", *Journal of Crystal Growth*, 219, (2000), 451-464.

Light Metals 2006 *Edited by Travis J. Galloway* *TMS (The Minerals, Metals & Materials Society), 2006*

PLANTWIDE REPLACEMENT OF THE EXISTING CONTROL EQUIPMENT BY A NEW DCS AT AOS

Mr. Joerg Ruester[1], Dr. Rolf Arpe[1], Mr. Holger Grotheer[1]
[1]Aluminium Oxid Stade GmbH, Johann-Rathje-Köser-Strasse, Stade 21683, Germany

Keywords: DCS, Automation, Control Room

Abstract

Since the start–up of the plant in 1973, AOS has been running with single loop process control equipment. The majority of the controllers need replacement, since there is no more support available. Therefore, AOS decided to install a modern DCS. Along with the implementation of the system, a reduction of staff is supposed to be achieved by means of a centralization of the current eleven control rooms and by a higher automation level.
This paper describes the experience with the hot change-over of the DCS in a running system, the opportunities of a DCS at AOS and the design of a new central control room.

Introduction

The alumina plant Aluminium Oxid Stade (AOS) GmbH was built in 1970 – 1973 and started up in 1973. The plant is located in the Northern region of Germany along the river Elbe close to Hamburg. The plant was equipped with single loop controllers. Within the past 30 years of operation, the controllers were partly replaced and expanded by state of the art controllers. Therefore, the current installation has a big variety of instrumentation; the whole history of control equipment from pneumatic thru modern SPS is present. For many of these instruments support is no longer available.

Fig. 1: Aluminium Oxid Stade GmbH

New functions and expansions of the existing plant have always meant a lot of wiring work in the old technology. Sophisticated control strategies are almost impossible to install, which limits the optimization of the plant. Furthermore, it is not that easy to keep this "old" knowledge or to get skilled people, since universities or technical schools do not provide this knowledge anymore.

The operators work in eleven permanently manned control rooms. Getting information from other parts of the plant is only possible by telephone. Therefore, the flow of information was insufficient. Moreover, in order to provide the process engineers and managers with information from the plant, almost every single data entry needed to be typed in manually thru an AOS-made software tool. This means a big gap of information which makes a proper analysis of the plant very difficult. Furthermore, the operators are busy with non-process relevant actions.

Therefore, the decision of the implementation of a DCS is a big opportunity for AOS.

1. Project

Scope of the Project

The scope of the project contains

- Change-over of all I/O's (15.000) to a DCS. AOS has chosen the Experion PKS from Honeywell.
- Implementation of a new IT-tool for the plant history. AOS has chosen PHD from Honeywell.
- Centralization of the eleven control rooms.
- Installation of an integrated video-system (21 cameras) which is the DVM-Manager from Honeywell.
- Training of 200 operators

Along with the DCS-Implementation AOS started simultaneously a revamping program of the whole control strategy in the plant in order to get the benefits of a DCS right away.

In order to get a better communication among the operators who run the process, AOS decided to centralize the existing control rooms. Due to the reduction of control rooms it is also possible to reduce the number of operators.

Since there will be fewer permanently manned control rooms than before, AOS installed a Video-system (DVM) to observe certain spots in the plant. Before, these spots were checked regularly by the control room operators. The DCS also allows the integration of pictures in the process graphics. This is a very helpful feature, which is very often used by the operators. Moreover, it makes the process safer, since the critical spots are observed permanently.

For each step of the project, AOS started a training program for involved operators. The training contains firstly a basic part (DCS structure, general basics of a DCS like calling up graphics, pop-ups, face plates, alarm page etc.) and secondly, an AOS-specific part (operating the specific plants at AOS including the changes and modification caused by the revamping). Altogether, almost 200 operators and service staff needed to be trained on the new system. This was done by means of internal seminars which were held by AOS-engineers.

Project team For the whole project AOS installed a project team consisting of seven core people, which can be seen in Fig. 2.

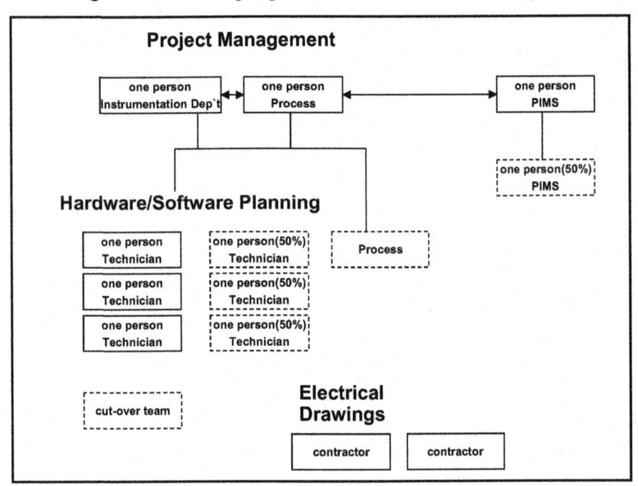

Fig. 2: Project team

The project management consists of three people from the instrumentation department, one process engineer and one of the IT-department. The process engineer coordinates the demands of the process, especially for the software, with the Control department and with the IT-department regarding PHD and does the software revamping. The instrumentation department is responsible for hardware and software planning which is done by three fulltime and three part-time technicians. The technicians are supported by two contractors who are mainly doing the documentation.

Each of the technicians is responsible for the change over of a part of the plant, which is done together with the cut-over-team. The actual change-over is carried out by experienced AOS-staff, since they know the plant details very well and are supported by one or two contractors.

The wiring, system set-up and the programming is carried out by contractors.

Project Schedule The project was supposed to be accomplished within five years and is subdivided in five steps. The schedule is:

- 2003 system set-up and milling
- 2004 digestion including surrounding plants.
- 2005 red mud separation
- 2006 liquor cooling, liquor cleaning, precipitation and filtration.
- 2007 calcination and boiler house

Prior to the start of the project, one and a half years were needed for the basic soft- and hardware specification of the DCS and for finding the best system and negotiations with the suppliers.

Each step consists of an AOS-planning section, a Honeywell programming section, a pre-Factory-Acceptance-Test (FAT) and the final FAT. After the FAT the system is shipped to AOS followed by the Side-Acceptance-Test, where the system set-up is checked at local conditions. Once this is done, the actual change-over takes place. AOS calculated about 10 tags/day to be switched over. Right now the project is on schedule.

Problems One of the main problem is the documentation of the existing plant. In areas of high safety and directly to production related facilities like digestion, the documentation is very good. In areas of lower focus like red mud filtration, the level of documentation is very poor. Therefore, almost two thirds of the time of the planning-team is used for updating the documentation.

Most of the plant could be cut over during turn-arounds. More effort has to be spent on facilities which have to be cut over while they are running. This can become very critical, since the whole plant flow is dependent on these parts of the plant. But due to the good preparation of the project and the fact that experienced AOS-staff was doing the cut-over, major failures could be avoided until now.

One major issue is the participation of the operators at the project. They were part of the planning at all steps of the project. Design of the graphics as well as some software modifications (compared to the operation prior the DCS) were always double checked with the operators.

2. Opportunities of the DCS

AOS has switched over more than half of the refinery. Now advantages of the DCS become quite obvious. The availability of information lead to a much better understanding of our process. Not only for operators, who can see the whole process and can therefore react earlier. It also can be seen from historic events and trends what "really" happened in the process besides that, what operators sometimes tend to tell.

We found a couple of some interesting relations which were not clear before the DCS was installed. E. g. variations of the tank level prior to the digestion feeding pumps strongly influence the molar ratio (a/c-ratio). A sophisticated tank level controller solved that problem.

The DCS enabled AOS to install online measurements for the molar ratio at the outlet of the tube digesters and a complex controller for this equipment. Now the molar ratios run much smoother and more stable than before. This was always the target in the past decades, but could not be achieved, since the technology for the required data treatment was not available.

The engineering in the existing programs like adding new control loops or interlocks became very simple compared to the wiring demand before. Optimizations could be worked out much quicker or are in the first place even possible.

Complex start ups and shut downs became easier now, since they have been automated. E. g. operation of mills as well as tube digesters are fully automated now. Once the plant is set up for operation, it is only a click in order to start a whole digestion line. The program also assists at steps, where manual work has to be done. This avoids human failures and helps operators to run the plant, since there is a set standard given by the DCS sequence.

Since most of the DCS-suppliers have Window-based systems, the design of the process graphics became very powerful and a big help as well. Operators can easily overview the process status with its actual data, since the possibility of arranging tanks, pumps, pipework etc. lead to a better understanding of the connected plants. Furthermore, by means of animated points or piping, hints and scripts the operator is support by the DCS, when he analyzes the process.

Operators at AOS accept the DCS and nobody wants to go back to the analog technique. The adaptation was easy even for operators, who usually do the heavy work in the plant.

3. Central Control Room

Part of the project is also the centralization of the existing eleven control rooms in order to share the information quicker and to reduce the staff.

Currently all control rooms are permanently manned as can be seen in Fig. 3.

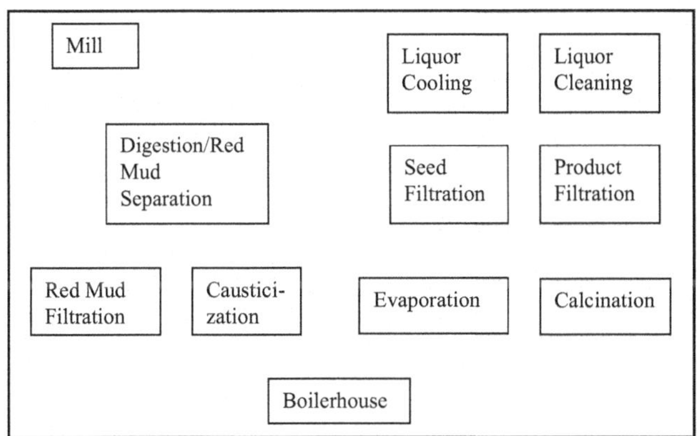

Fig. 3: "Old" control rooms

In the future just the central control room will remain permanently manned. In the other control room stations with access to the DCS will be installed for start up/shut down and trouble shooting purposes. An exception will be the control room for the seed- and product-filtration. This one will remain temporarily manned, since the startup- and shut downs of the filters cannot be done from the Central Control Room.

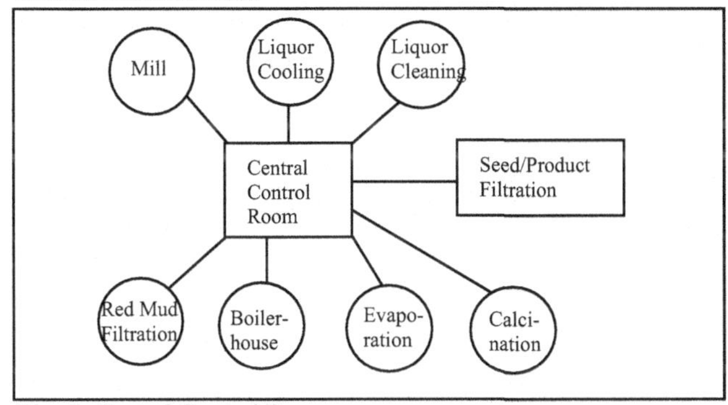

Fig. 4: Central Control Room

In order to get the Central Control Room, an existing room was redesigned. The existing room was the control room for digestion and red mud separation. The original design of the plant was considered to have triple of the current capacity. Therefore, the existing control room has sufficient space for the redesign.

The new central control room consists of four compartments. The biggest one contains the actual control center. There are three operators located, who run the whole process. Another compartment is the configuration room, where all engineering takes place. A third room is for trouble shooting and acts also as a sluice for operators who come from outside. The last room is a room for visitors.

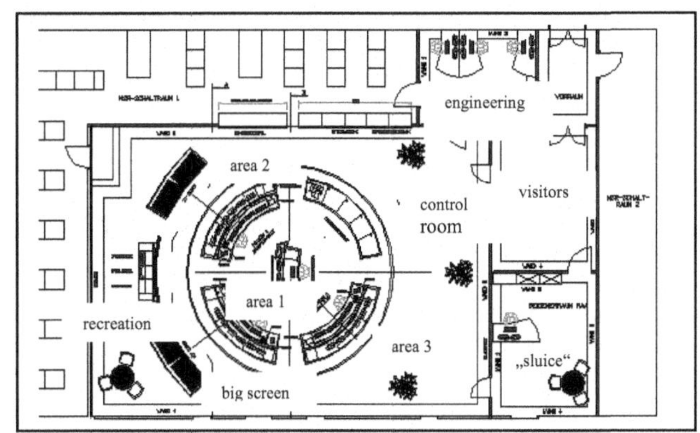

Fig 5: Design of the Central Control room

The idea was to have a control room with as low traffic as possible, so that the operators can concentrate on the essential things, which is controlling the process. There are three main operators located in the Central Control Room. They are arranged in a circle in order to get all three of them as close as possible. The coordination of the Bayer circuit becomes much more effective.

Fig. 6: picture of the Central Control Room

AOS also installed two big screens. These screens allow a very good process overview of a whole area, since it is not possible having all related process information combined in one graphic, which is very helpful during trouble shooting. The screens can be operated from any of the process stations in the control room.

The new control room has been widely accepted by the AOS-staff. The extraordinary design creates a very good working atmosphere and work motivation.

Summary

The DCS project was approved in early 2003, since there was no longer support available for the majority of the controllers. The project is done mainly by AOS-staff and is more than half accomplished. Until now the experiences with the DCS are very good. No major failure of any plant occurred during the cut-over. Lots of benefits came along with the implementation, even benefits that AOS did not expect. The Central Control Room became the heart of the process operation and is a big help, since information can be exchanged directly among the operators, who control the process.

Acknowledgements

The authors would like to express many thanks to Victor Phillip M. Dahdaleh for his generous support and his great efforts for the approval of the project. The authors are also grateful to the good co-operation with Honeywell, who never hesitated to transpose the ideas of the AOS-team.

Precipitation Fundamentals

SESSION CHAIR

Songqing Gu
Zhengzhou Research Institute of Chalco
Zhengzhou, Henan, China

Light Metals 2006 *Edited by Travis J. Galloway* ***TMS (The Minerals, Metals & Materials Society), 2006***

INFLUENCE OF ALUMINA MORPHOLOGY AND MICROSTRUCTURE ON ITS STRENGTH

Gu Songqing, Qi Lijuan, Yang Zhimin, Yin Zhonglin
Zhengzhou Light Metal Research Institute, Chalco
No.82 Jiyuan Rd., Shangjie District, Zhengzhou, Henan, China 450041

Keywords: alumina, hydrate, strength, attrition index, morphology, microstructure, agglomeration, calcination

Abstract

A study on the influences of alumina morphology and microstructure on its strength has been carried out. The alumina hydrate grain growth types are classified by SEM observations based on its microstructures. It is found that agglomerated alumina hydrate particles have much more opportunities to be strengthened than those particles which grow individually. The strength of the agglomerated alumina particles will be kept higher because the water vapor can be released gradually through the gaps among the agglomerated hydrate particles during calcinations without damage of the whole grain microstructure of the particles. The better control of agglomeration and growth processes in precipitation and smoother calcinations are the key solutions for obtaining stronger alumina product.

Introduction

There exists a huge demand for alumina in China with great development of Chinese aluminum industry in the recent years.

The high amperage aluminum smelting cell technology is used widely and most of the smelters equipped with Soderberg cells have been shut down because of the environment protection issue and lower current efficiency, which brings about the requirement for high quality sandy alumina. Alumina quality is becoming more and more important for improving smelting cell performance and the environment in both smelting workshops and surroundings.

Chinese alumina industry is growing rapidly in the past years and total domestic alumina yield produced mainly by six Chinese alumina refineries of Chalco will reach more than 7.6 million tons in 2005, which is a little more than half of alumina consumption in China.

As well known, the lower grade of diasporic bauxite is the only raw material resource for alumina production in China, which leads to the great difficulty of sandy alumina production due to the following issues:

(1) The process becomes very complicated because of including both Bayer process and Sintering process, and so-called Combined process as well. Both of Bayer precipitation and liquor carbonation in sintering process should be considered for alumina quality improvement.

(2) The worse digestion performance of diasporic bauxite at higher caustic concentration and temperature results in lower supersaturation and higher caustic concentration in the precipitation.

(3) The carbonation process of the pregnant liquor for hydrate precipitation in sintering process is difficult to control due to too quick neutralization of liquor by CO2 . About 40% of Chinese domestic alumina production is from the carbonation process.

Two important properties of sandy alumina should be paid great attention. One is its size distribution, the other is the alumina strength. Alumina size distribution can be controlled by changing parameters in the precipitation and calcinations processes, such as temperature reduction gradient during whole process, seed additions and even concentrations and molar ratio of liquor. A series of process conditions have great and complicated influences on alumina strength, some of which should be further studied.

This paper deals with the following study goals:

(1) Classifying different types of alumina microstructure;

(2) Studying the relationship between alumina strength and its microstructure;

(3) Finding out the solutions to improve alumina strength;

The Description of Alumina Strength

A study on the strength of different kinds of alumina samples was carried out by a series of testing ways. It has been found that alumina strength is closely related with its morphology and grain microstructure observed by SEM (scanning electronic microscope).

Various alumina and hydrate samples are taken from Bayer precipitation and carbonation in sintering process of different Chinese refineries. Some alumina samples from the refineries abroad are tested at the same time for comparison.

So-called Attrition Index is applied for alumina strength measurement. The testing is carried out by a special test instrument.

The size distribution of alumina sample can be tested by laser size analysis and mesh analysis respectively. Then the same sample is put into attrition index analyzer for attrition testing. A certain gas stream is blowing the sample from a definite diameter orifice for some time. The sample after attrition testing is analyzed for size distribution again.

The attrition index of the alumina sample can be calculated by both size distribution results before and after attrition testing as following equation:

AI = 100*(Pa-Pb)/(1-Pb)

in which Pa is the percentage of the portion in weight less than 45 μ m in the sample after attrition testing, Pb is the percentage of the portion in weight less than 45 μ m in the sample before attrition testing.

Generally, the attrition index represented by AI is ranged in 5-40. The alumina with AI of 5-20 can be considered as sandy alumina grade in strength. The lower the AI is, the stronger the alumina will be.

Compared with alumina the hydrate is much stronger with lower AI. And there is no strict correspondence of AI between both alumina and hydrate. And it is obvious that alumina becomes weaker owing to grain microstructure breakage during calcination. So the calcination process has a great influence on alumina strength in addition of the hydrate strength factor.

Different Types of Alumina Microstructure

Different alumina morphology and microstructures can be observed by SEM and classified as following growth types:

Single Crystal Growth Type (SCG Type)

Single crystal growth type is that a single hydrate grain is growing gradually during precipitation without evident fine seed agglomeration on its surface, as shown in Fig.1. This kind of hydrate grain usually is produced by keeping relatively low temperature growth without agglomeration and cycling many times in seeded precipitation.

Strength of the alumina calcinated from single grain growth hydrate is very low because water release damage its internal microstructure during calcination due to longer water penetrating distance, as a result of which some big gaps vertical to the grain axis will form, as shown in Fig. 2.

Fig.1 SCG Type Hydrate Fig.2 SCG Type Alumina

Agglomerated Mosaic Growth Type (AMG Type)

Agglomerated mosaic growth type is that a big hydrate grain is made of a number of very fine and different size hydrate grains, as shown in Fig.3. This kind of grains is formed by high temperature agglomeration from fine seeds in seeded precipitation and then grows to be the mosaic morphology.

The alumina with agglomerated mosaic growth type has the strongest microstructure and very low AI. The sandy alumina usually has the morphology like this.

The reason why agglomerated mosaic alumina is the strongest is the water released during calcinations can easily escape through the small gaps between agglomerated fines without damage of the crystal microstructure. It also can be observed under SEM that a big grain usually consists of lots of fines with various sizes, which is reasonable for a high strength like concrete. Fig.4 shows the shape and morphology of sandy alumina after AI test without visible change.

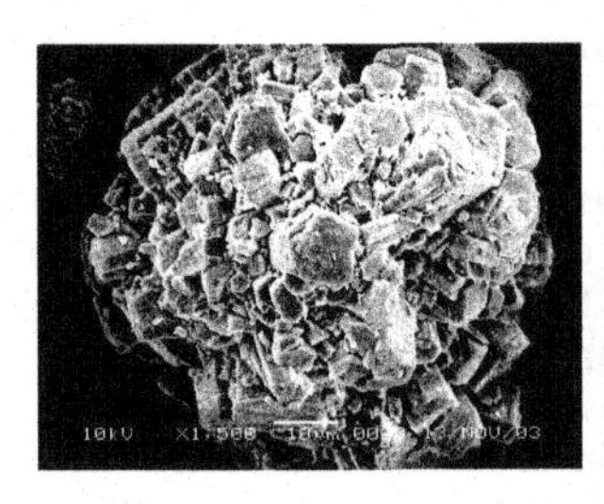

Fig.3 AMG Type Fig.4 Sandy Alumina after AI Test

Scattering Pillar Growth Type (SPG Type)

Scattering pillar growth type is that lots of short pillars are growing based on the grain with a scattering directions, as shown in Fig.5. Very rapid neutralization reaction of liquor by CO2 in carbonation of sintering process without suitable control leads to formation of this kind of grains.

It is obvious that this kind of grains is easily broken by attrition and collision so that its AI will be very high. It can be found that a huge number of fine particles appear after AI test of this kind of grains.

Mosaic Aggregate Growth Type (MAG Type)

Mosaic growth type is that lots of relatively bigger hydrate grains get together loosely and grow gradually to form mosaic shape as shown in Fig.6. Both aggregation and the following growth promote formation of mosaic type growth.

This type of grains is popular in the Chinese refineries due to very low supersaturation in Bayer precipitation. And it is not strong enough, but has acceptable medium strength.

Fig.5 SPG Type Fig.6 MAG Type

Agglomerated Radial Growth Type (ARG Type)

Agglomerated radial growth type is that all the fine grains are agglomerated closely to form a spherical shape and then growing by the radial directions as shown in Fig.7 . This kind of grains can be obtained in the well controlled carbonation of sintering process.

The very high strength of this kind of grains comes from their tough microstructure and round shape. There are many fines in the gaps of grains for a strong cementation.

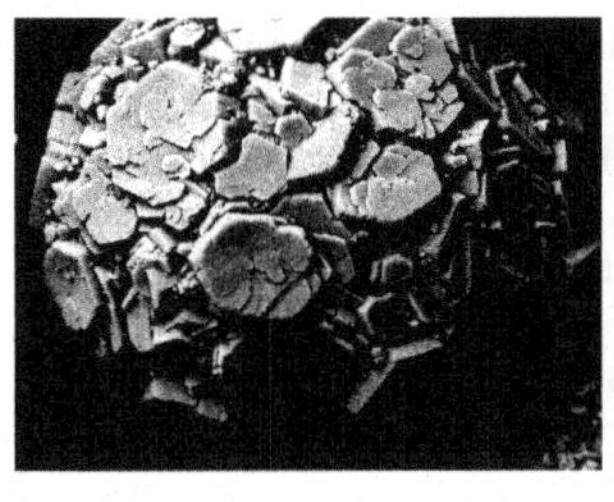

Fig.7 ARG Type   Fig.8 Deep Gaps on Alumina Surface

Relation Between Alumina Strength and Its Microstructure

A lot of study on the morphology and microstructure change from hydrate to alumina by SEM has proved that alumina after calcination keeps almost the same shape and morphology with hydrate.

An obvious change might happen that some new parallel gaps vertical to crystal axes appear during calcination on the grain surface in some cases, especially for SCG and MAG types of hydrate, as shown in Fig.8. And these gaps in alumina grains will damage their microstructures and strength as well.

The attrition index of various types of alumina is quite different because of their internal microstructure characteristics. Table I shows the possible attrition index ranges for different types of hydrate. It is clear that the agglomerated mosaic growth type and agglomerated radial growth type of hydrate are preferred to form during hydrate precipitation for the stronger product.

Table I. Attrition Index Ranges of Different Grain Types

Grain type	SCG	AMG	SPG	MAG	ARG
AI, %	30-50	5-15	30-60	15-30	5-15

There are two kinds of possible changes for the grains in AI tests: one is that only surface erosion happens without damage of grain microstructure, the other is the grains are broken into some fragments.

The internal microstructure of hydrates is very complicated. It is shown in the picture of the alumina grains broken after attrition index tests in Fig.9 that the internal microstructure of some grains is likely radial, which indicates the crystal growth directions, and the another shows a layer upon layer microstructure (Fig.10).

The SEM observations on some spot surface areas on the big grains of sandy alumina and non-sandy alumina after AI tests respectively revealed that there exist a quite rough surface on the sandy alumina grain and lots of fines filling in the grain gaps, while there are smoother surface and less fines in the gaps for the non-sandy alumina grain.

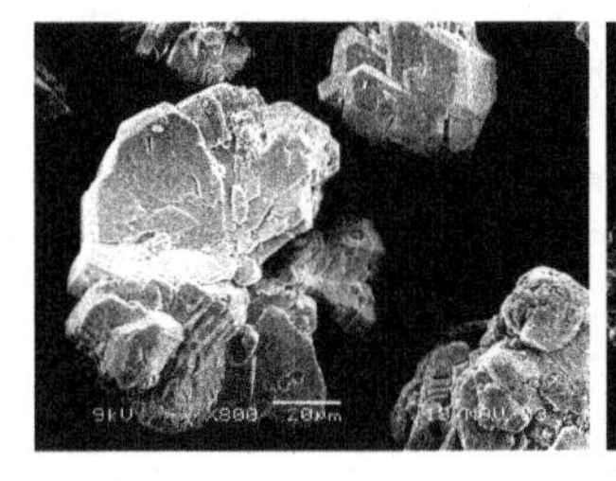
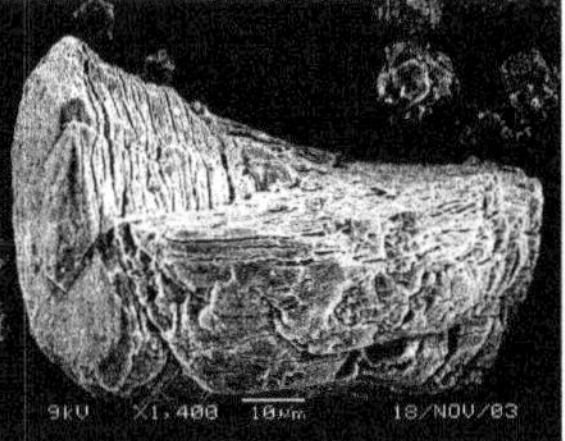

Fig.9 Broken Alumina with Radial Microstructure   Fig.10 Layer upon Layer Internal Microstructure

The high magnification SEM tests have further clearly revealed microstructure differences of different types of alumina as seen in Fig.11 and Fig.12.

Fig.11 Weaker Hydrate Grain with Loose Microstructure

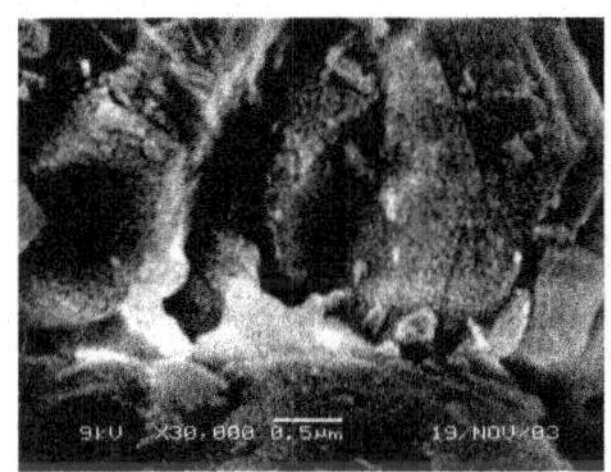

Fig.12 Stronger Hydrate Grain with Tight Microstructure

Improvement of Alumina Strength

It is concluded from the study on the relationship between alumina strength and its microstructure above mentioned that the alumina strength mainly depends on its internal microstructure including size distribution of the fines composing hydrate grain, microstructure types, grain cementation firmness, and microstructure damages by the gaps forming in calcination as well.

Suitable size distribution of the fines in alumina grains, AMG and SAM types of hydrate and stronger cementation between the fines in the grain are preferred to produce stronger sandy alumina.

An agglomeration microstructure and tough cementation of fine seeds with a suitable size distribution are the basis for the strong hydrate formation. The further filling the gaps of the agglomerated grain by precipitated hydrate during the following growth stage is very important for hydrate grain strengthening.

It is concluded that some small hydrate grains should be agglomerated in the first step and further strengthened by the following growth in the relatively high supersaturation in order to obtain alumina with high strength.

The gap formation and microstructure stress can be reduced by gradually temperature raising and not too strong erosion and collision during calcinations.

The following measures can be taken to improve alumina strength:

Intensifying Agglomeration

The agglomeration conditions should be well controlled, such as agglomeration temperatures, seed content and size distribution, duration and pregnant liquor supersaturation. These conditions sometimes impact on each other so that a systematic optimization of the process should be a key technology.

Improving the Precipitation Filling at the Early Growth Stage

the agglomerates produced in the agglomeration stage are quite weak and should be strengthened in the following growth stage. So a little higher supersaturation at the early growth stage will provide better condition to make the grain gaps to be filled better and become stronger, for that the molar ratio of liquor at the end of agglomeration stage should be controlled.

Control of Seed Content and Size Distribution in Precipitation System

Seed content in all the precipitation process stages and its size distribution sometimes play a key role to systematic size change and strength trend of alumina, which greatly impacts on agglomeration microstructure and strength, precipitation rate and the ratio of liquor supersaturation to seed surface area.

CGM Addition

CGM can be used for agglomeration and growth rate control for the optimized process conditions.

Application of Gas Suspension and Fluidized Calcination Process and Relevant Equipment

There are quite different testing results of size distribution, strength, alpha phase content, surface area and repose angle etc. for the alumina samples produced by rotary kiln calciner and gas suspension calciner respectively.

Application of gas suspension and fluidized calciners makes the calcinations process less breakage of alumina, less overheating of the material, slower temperature raising and more homogeneous water release. The temperature and flow speed control at various calcination stages of gas suspension or fluidized calcinations is still a key problem for alumina quality.

Conclusion

1. There are at least 5 types of morphology and microstructure of the hydrate precipitated from liquor.

2. Alumina strength mainly depends on the precipitation process, which greatly impacts on its grain microstructure and cementation toughness of hydrates, and transformation process from hydrate to alumina, which might have a great influence on alumina microstructure damage and gap formation in the grains.

3. The better control of agglomeration and growth processes in precipitation and smoother calcinations are the key solutions for obtaining stronger alumina product.

References

1. J.Tan et al.," Microstructure Analysis of Aluminum Hydroxide and Alumina", *Light Metals 2005*, 111-114.

2. D. Olsen et al., "Effects of Morphology on Alumina Strength and Dustiness", *Proceedings of the 6th International Alumina Quality Workshop 2002*, 1-9.

3. E. Rodriguez, R. Mendoza, "Evaluation of the Precipitation Circuit Operation Parameters to Control the Alumina Attrition Index", *Light Metals 2002*, 145-148.

4. J. Eduardo Lopez, Ingemar Quintero, "Evaluation of the Agglomeration Stage Conditions to Control Alumina and Hydrate Particle Breakage", *Light Metals 1992* 199-202.

Light Metals 2006 *Edited by Travis J. Galloway* *TMS (The Minerals, Metals & Materials Society), 2006*

STUDY ON THE RATE OF CRYSTAL GROWTH AND THE PHENOMENA OF TEMPLATE CRYSTALLIZATION DURING SEEDED PRECIPITATION FROM SODIUM ALUMINATE LIQUOR

Bai Wanquan, Yin Zhonglin, Li Wangxing, Qi Lijiuan, Yang Qiaofang, Wu Guobao
Zhengzhou Research Institute Aluminium Corporation of China Limited
Shangjie, Zhengzhou, Henan, 450041, China

Keywords: sodium aluminate solutions, crystal morphology, template crystallization

Abstract

Seeded precipitation from supersaturated sodium aluminate liquor with high concentration of diasporic bauxite was studied by using a laboratory batch crystallizer. It was found that the agglomeration rate of the particles can reach 7-8μm/hour, and it's a major part of particle size growth. Some special particles also have been found in experiments, which had relatively large particle size and high level of sphericity. This phenomenon may be due to some kind of crystal templates, where Al $(OH)_3$ is precipitated layer by layer on the surface of the template.

Introduction

In aluminum hydroxide precipitation process, caustic sodium concentration and supersaturation of input sodium aluminate solutions, temperature, seed charge and precipitation retention time are among the key parameters that determine the yield and quality of precipitated hydrate. In order to produce sandy alumina, it is very important to study the effects of these parameters on precipitation process. The precipitation process is that solid aluminum hydroxide precipitate from sodium aluminate solutions, the essential reaction is represented by the following equation:

$$NaAl(OH)_4 \rightarrow Al(OH)_3 + NaOH \quad (1)$$

However, precipitation of aluminum hydroxide from super saturated sodium aluminate solutions is different from other precipitation of inorganic salt, it involves (a) secondary nucleation, (b) agglomeration, (c) crystal growth and (d) attrition [1]. And some of these mechanisms will be dominate under some certain conditions. Agglomeration and crystal growth are the main mechanisms for producing coarse aluminum hydroxide.

Agglomeration and agglomeration mechanism have been studied by a lot of scholars[2-5]. As found by previous works[6], the growth of fine crystal is extended in axis direction and radial direction of hexagonal prism, and pseudo-hexagonal lozenges grow on the {001} face of seed crystal. Aluminum hydroxide crystallization activation energies and the second order dependence of growth rate on solution Al(III) supersaturation indicate that the crystallization process is reaction rate controlled rather than diffusion rate controlled[7,8].

Few reports are studied on the rate of increasing particle size for agglomeration and crystal growth in precipitation process from pregnant liquor with high concentration from diasporic bauxite in China. The SEM morphology that has been reported mainly consists of crystal particles with agglomeration or crystal radial growth. But template crystallization has been seldom reported.

Experiment

Agglomeration Experiment

The agglomeration experiment is done in a water bath precipitation tanks of 2.5L, which are controlled by an automatic temperature controller. The sodium aluminate liquor with a caustic to Al (III) molar ratio of 1.5 and caustic concentration of 145g/L is from Bayer plant.

Seed: seed ratio is 0.3 and seed comes from laboratory product and particle size is lower than 30μm. The agglomeration experiment is carried out five times in turn. The next experiment seed is from the last experiment product.

Precipitation time: 2hours; Temperature system: temperature is continually varied from 76°C to 74°C.

The solution is analyzed by titration method, and the solid samples washed and dried for particle size analysis with Laser Light PS Instrument.

Crystal Growth Experiment

The precipitator is the same as that used in the agglomeration experiment. The sodium aluminate liquor with a caustic to Al (III) molar ratio of 1.5 and caustic concentration of 155g/L is also from Bayer plant.

Seed: seed ratio is 4.1 and seed comes from coarser seed in Bayer plant. The experiment is carried out four times in turn. The next experiment seed is from the last experiment product.

Precipitation time: 40hours; Temperature system: temperature is varied from 61°C to 51°C.

The solid aluminum hydroxide washed and dried for particle size analysis by Elzone-5382 micromeritics made from USA.

Circulation Experiment

The agglomeration temperature is fallen from 76°C to 74°C during the 3hours running period, then rapidly drop to 61°C and turned next phase named crystal growth. In growth phase, the temperature continually drop to 51°C at holding time of 40 hours. The seed is classified and the coarse fraction of the produced hydrate is then entered in the growth section as seed, and the fine fraction is used as seed in the agglomeration section. A part of pregnant liquor is added into the growth section to increase supersaturation to fill the agglomerates produced from the agglomeration section.

The production aluminum hydroxide separated was washed and

dried and then calcined to alumina. The particles of aluminum hydroxide are analyzed by SEM to study the crystal morphology.

Results and Discussions

The Particle Growth Rate by Agglomeration

The particle size distributions of seed and agglomeration product are shown in table1. The percentage of fine particles under 10μm significantly decreased and the percentage of particles over 10μm(especially range from 10 to 16μm) increased in the first agglomeration by agglomeration. It is demonstrated in the 2nd experiment that the percentage of particles under 16μm decrease with the increasing of particles over 16μm (especial range from 20 to 30μm). Then in the 3rd experiment, the size of agglomerating particles reaches the range from 16μm to 20μm and the percentage of particles between 20-50μm increases accordingly. It is shown in the PSD of 4th product that, if the percentage of particles under 20 μm is less, the percentage of 20-30μm particles will decrease with the increasing of percentage of particles over 30μm by agglomeration of the particles(from 20μm to 30μm). In the 5th experiment, the particles between 20-30 μ m continuously agglomerate while the particles over 30μm do not agglomerate because of the percentage of 30-40μm particles not decreasing.

The figures in table1 also indicates that the percentage of particles under 10μm decrease with the increasing of the times of agglomeration. However, the particles under 10μm are not entirely disappeared, which is explained by the four mechanism existing together in the procedure of seeded precipitation. It is shown from the variations of average particle size (D50) that the particle size of product evidently increases by particles agglomeration. For example, the average size is increased by 7-8μm per hours from 2nd to 3rd experiment.

Figure1, Figure2 and Figure3 show the SEM photographs of hydrated of 1st , 2nd and 3rd agglomeration, respectively. It can be seen that agglomerate changes to firm by filling of aluminum hydroxide precipitated and the globe of the agglomerate is improved by some times of agglomeration.

Figure 1 SEM of hydrated obtained by 1st agglomeration

Figure2 SEM of hydrated obtained by2st agglomeration

Figure3 SEM of hydrated obtained by 3st agglomeratio

Table1: Particle Size Distribution (PSD) in Seed and Agglomeration Product (Weight Percentage/%)

PS/μm	2	2-5	5-10	10-16	16-20	20-30	30-40	40-50	D50
seed	7.72	23.25	51.12	16.28	1.51				6.59
1st	4.68	6.27	34.47	38.63	10.75	4.87			10.57
2nd	1.64	1.41	7.24	30.21	12.97	34.68	9.56	2.17	19.28
3rd	0.19	0.41	1.13	2.6	4.71	27.59	31.45	20.48	34.02
4th	0.11	0.32	0.52	1.86	3.02	20.56	38.41	21.3	45.25
5th	0.1	0.28	0.50	1.31	1.45	16.02	38.66	23.12	49.98

The Crystal Growth Rate

The principle of Elzone-5382 is of the following, when a particle suspended in electrolyte solution pass through the orifice, the voltage between the both sides of the orifice will change at the same time. The particles number of per gram in the four crystal growth experiments was measured, as being shown below in Figure 4 to Figure 7, in which the vertical axis is the particles number per gram while horizontal axis is logarithm value of particle size.

Figure4 and Figure5 show the particle size distribution for under 20μm(logarithm value 1.3). There is a peak in every curve and the peak moves right with experiment times, it indicate that the result is dominated by crystal growth rather than agglomeration. The peak of 1st locate near 0.74(5.5μm), the peak of 2nd locate near 0.94(8.7μm), the peak of 3rd locate near 1.02(10.5μm), and 4th peak is near 1.1(12.6μm). The average rate of crystal growth every experiment is about 2.4μm which is between 2.1μm and 3.2μm.

Figure6 and Figure7 show the particle size distribution for over 20 μm. The peaks located near 1.75-1.8(about60μm) show there are many of particles about 60μm. The peak moves from 1.74(58.3μm) to 1.8(62.6μm) at the first two experiments and the peaks of the last two experiments also locate about 62.6μm, which indicate the crystal growth rate is very slow.

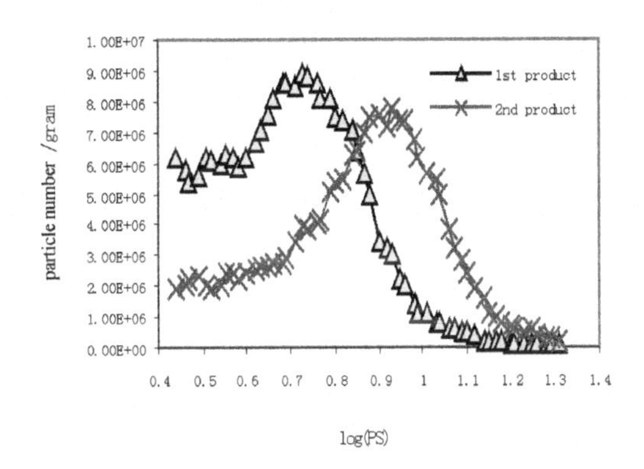

Figure 4.PSD of -20μm of 1st and 2nd

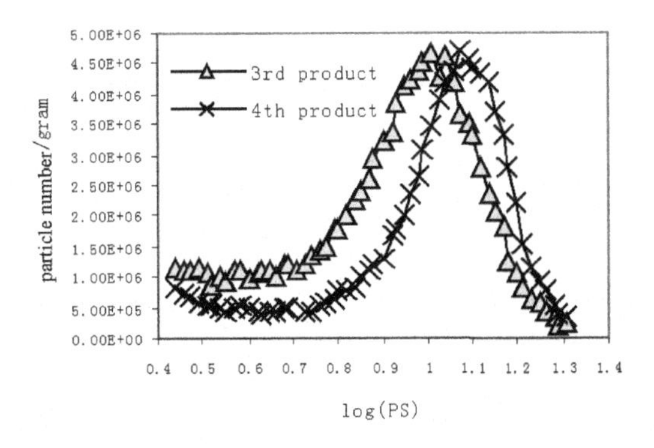

Figure 5.PSD of -20μm of 3rd and 4th

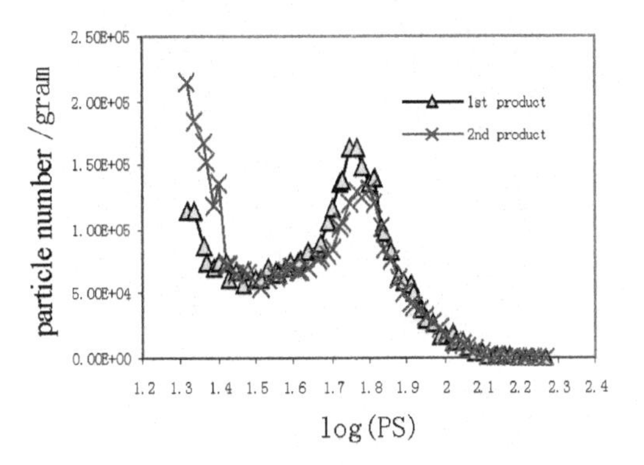

Figure 6. PSD of +20μm of 1st and 2nd

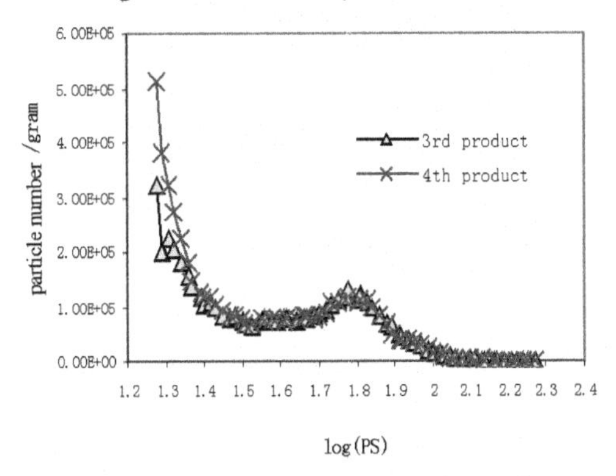

Figure 7. PSD of +20μm of 3rd and 4th

The Phenomena of Template Crystallization

The circulate experiment work presents the SEM morphology of aluminum hydroxide. Figure8 shows a particle of aluminum hydroxide, whose particle size is larger and shape is globe, the crystallites size on the surface is relatively uniform(under10μm). The morphology is much different from that of crystal growth or agglomeration particles.

It is conjectured that the formation of these particles is based upon some organic or inorganic salt crystal templates, rather than aluminum hydroxide. Then, $Al(OH)_3$ is precipitated layer by layer on the surface of the template. This organic or inorganic salt template is formed by partial supersaturation in solution under changing for parameter, such as falling rapidly at temperature. The salt template is high activation surface of $Al(OH)_3$ precipitation. And the fine aluminum hydroxide particles precipitated on the salt surface are the activation seed of $Al(OH)_3$ precipitation, so, step by step, the globular $Al(OH)_3$ particle is achieved.

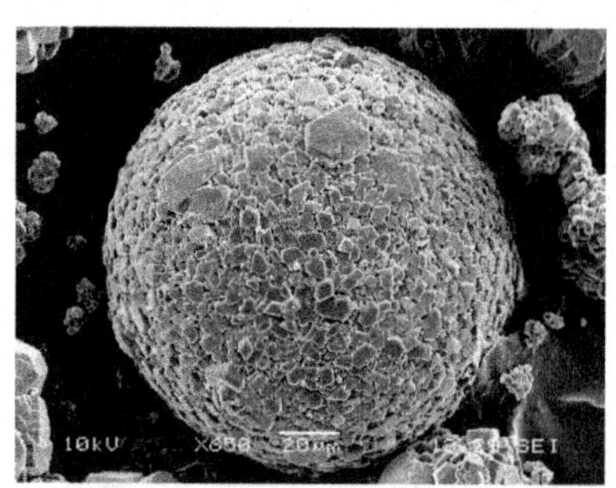

(a) amplifier coefficient:650

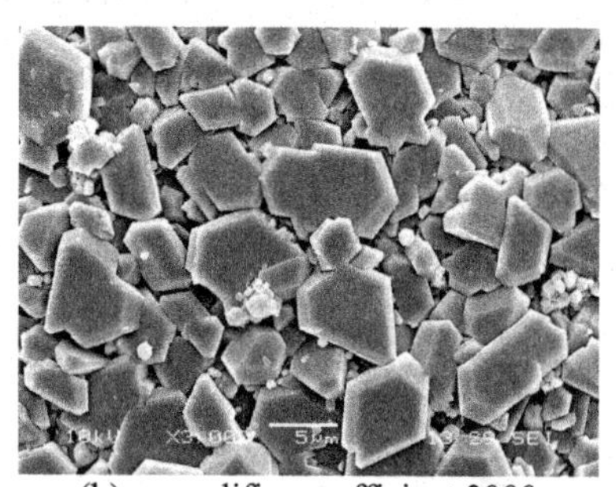

(b) amplifier coefficient:3000

Figure 8 SEM of hydrated obtained by Circulate Experiment

Conclusion

In order to obtain coarse aluminum hydroxide and sandy alumina, it is necessary that the parameters in the precipitation process should be adjusted to promote the agglomeration of fine particles within the appropriate values. The highest particle growth rate reaches 7-8μm per hours under the current condition, which is faster than the particle growth rate of crystal growth with about 2.4 μm per experiment.

Some special particles based on some organic or inorganic salt crystal templates indicate that the phenomena and mechanism of the template crystallization exist in the seeded precipitation from aluminate solutions.

Acknowledgement

This research is supported by "China National Key Fundamental Research Development Project"(973), No. 2005CB623702 and Center of Research Development Aluminum Corporation of China Limited.

References

1. Yang Zhongyu. Technology of Alumina Production. Beijing: metallurgical Industry Press, 1993
2. Sakamoto K., Kanehara M., Matsushita K. Agglomeration of Crystalline Particles of Gibbsite during the Precipitation in sodium Aluminate Solution. Light Metals,1976,2:149-162
3. Steemson M.L., White E.T., Mrashall R.J. Mathematical Model of Precipitation Section of a Bayer Plant. Light Metals,1984, 237-253
4. Sang J.V. Fines Digestion and Agglomeration at High Ratio in Bayer Precipitation. Light Metals,1989:33-39
5. Zhou Huifang. Study on Crystal Agglomeration Mechanism in Sodium Aluminate Solution. Non-Ferrous Metal, 1994,46(4), 54-57
6. Zhao Jihua, Chen Qiyuan. Advances in Research on Intensification of Precipitation Process from Aluminate solution. Light Metals,(2000),4:29-31
7. White E.T., Bateman S.H. Effect of Caustic Concentration on the Groth Rate of Al(OH)3 Particles. Light Metals, (1988) ,157-162
8. W.R. King. Some Studies in Alimina Trihydrate Precipitation Kinetics. Light Metals, (1973) ,551-563

EFFECTS OF MONOHYDROXY-ALCOHOL ADDITIVES ON THE SEEDED AGGLOMERATION OF SODIUM ALUMINATE LIQUORS

Jianguo Yin[1], Jie Li[1], Yanli Zhang[2], Qiyuan Chen[1], Zhoulan Yin[1]
[1]School of Chemistry and Chemical Engineering, Central South University, Changsha 410083, China
[2]Zhengzhou Research Institute of Chalco, Zhengzhou, Henan, 450041, China

Keywords: monohydroxy-alcohol additives, seeded agglomeration, sodium aluminate liquors, precipitation ratio, particle size distribution

Abstract

A series of monohydroxy-alcohol additives were adopted in Bayer process. The effects of dosage and carbon chain length on precipitation ratio of sodium aluminate liquors and particle size distribution (PSD) of gibbsite were investigated. Experimental results indicated that: all three monohydroxy-alcohols can increase precipitation ratio of sodium aluminate liquors at the proper dosage. They can accelerate precipitation process obviously at lower dosage at the beginning and yet last longer effects at higher dosage. Particle size of gibbsite products is enlarged when 1-Octadecanol is added at low dosage, yet the product fines under all other conditions. It can be concluded that: long chain additives with low dosage or short chain additives with high dosage is favorable for the seeded agglomeration of sodium aluminate liquors. The relationship between carbon chain length and its dosage may be complementary. Action mechanism among monohydroxy-alcohol additives, sodium aluminate liquors and gibbsite crystals will be further investigated based on the information mentioned above.

Introduction

Seeded precipitation of sodium aluminate liquors is an important and also a most time-consuming step in Bayer process for the production of alumina, which has a long precipitation period (45-72 hours) and low precipitation ratio (less than 55%) (1). All kinds of enhancing methods aiming at seeded precipitation process have been caused great concern, for example, optimizing technological parameters in seeded precipitation process, adopting activation crystal seeds, using additives, applying outer fields and so on (2). Seeking additives which can enhance precipitation ratio of sodium aluminate liquors and improve particle size and intensity of alumina products is becoming a focus, which is concerned by alumina industry in the world (3-15). Some countries have owned matured additive products, but those additives are designed for digestion liquors of gibbsite bauxite and they can't work well for digestion liquors of diaspore bauxite. Selection of additives in China is a little random and based basically on experiences. Great progress of enhancing seeded precipitation of sodium aluminate liquors by additives hasn't been made in China. So it is significant for us to explore action mechanisms among additive molecules, gibbsite crystals and supersaturated sodium aluminate liquors. It is also necessary to have a molecular design of additives based on mechanisms in order to enhance seeded precipitation process. A series of monohydroxy-alcohol additives were adopted in this paper and their effects of dosage and carbon chain length on precipitation ratio of sodium aluminate liquors and PSD of gibbsite products were investigated. The experimental results are helpful for further studies on action mechanisms among additive molecules, gibbsite crystals and supersaturated sodium aluminate liquors.

Experimental

Experimental methods

A certain volume and concentration of sodium aluminate liquors were added into precipitation tank which was preheated to a constant temperature, additives of precise dosage were added as the liquors reached a constant temperature. Crystal seeds were added and reaction time was recorded after 20 minutes' isotherm process. Samples were taken periodically from precipitation tank and centrifugated. Clear liquors were used to analyse the content of alumina and alkali, and the solid samples were washed for three times in hot distilled water for the PSD determination. Metallurgical industry standard of YB-817-75 was adopted for the analysis of chemical components, and particle size analyzer of Mastersizer 2000 was used for the PSD determination. Blank experiment was performed at the same time.

Experimental apparatus

Precipitation tank, with a volume to be 1L and valid volume 0.8 L, is a stainless steel vessel. The tank is set in an isothermal bath.

Liquors and seeds

The liquors were prepared from industrial alkali and aluminum hydroxide. Industrial sheet alkali was dissolved in half of the required volume of distilled water in a stainless steel vessel. Aluminum hydroxide from Pilot Alumina Plant of Zhengzhou Research Institute of Chalco was dissolved slowly in hot alkali liquors with agitation. Then the liquors were diluted to a certain volume with hot distilled water and filtered twice for experiments. Crystal seeds from Pilot Alumina Plant of Zhengzhou Research Institute of Chalco were ball milled, dried and then passed through sieve of 45 μm. Minus mesh seeds were mixed homogeneously for experiments.

Additives

1-Octanol, 1-Dodecanol and 1-Octadecanol of chemical grade were adopted in our experiments.

Conditions of seeded precipitation

Experiments were performed under the conditions of alkali (Na_2O_k) concentration 147g·l^{-1}, initial molecular ratio (α_k , Na_2O_k to Al_2O_3) 1.47, temperature 74.6℃, agitation rate 100r·min^{-1}, seeds mass ratio (*Ks*) 0.25, precipitation time 10 hours.

Results and discussion

Effect of monohydroxy-alcohols dosage on precipitation ratio

Effect of 1-Octanol dosage on the precipitation ratio is shown in Fig.1. It can be seen that precipitation ratio increase stably with time and it reaches 24.2% at the 10th hour for the blank. Addition of 1-Octanol obviously enhances precipitation ratio of sodium aluminate liquors and final precipitation ratio is increased 2.25% at the dosage of 7.40×10^{-4} mol/l compared with the blank. Precipitation ratio at the dosage of 1.85×10^{-4}mol/l is always the best in the former 6 hours. Yet precipitation ratio at the dosage of 7.40×10^{-4}mol/l is the highest in 6-10 hours. To 1-Octanol, dosage at 1.85×10^{-4}mol/l may be more favorable for the precipitation of sodium aluminate liquors, its initial precipitation ratio is larger. Although dosage at 7.40×10^{-4}mol/l is higher than the favorable one, it can act as a reservoir. Its residue dosage can still work although it is also consumed in the process. So it has a longer actuation duration than others.

Effect of 1-Dodecanol dosage on the precipitation ratio is shown in Fig.2, which shows that addition of 1-Dodecanol can also increase precipitation ratio at the selected dosages compared with the blank. Precipitation ratio at the dosage of 1.85×10^{-4}mol/l and 3.70×10^{-4}mol/l is higher in the initial 6 hours. Yet the difference of precipitation ratio is becoming less and less as experiments go on. For 1-Dodecano of high dosage, continuous improvement of precipitation ratio is appearing. Final precipitation ratios are 25.7%, 26.0% and 25.9% respectively when the dosages of additives are 1.85×10^{-4} mol/l, 3.70×10^{-4} mol/l and 7.40×10^{-4} mol/l at the end of 10th hour.

Effect of 1-Octadecanol dosage on the precipitation ratio is shown in Fig.3. It is shown that dosage of 1-Octadecanol has great influence on precipitation ratio. Compared with the blank, final precipitation ratios are increased by 2.8%, 2.0% and 1.0% respectively when the dosages are 7.40×10^{-4}mol/l, 1.85×10^{-4}mol/l and 3.70×10^{-4}mol/l.

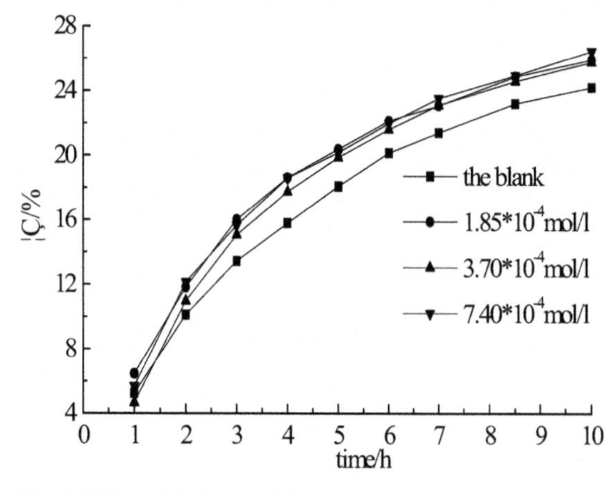

Fig.1 Effect of 1-Octanol dosage on the precipitation ratio

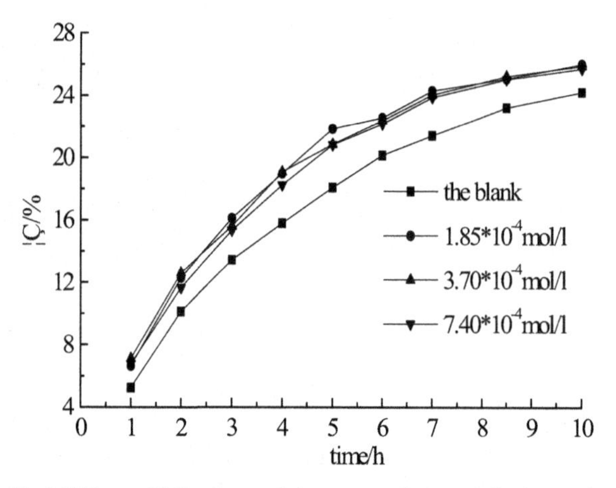

Fig.2 Effect of 1-Dodecanol dosage on the precipitation ratio

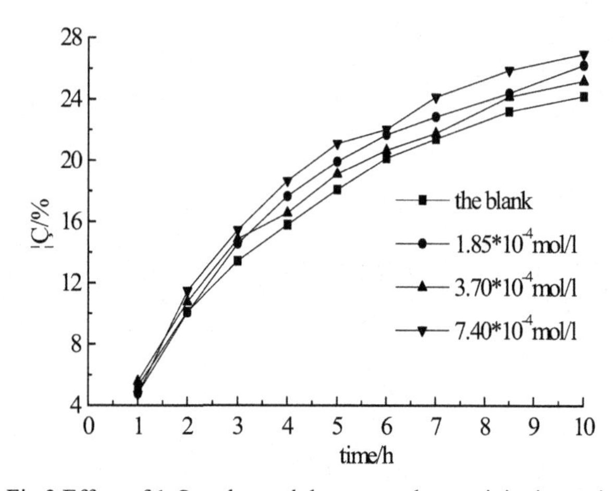

Fig.3 Effect of 1-Octadecanol dosage on the precipitation ratio

Effect of carbon chain length of monohydroxy-alcohols on precipitation ratio

Effects of carbon chain length of monohydroxy-alcohols at the dosage of 1.85×10^{-4}mol/l, 3.70×10^{-4}mol/l and 7.40×10^{-4}mol/l are shown in Fig.4-6. Alcoholic additives of short carbon chain improve precipitation ratio more in the initial 1-8.5 hours at the dosage of 1.85×10^{-4}mol/l. Their relative activity are 1-Dodecano >1-Octanol>1-Octadecanol. But the differences of precipitation ratio are becoming less and less in the precipitation process and the precipitation ratios are almost the same at the end of experiments. It is similar at the dosage of 3.70×10^{-4}mol/l. Additives of shorter carbon chain have better precipitation ratio at the initial 0-3 hours at the dosage of 7.40×10^{-4}mol/l. Yet additives of longer carbon chain have better continuous enhancement of precipitation ratio at the 8.5-10 hours. Final activation results are 1-Octadecanol>1-Dodecano>1-Octanol.

It can be concluded from the above experimental results that all three additives can improve precipitation ratio under the above conditions. Precipitation ratio of sodium aluminate liquors can be improved more at low dosage of additives at the beginning. Yet continuous improvement effects are displayed as experiments go on and final precipitation ratios are better for additives of larger dosage.

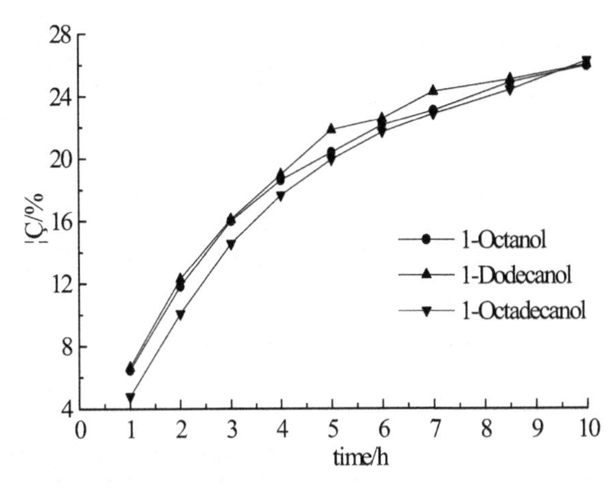

Fig.4 Effect of carbon chain length of monohydroxy-alcohols on the precipitation ratio at the dosage of 1.85×10^{-4}mol/l

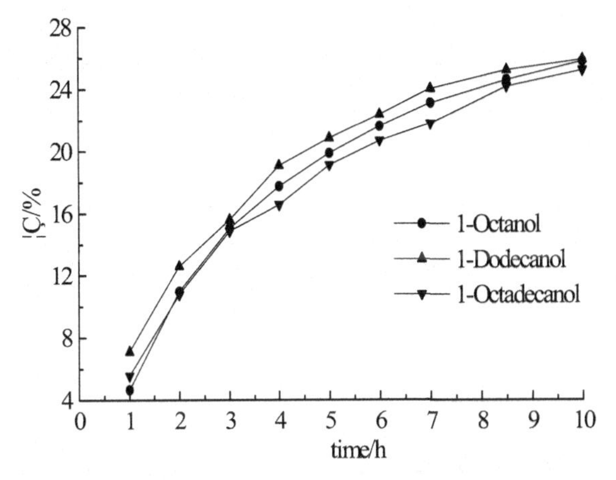

Fig.5 Effect of carbon chain length of monohydroxy-alcohols on the precipitation ratio at the dosage of 3.70×10^{-4}mol/l

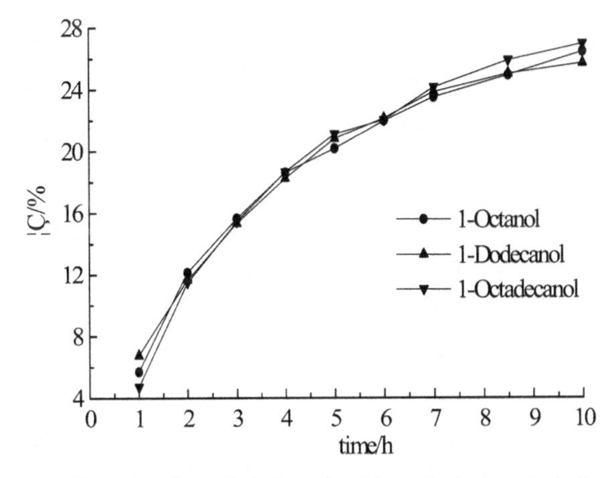

Fig.6 Effect of carbon chain length of monohydroxy-alcohols on the precipitation ratio at the dosage of 7.40×10^{-4}mol/l

To monohydroxy-alcohol additives, small dosage may be more favorable for the precipitation of sodium aluminate liquors, so its initial precipitation ratio is larger. Although large dosage is higher than the favorable dosage, it can act as a reservoir. Its residue dosage can still work although it is also consumed in the process. So it has a longer actuation duration than others.

Monohydroxy-alcohols of shorter carbon chain are better at the beginning. Yet the differences of precipitation ratio are diminishing, and additives with long chain have better continuous improvement of precipitation ratio as experiments go on.

Carbon chain belongs to hydrophobic group and its length has relationship with dissolvability in sodium aluminate liquors. So carbon chain length may has the same effect as the dosage.

Effect of monohydroxy-alcohols' dosage on PSD of gibbsite products

Effect of 1-Octanol's dosage on PSD of gibbsite products is shown in table 1. It can be seen that particles less than 20 μm decrease obviously and particles over 20 μm increase in the gibbsite products compared with crystal seeds, which demonstrates that particles less than 20 μm are agglomerated in the precipitation process. It is similar when using other additives. All these experimental results illustrated that our experimental conditions are favorable for the agglomeration of little particles. Addition of 1-Octanol increases particles less than 30 μm and makes gibbsite products fine.

Table 1 Effect of 1-Octanol's dosage on PSD of gibbsite products (volume percentage (VP)/%)

1-Octanol dosage /mol.l^{-1}	PSD/%					
	0~ 10μm	10~ 20μm	20~ 30 μm	30~ 45μm	+45 μm	d_{50} μm
Seeds	26.27	25.23	21.06	18.26	9.18	19.40
0	2.34	17.25	26.92	30.57	22.92	31.39
1.85×10^{-4}	2.84	18.64	27.39	29.91	21.22	30.45
3.70×10^{-4}	3.04	19.33	27.45	29.53	20.65	30.08
7.40×10^{-4}	2.55	17.63	27.14	30.42	22.26	31.06

Effect of 1-Dodecanol's dosage on PSD of gibbsite products is listed in table 2. It can be learned that products fines with the addition of 1-Dodecanol compared with the blank. Particles less than 10 μm and particles of 20-45 μm increase, yet particles of 10-20 μm decrease with the addition of 1-Dodecanol. Particles less than 30 μm in gibbsite products increase more with the further addition of 1-Dodecanol. Particles less than 30 μm relatively decrease at the dosage of 7.40×10^{-4} mol.l^{-1}. Refinement of products is not obvious at the dosage of 1.85×10^{-4} mol.l^{-1}.

Effect of 1-Octadecanol's dosage on PSD of gibbsite products is shown in table 3. It can be seen that 1-Octadecanol's dosage has great effect on PSD of gibbsite products. Particles less than 30 μm in the products decrease with the addition of 1-Octadecanol, then it increase with the further addition of additives. Particles of 0-10 μm and 20-45 μm increase yet particles of 10-20 μm decrease at the dosage of 7.40×10^{-4} mol.l^{-1}. Particle size of products are enlarged by decreasing fine particles and adding +30 μm particles at the dosage of 1.85×10^{-4} mol.l^{-1}. Products become fine at the dosage of 3.70×10^{-4} mol.l^{-1} and 7.40×10^{-4} mol.l^{-1}.

Table 2 Effect of 1-Dodecanol's dosage on PSD of gibbsite products (VP/%)

1-Dodecanol Dosage /mol.l^{-1}	PSD/%					
	0~ 10μm	10~ 20μm	20~ 30μm	30~ 45μm	+45 μm	d_{50} μm
0	2. 34	17. 25	26. 92	30. 57	22. 92	31.39
1.85×10^{-4}	2. 38	17. 22	27. 10	30. 67	21. 63	31.30
3.70×10^{-4}	2. 59	18. 41	27. 61	30. 21	21. 18	30.55
7.40×10^{-4}	2. 58	17. 90	27. 39	30. 40	21. 73	30.84

Table 3 Effect of 1-Octadecanol's dosage on PSD of gibbsite products (VP/%)

1-Octadecanol dosage /mol.l^{-1}	PSD/%					
	0~ 1μm	10~ 20μm	20~ 30μm	30~ 45μm	+45 μm	d_{50} μm
0	2. 34	17. 25	26. 92	30. 57	22. 92	31.39
1.85×10^{-4}	1. 32	13. 60	25. 81	32. 16	27. 11	33.70
3.70×10^{-4}	2. 88	18. 62	27. 37	29. 90	21. 23	30.45
7.40×10^{-4}	2. 97	18. 51	27. 38	29. 92	21. 22	30.45

Effect of carbon chain length of monohydroxy-alcohols on PSD of gibbsite products

Effect of carbon chain length of monohydroxy-alcohols on PSD of gibbsite products at the dosage of 1.85×10^{-4} mol.l^{-1} is shown in table 4. It can be seen that particles less than 30μm decrease gradually, yet particles larger than 30μm and d_{50} gradually increase with the additives' carbon chain. It demonstrates that particles less than 30μm are agglomerated easily and particle size of products are enlarged by using additives of longer carbon chain at the dosage of 1.85×10^{-4} mol.l^{-1}.

Effect of alcoholic species on PSD of gibbsite products at the dosage of 3.70×10^{-4} mol.l^{-1} is shown in table 5. It shows that average particle size of products are all lower than the blank with the addition of additives. Yet 1-Dodecanol slows down the diminution of products by reducing particles of less than 20μm. Capacity of enlarging particle size are 1-Dodecanol >1-Octadecanol >1-Octanol.

Effect of alcoholic species on PSD of gibbsite products at the dosage of 7.40×10^{-4} mol.l^{-1} is shown in table 6. It can be seen that, with the increase of additives' carbon chain particles less than 30 μm increase and products fine. Capacity of enlarging particle size are 1-Octanol >1-Dodecanol > 1-Octadecanol.

It can be concluded from the above experimental results that monohydroxy-alcohols easily make products fine at relatively high dosage. Long chain additives with low dosage, median chain additives with median dosage and short chain additives with high dosage are favorable to enlarge particle size of gibbsite products.

Table 4 Effect of carbon chain length of monohydroxy-alcohols on PSD of gibbsite products at the dosage of 1.85×10^{-4} mol.l^{-1} (VP/%)

Additive Species	PSD/%					
	0~ 10μm	10~ 20μm	20~ 30μm	30~ 45μm	+45 μm	d_{50} μm
1-Octanol	2.84	18.64	27.39	29.91	21.22	30.45
1-Dodecanol	2.38	17.22	27.10	30.67	21.63	31.30
1-Octadecanol	1.32	13.60	25.81	32.16	27.11	33.70

Table 5 Effect of carbon chain length of monohydroxy-alcohols on PSD of gibbsite products at the dosage of 3.70×10^{-4} mol.l^{-1} (VP/%)

Additive Species	PSD/%					
	0~ 10μm	10~ 20μm	20~ 30μm	30~ 45μm	+45 μm	d_{50} μm
1-Octanol	3.04	19.33	27.45	29.53	20.65	30.08
1-Dodecanol	2.59	18.41	27.61	30.21	21.18	30.55
1-Octadecanol	2.88	18.62	27.37	29.90	21.23	30.45

Table 6 Effect of carbon chain length of monohydroxy-alcohols on PSD of gibbsite products at the dosage of 7.40×10^{-4} mol.l^{-1} (VP/%)

Additive Species	PSD/%					
	0~ 10μm	10~ 20μm	20~ 30μm	30~ 45μm	+45 μm	d_{50} μm
1-Octanol	2.55	17.63	27.14	30.42	22.26	31.06
1-Dodecanol	2.58	17.90	27.39	30.40	21.73	30.84
1-Octadecanol	2.97	18.51	27.38	29.92	21.22	30.45

Conclusion

Effects of dosage and carbon chain length on precipitation ratio and PSD of gibbsite products for three monohydroxy-alcohol additives were investigated under the conditions of alkali (Na_2O_k) dosage 147g·l^{-1}, initial molecular ratio (α_k, Na_2O_k to Al_2O_3) of 1.47, temperature 74.6 □, agitation ratio 100r·min^{-1}, seeds mass ratio (*Ks*) 0.25, precipitation time 10 hours. Based on experimental results, conclusions can be drawn as follows:

1 All three monohydroxy-alcohols can increase the precipitation ratio of sodium aluminate liquors at the adopted dosages. They can raise precipitation ratio larger at lower dosage at the beginning , yet last longer at higher dosage.

2 Monohydroxy-alcohols with short carbon chain raise precipitation ratio larger at beginning stage of experiments yet monohydroxy-alcohols of long carbon chain have better continuous improvement effect at the same dosage.

3 Particle size of gibbsite products is enlarged when 1-Octadecanol is added at low dosage, yet products are all fined under the other conditions.

4 Long chain additives with low dosage or short chain additives with high dosage is favorable for the precipitation of sodium aluminate liquors. The relationship between carbon chain length and its dosage may be complementary.

Acknowledgement

This work is supported by National Natural Science Fund of China projects "Molecular design of additives for the seeded precipitation of sodium aluminate liquors during the alumina production" (project number:20476107) and National Priority Development Project Fundamental Research in China ('973' project) "fundamental research on extracting alumina from concentrated bauxite with high efficiency" (project number : 2005CB623702)

References

1. Xie Yanli, Bi Shiwen, and Ren Wencai. "Application of additives in precipitation of Bayer sodium aluminate liquors," Light Metals of China, 1(2000), 25-26, 49.

2. Chen Guohui et al, "Development of gibbsite consolidation precipitation from caustic aluminate liquors with seeds," Hunan Metallurgy, 31(1)(2003), 3-6.

3. Xue Hong et al, "Enhancing precipitation of sodium aluminate liquors in the Bayer process by additives," The Chinese Journal of Nonferrous Metals, 8(2)(1998), 415-417.

4. Chen Feng et al, "Effect of anioncs-oily additive on seed precipitation from sodium aluminate liquors," Journal of Northeastern University (Natural Science), 25(6)(2004), 606-609.

5. Zhao Su et al, "Effect of additives on particle size and strength of aluminum hydroxide decomposed with seed crystal used," Journal of Northeastern University (Natural Science), 24(10)(2003), 939-941.

6. Zhao Su et al, "Effect of anionic surfactant on decomposition of sodium aluminate liquors," Journal of Northeastern University (Natural Science), 25(2)(2004), 139-141.

7. Li Haipu et al, "Function of sodium polyacrylate in seeded precipitation of sodium aluminate liquors," J. Cent. South. Univ.(Science and technology), 35(1)(2004), 138-142.

8. Yang Zhenghui, Chen Wenmi, and Yang Zhengbao, "Study on the enhancing seed precipitation process of sodium aluminate liquors by crystallization additive," Hunan Metallurgy, 31(1)(2003), 16-20.

9. Chen Wenmi et al, "Study on enhancing seed precipitation process of sodium aluminate liquors by crystallization additive," Hunan Nonferrous Metals, 18(6)(2002), 23-26.

10. Zhang Bin, Chen Guohui, and Chen Qiyuan, "Study on organic surfactants improving the aluminum hydratio seeds' particles size precipitated from sodium aluminate liquors," Nonferrous Metals (Metallurgy), 5 (2005), 28-31.

11. I. Seyssiecq et al, "The influence of additives on the crystal habit of gibbsite," Journal of Crystal Growth, 196(1999), 174-180.

12. A. M. Paulaime et al, "The influence of organic additives on the crystallization and agglomeration of gibbsite," Powder Technology, 130(2003), 345-351.

13. P. G. Smith et al, "The effects of model organic compounds on gibbsite crystallization from alkaline aluminate liquors: polyols," Colloids and Surfaces A: physico-chemical and Engineering Aspects, 111(1996), 119-130.

14. H. Watling, J. Loh, anf H. Gatter, "Gibbsite crystallization inhibition 1. effects of sodium gluconate on nucleation, agglomeration and growth," Hydrometallurgy, 55(2000), 275-288.

15. A. R. Hind, S. K. Bhargava, and S. C. Grocott, "The surface chemistry of Bayer process solids: a review," Collides and Surfaces A: Physicochemical and Engineering Aspects, 146 (1999), 359-374.

Light Metals 2006 *Edited by Travis J. Galloway* ***TMS (The Minerals, Metals & Materials Society), 2006***

STUDY ON THE EFFECT OF K_2O ON SEED PRECIPITATION IN SODIUM ALUMINATE LIQUORS

Yanli Xie[1,2], Qun Zhao[3], Zhenan Jin[2], Zijian Lu[2], , Shiwen Bi[2]
[1]School of Materials Science and Engineering, Georgia Institute of technology
771 Ferst Dr, Atlanta, GA30332-0245

[2]School of Materials and Metallurgy, Northeastern University,
No. 11, Lane 3, Wenhua Road, Heping District, Shenyang, Liaoning, 110004

[3]Zhengzhou Research Institute of CHALCO (Aluminum Corporation of China Limited)
No.82 Jiyuan Road, Shangjie, Zhengzhou, Henan, 450041, China

Keywords: precipitation, potassium aluminate solution, aluminum hydroxide

Abstract

The potassium oxide accumulates during the liquor recycling and its concentration can up to 55g/l in pregnant liquor with the total soda 178.4g/l, which bring negative effect on the seed precipitation procedure. This paper studied the effect of K_2O on precipitation yield and hydrate particle size distribution when the concentration of K_2O is on the range of 0-60g/l from 160g/l caustic soda. The results demonstrated that the potassium contaminated sodium aluminate liquor has higher precipitating ratio, but its effect on particle size distribution is complicated, which varied with time and potassium concentration. The mechanism on why potassium affects precipitation procedure is also studied in this paper.

Introduction

The 90% alumina in this world is produced by Bayer process, and precipitation is the slowest but most important procedure in Bayer recycling, which determines the quality of product alumina and the liquid yield. As a result, ample researches on precipitation in sodium aluminate liquors have been done in the past several decades. However, little research on precipitation from potassium aluminate liquor, especially from potassium contaminated sodium aluminate solution has ever been done. The interests in effect of potassium on precipitation increase recently due to the necessity to produce sandy alumina.

The alkali ion Na^+ and K^+ has similar electric structure and property, but their effects on the precipitation are distinct [1]. The sodium is consumable during the process and need adding certain amount of caustic to compensate for the lost to guarantee the success of next Bayer recycle. However, the potassium accumulates in the procedure. The amount of K_2O impurity from bauxite containing illite is only ~0.001% during each Bayer recycling, but it accumulate in Bayer solution and the content can exceed 55g/l in a certain Chinese refinery [2].

The potassium in pregnant liquors can change the viscosity, density, rheological properties of aluminate solution, and the interaction between gibbsite and solution, hence affecting the precipitation procedure. Typically the potassium in aluminate solutions decreases the particle size of aluminum hydrate product, elongates the gibbsite crystal and results in fine particles with weak strength. However, the results on whether potassium influences precipitation ratio are contradict. Some researchers find the K_2O in aluminate liquors decrease the liquid yield, but some others find it increases the precipitation ratio or has no effect because they use different precipitation conditions.

Experimental

Precipitation experiment

The batch precipitation experiment is done in eight stainless thickeners with capacity of 100ml, which are put in an isothermal oil bath controlled by an automatic temperature controller.

The sodium aluminate liquors with different concentration of potassium oxide were prepared by dissolving gibbsite to sodium hydroxide and potassium hydroxide. The precipitation condition is as follows:

Temperature system: constant at 60°C;
Initial solid content: 550 g/l, and the seed comes from plant but is washed by hot distilled water and then dried at 110°C for 2 hours;
Precipitation time: 24 hours;
Solution condition: molecular ratio of caustic concentration to alumina concentration is 1.55, caustic concentration is 160g/l.

The concentrations of alkali and alumina are determined by titration, while the solid sample is washed by hot water and dried in oven for particle size analysis and SEM.

Particle size analysis

The particle size distribution of aluminum hydroxide is analyzed by LS 13 320 laser diffraction particle size analyzer.

The SEM photos are taken by Hitachi S-800.

Effect of potassium on precipitation ratio

The effect of potassium on precipitation ratio varied with different decomposition condition, such as caustic concentration, A/C ratio and temperature. In this experiment, the precipitation was done at 60°C with caustic molecular ratio of 1.55 and caustic concentration of 160g/l. The plot of precipitation ratio to time at different potassium concentration (0g/l, 15g/l, 30g/l, 40g/l, 50g/l and 60g/l) was shown in Figure 1. It illustrated the precipitation ratio increased with the rising of potassium concentration, but the difference decreased when potassium content higher than 40g/l. The low potassium concentration (15g/l) had little effect on the precipitation rate.

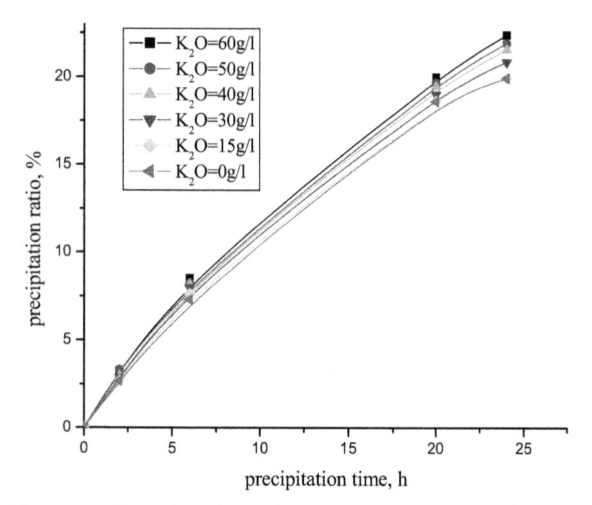

Figure-1 Effect of K_2O concentration on precipitation ratio

The reason why potassium can enhance precipitation was the K_2O in aluminate liquors changed the precipitation mechanism and nucleation predominant the whole process. As a result, there were much more fine particles with bigger surface area and more active, hence increasing the precipitation rate.

Effect of potassium on particle size distribution of gibbsite

The potassium in aluminate solution not only resulted in strong nucleation, but inhibited agglomeration between fine particles [3, 4, 5]. As a result, the gibbsite produced from potassium contaminated sodium aluminate solutions always has smaller size than the one from pure sodium aluminate liquors.

The particle size distribution of aluminum hydroxide obtained from pure sodium aluminate liquors and from solution contained 30g/l, 60g/l K_2O impurity was illustrated in figure 2 ~ figure 4 when precipitate for 2h, 6h and 24h respectively.

Figure 2 (precipitate 2h) showed that the particle size distribution (PSD) curves for three samples were almost overlapped at fine particle size region, while the height of PSD peaks decreased in the sequence from 0g/l to 60g/l and the peak position also shifted to left a little at same sequence. It illustrated that particle size of alumina hydrate from solution with higher potassium content was smaller than the one with lower content, but the difference wasn't very big.

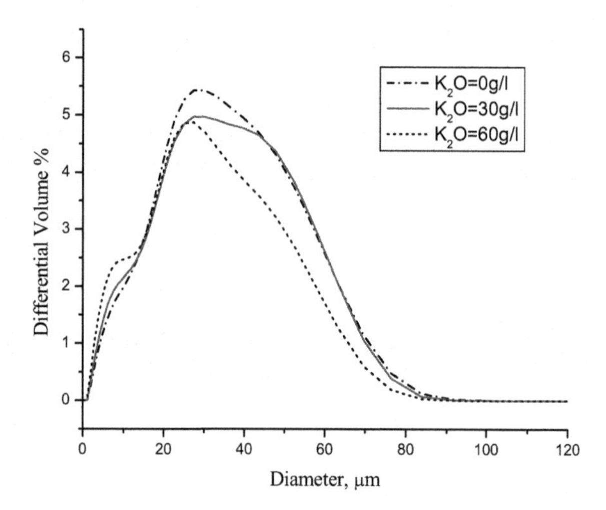

Figure-2 Particle size distribution of alumina hydrate after precipitating for 2hrs

However, the peaks of particles from potassium contaminated sodium aluminate liquors shifted left significantly to small particle size when precipitate for 6h, as shown in figure 3. The higher the concentration of K_2O, the more the peak position shifted to small size, which means the particle size decreased with increasing of potassium content. The volume percent of fine particles in product increased dramatically with rising of potassium concentration and precipitation time.

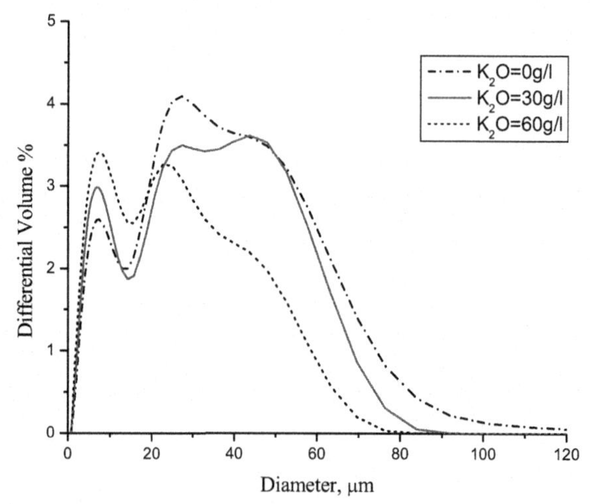

Figure-3 Particle size distribution of alumina hydrate after precipitating for 6hrs

The differences among the PSD curves become larger with precipitation processing as shown in figure 4. The PSD shape also became different after precipitating for 24h. There are two peaks for particles from pure sodium aluminate solutions (K_2O=0g/l) and from solution with relatively lower potassium content (K_2O=30g/l), but the peak height of K_2O=30g/l is higher than the one of blank at fine particle region, while its height lower than the blank at big particle size range. Moreover, the PSD curve for the particles from aluminate solution with higher potassium concentration (K_2O=60g/l) had only one peak at small size region and the height of this peak is the highest, that is, the fine particles dominate the hydrate products.

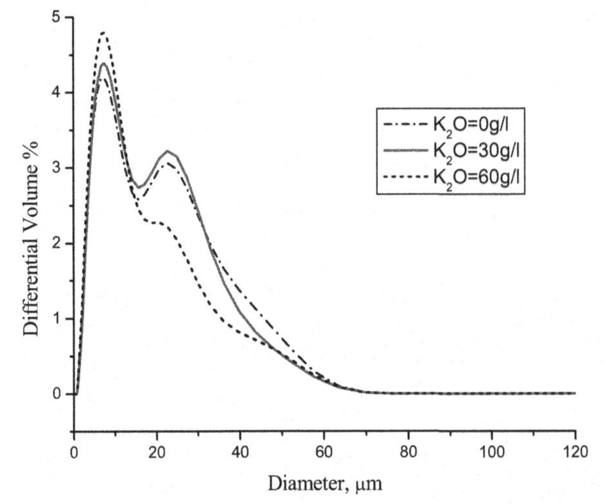

Figure-4 Particle size distribution of alumina hydrate after precipitating for 24hrs

Generally speaking, the particle size of hydrate prepared from potassium contaminated sodium aluminate liquors was smaller than the one from pure sodium-based solutions.

The potassium in solutions inhibited the agglomeration of gibbsite because the repulsion between particles in sodium aluminate solutions decreased faster than in potassium-based solutions upon aging, thus resulting in the higher interparticle cohesion and growth rate in sodium solutions, hence leading to the agglomeration in sodium liquors [3]. However, the strong repulsion between particles in potassium solutions prohibited the adhesion of gibbsite, which made agglomeration difficult in potassium solutions.

Effect of potassium on morphology of gibbsite

The K_2O in sodium aluminate liquors significantly influenced the morphology of gibbsite. It elongated the crystal of gibbsite and prohibited the agglomeration between particles, as shown from figure 5 to figure 10, which were SEM photos of hydrate seed and gibbsite prepared from sodium aluminate solutions with different concentration of K2O: 0g/l, 15g/l, 50g/l and 60g/l respectively.

Figure 5 showed the typical morphology of gibbsite prepared from industrial sodium aluminate solutions, which was spherical polycrystalline composed of many hexagonal plates or blocks, while the pores between big particles was inlet by fine hydrates.

There were almost no variations on the shape of gibbsite (hexagonal crystal) except the edges were smoothed by strong agitation after precipitating in pure sodium aluminate solutions for 24 hours (shown in Figure 6). Some nuclei were observed on the tilt plane or on the boundary between particles. The relative supersaturation of sodium aluminate liquors we used is ~0.6, the crystal should growth according to mixture of step growth and nucleation and spread mechanism [5]. However, the strong agitation facilitate the diffusion of Al $(OH)_4^-$ and OH^- ions, hence enhancing the nucleation and resulting more nuclei.

However, the phenomena were different for gibbsite prepared from potassium contaminated solutions. The potassium in sodium aluminate solutions changed the crystal growth mechanism, leading to prismatic crystal (figure 7-figure 10). The percentage of crystal being elongated was proportional to the concentration of potassium in sodium aluminate liquors.

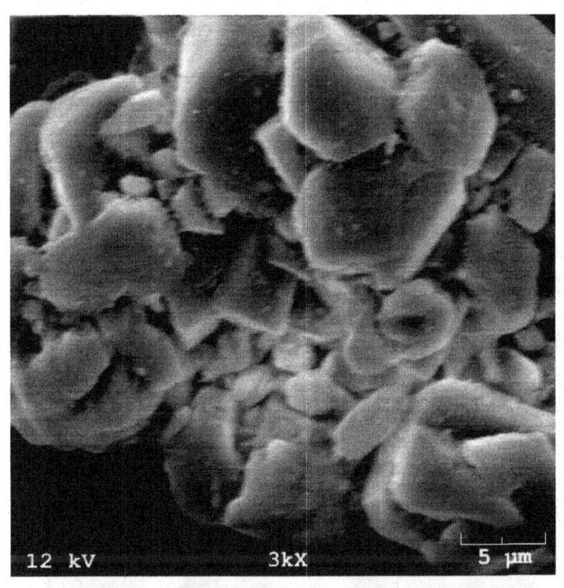

Figure-5 SEM photo of alumina hydrate seed

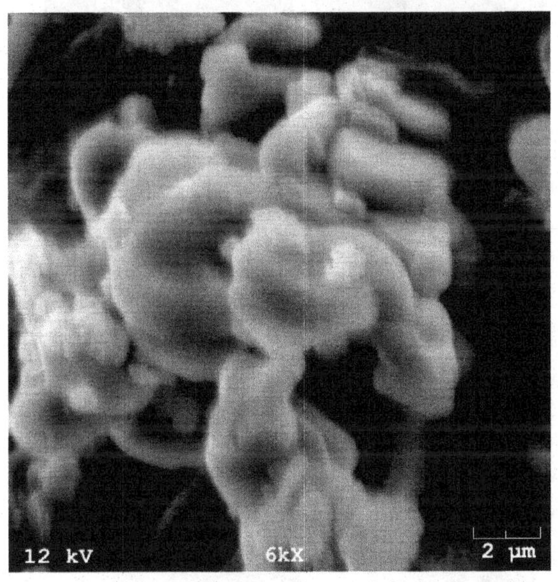

Figure-6 SEM photo of gibbsite after precipitating for 24hrs with K_2O=0g/l in sodium aluminate liquor

Only a few elongated crystals were observed (figure 7) in the product hydrate prepared from solutions with lower potassium content (15g/l). The morphology of gibbsite was analogical to the one from sodium solutions, a polycrystalline with round edge.

Unfortunately, the structure was destroyed at higher concentration of potassium ($K_2O \geq 50g/l$). The morphology of gibbsite was still polycrystalline at potassium content of 50g/l, but the crystals were needle-like which extruded out and formed granules with rough surface (figure 8). It worsened when increasing potassium content to 60g/l. All of the crystals are prismatic which separated from each other (figure 9)-agglomeration was difficult at high potassium concentration, and even original aggregates were destroyed by K^+.

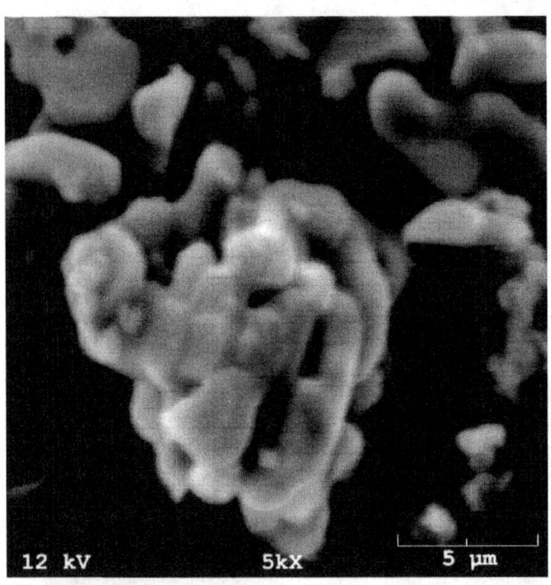

Figure-7 SEM photo of gibbsite after precipitate for 24hrs with K_2O=15g/l in sodium aluminate liquor

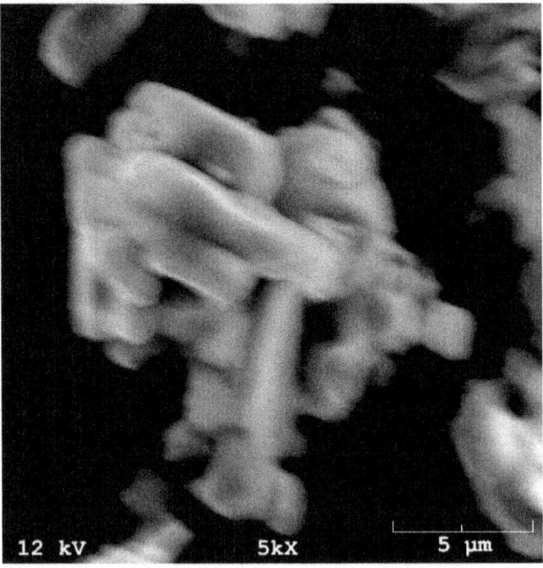

Figure-8 SEM photo of gibbsite after precipitate for 24hrs with K_2O=50g/l in sodium aluminate liquor

The erosion of K^+ on gibbsite agglomerates was processed gradually with time. Figure 10 gave the information about gibbsite prepared from sodium solutions with 60g/l K_2O after precipitating for 6h. The morphology was similar to seed, polycrystalline composed of hexagonal crystals, which hadn't been affected by K^+. However, the structure was destroyed gradually with time, and finally formed separated needle-like single crystals.

The difference in morphology of gibbsite from pure sodium-based solutions and from potassium contaminated solution was due to the distinction of species and strength of ion-pairing in the solutions [7, 8]. As mentioned above, nucleation and spread was the dominant mechanism under our precipitate condition. However, the small nuclei appeared in sodium solutions was hexagonal, while chains of elongated needles along the prismatic faces predominant the nucleation from potassium contaminated aluminate solutions.

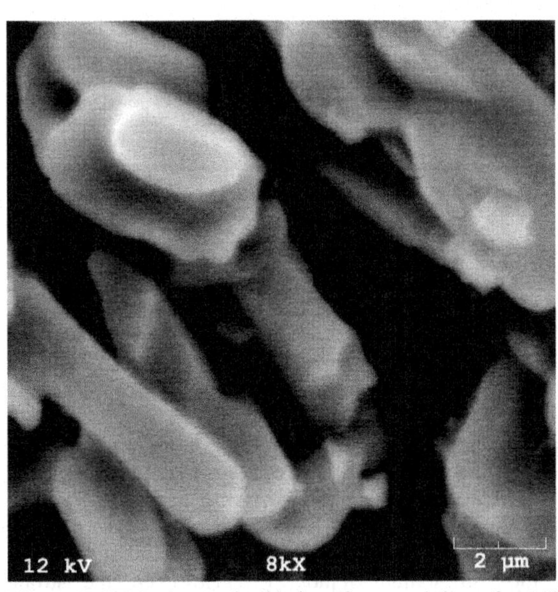

Figure-9 SEM photo of gibbsite after precipitate for 24hrs with K_2O=60g/l in sodium aluminate liquor

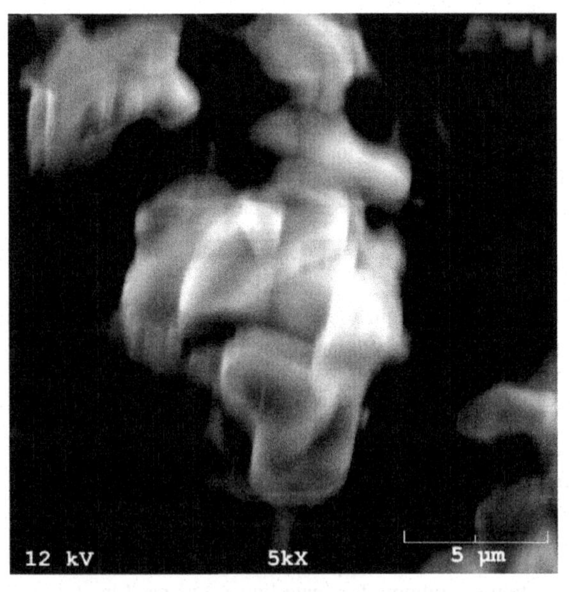

Figure-10 SEM photo of gibbsite after precipitate for 6hrs with K_2O=60g/l in sodium aluminate liquor

Conclusion

The potassium in sodium aluminate solutions had significantly affected precipitation procedure and changed the precipitation mechanism.

1) K^+ had little effect on the morphology of gibbsite at low concentration (15g/l), but worsened quickly with increasing of K^+ content. The crystals were elongated from hexagonal shape to needle shape, and finally formed separated prismatic single crystal of gibbsite at high potassium content of ~60g/l.
2) Potassium prohibited agglomeration between hydrate particles, and even destroyed original structure of gibbsite, thus resulting in fine particles.
3) The potassium in sodium aluminate solutions enhanced precipitation procedure. The precipitation ratio increased with rising of K^+ concentration.

Acknowledgements

The Sponsor for this program is Nature Science Funds of China, "Mechanism Research of Negative Effect of K_2O on Precipitation in Sodium Aluminate Solution" No.50304006.

Reference

1. J.S.L. Loh, H.R. Walting and G.M. Parkinson, "Alkali cations-role and effect in gibbsite crystallization", *Light Metals* 2002, 127-133

2. Z. Yin, S. Gu and H. Huang, "The influence of K_2O in spent liquor on bayer process of diasporic bauxite", *Light Metals* 2003, 173~176.

3. H.R. Watling, S.D. Fleming, W. van Bronswijk and A.L. Rohl, "Ionic structure in caustic aluminate solutions and the precipitation of gibbsite", *J. Chem. Soc. Dalton*, 1998, 3911-3917.

4. J. Addai-Mensah, C.A. Presidge and J. Ralston, "Interparticle forces, interfacial structure development and agglomeration of gibbsite particles in synthetic Bayer liquors", *Minerals. Eng.*, 1999, 12 (6), 655-

5. J. Addai-Mensah, C.A Prestidge, I. Ametov and J. Ralston, "Interactions between gibbsite crystals in supersaturated caustic aluminate solutions", *Light Metals* 1998, 159-

6. J.F. Sawsan, G.M. Parkinson and M.M. Reyhani, "Atomic force microscopy study of the growth mechanism of gibbsite crystal", *Phys. Chem. Chem. Phys.,* 2004 (6), 1049-1055.

7. J.S.C. Loh, H.R. Walting and G.M. Parkinson, "The effects of sodium, potassium and cesium on gibbsite precipitation from Bayor solutions", *International Symposium on Industrial Crystallization*, 14th, Cambridge, UK, Sept. 12-16, 1999, 706-716.

8. J.S.C. Loh, A.M. Fogg, H.R. Walting, G.M. Parkinson and D.O. Hare, "A kinetic investigation of gibbsite precipitation using in situ time resolved energy dispersive X-ray diffraction", *Phys. Chem. Chem. Phys.,* 2002 (2), 3597-3604.

Light Metals 2006 *Edited by Travis J. Galloway* ***TMS (The Minerals, Metals & Materials Society), 2006***

Influence of ions on the aggregation of Hydroxide Aluminum Seed

Lijuan Qi [1, 2] Songqing Gu[2] Jianli Wang [2,3] Qingwei Wang[2]
[1]Zhengzhou University Zhengzhou City, Henan Province,China 450001
[2]Zhengzhou Research Institute of Aluminum Corporation,China
No.82 Jiyuan Rd., Shangjie District, Zhengzhou, Henan, China 450041
[3]Central South University Changsha City,Hunan Province,China 410006

Key words: Hydroxide Aluminum Aggregation Zeta Potential impurities

Abstract

In the process of seeded precipitation to produce hydroxide aluminum, the aggregation of fine seed plays an important role for the production of hydroxide aluminum with large size. It is generally thought that the process of aggregation between particles is divided into three steps: firstly collision, secondly interaction, and thirdly cohesion. Among these three steps interaction is very important, which effective collision depends on. Interaction between fine seeds mainly lies in the static force, which can be measured by Zeta Potential on the surface of seed. In this paper investigated thoroughly are influences of ions in the liquor on the Zeta potential of particles surface, and concluded are the effects of them on the size distribution of hydroxide aluminum.

Introduction

In the process of seeded precipitation to produce hydroxide aluminum, secondary nuclear, aggregation and growth of seeds exist simultaneously, among which aggregation are the main factors to produce hydroxide aluminum with large size.

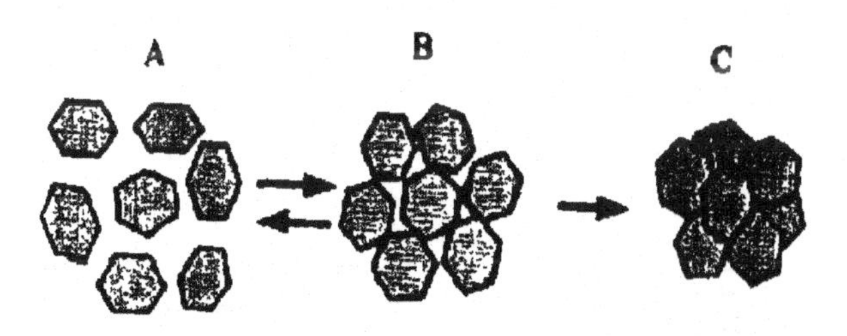

Figure1 The Process of Aggregation Between Particles of Hydroxide Aluminum

Aggregation can be divided into three steps: collision, interaction, and cohesion[1].Figure 1 shows this process. Among these three steps interaction is very important, which effective collision depends on. Interaction between fine seeds mainly lies in the static force, which can be shown through Zeta potential on the surface of seeds. Figure2 shows the double static layer on the surface of hydroxide aluminum with negative charges. The higher the absolute Value of Zeta potential on the surface of particles, more repulsion force between them, and more difficult cohesion. From figure 3 we can see this point.

There are so many factors which affect the zeta potential of hydroxide aluminum, such as surfacial structure, external field and internal field and so on[2]. Impurities belong to one of the factors among the internal field.

In the sodium aluminate liquor, there are some ions (cation or anion) adsorbed on the particles surface, which affects the double static layer of particles surface

In this study, the Zeta potential on the surface of seeds with existing ions of SO_4^{2-}, CO_3^{2-},$C_2O_4^{2-}$, $Cl^-$and k^+ in the liquor have been measured, and the influences of them on the size distribution of hydroxide aluminum have been explained.

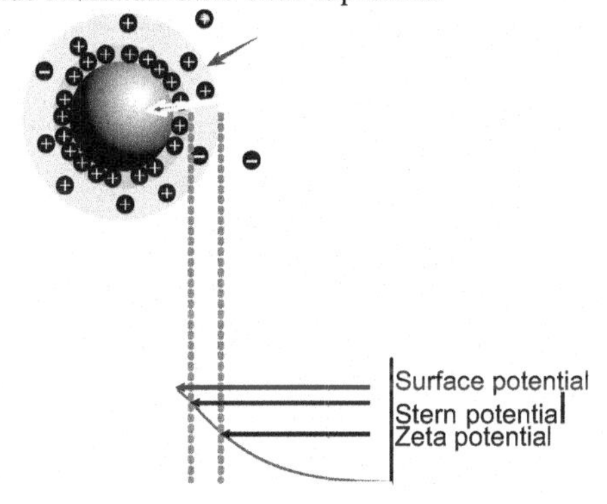

Figure2 Hydroxide aluminum with negative charges

Experiments

Hydroxide aluminum used in the test is from an Aluminary plant of Chalco. The average diameter of it is about 1 micron meters. The figure of its SEM (Scanning Electric Micrograph) is shown in Figure 4. The chemical composition of it sees table 1.

Zeta potential of particles surfaces is measured on the Zeta potential detector——Nano ZS90, manufactured by Malvon in England. It is based on the Henry equation:

$$U_E = 2\, \varepsilon\, z\, f(k\, a)/3\eta \qquad (1)$$

In the formula, U_E stands for electrophoretic mobility, and Z means Zeta potential, and ε is dielectric constant, and η. is viscosity, and f(ka) is Henrys function.

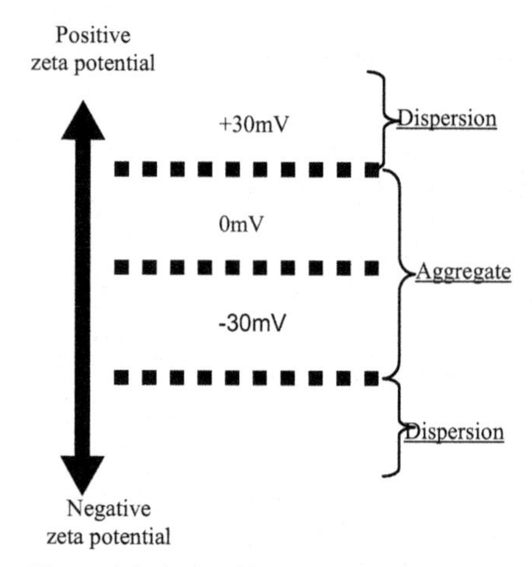

Figure 3 Relationship Between Zeta Potential and Stability of Hydroxide Aluminum

Table 1 Chemical Composition of Hydroxide Aluminum

SiO_2 (%)	Fe_2O_3 (%)	Al_2O_3 (%)	CaO (%)	MgO (%)	TiO_2 (%)	LOI (%)
0.037	0.0057	63.80	0.017	0.027	0.023	34.6

Figure4 SEM of Hydroxide Aluminum

Results

Influences of Cation and Anion ions on Zeta Potential of Hydroxide Aluminum in pure water

In the following figure, it can be seen that most of anions in the liquor have effect on the Zeta potential value of hydroxide aluminum. In pure water, due to adsorption of hydrogen ion, hydroxide aluminum particles bear positive charges. When the above impurities added; some anions enter the double static layer and reduced the layer thickness, and even change the bearing charges, enabling particles with negative charges. It shows that the above impurities influence the Zeta potential differently, their consequences are as following:

$OH^- > SiO_3^{2-} > CO_3^{2-} > C_2O_4^{2-} > SO_4^{2-} > Cl^-$

The results have been obtained when all impurities used are at the same level of ion concentration, that is 0.1mol/l. under this condition, It is obvious that Compared with OH^-、SiO_3^{2-}、CO_3^{2-}、$C_2O_4^{2-}$ and SO_4^{2-}, Cl^- ion is more difficultly affecting the Zeta potential, while OH^-、SiO_3^{2-}、CO_3^{2-}、$C_2O_4^{2-}$ and SO_4^{2-} influence greatly on the Zeta Potential of hydroxide aluminum.

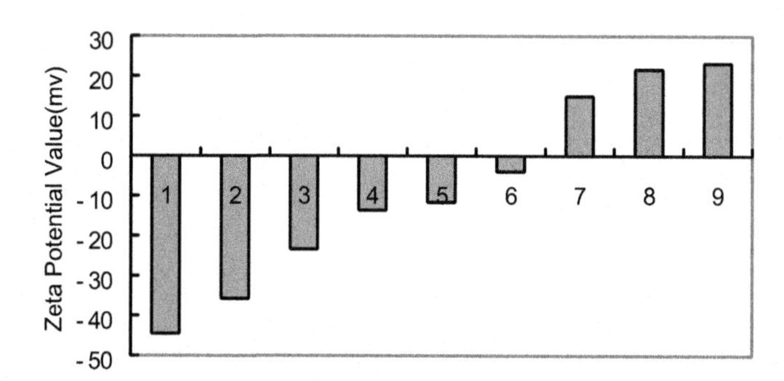

Figure 5 Influences of Impurities on Zeta Potential Value of Hydroxide Aluminum

1-NaOH 2-Na_2SiO_3 3-Na_2CO_3 4-$Na_2C_2O_4$ 5-Na_2SO_4
6-NaCl 7-KCl 8-H2O 9-$CaCl_2$

For cation ions, Ca^{2+}, K^+, and Na^+, they all try to enter the double static layer and have the tendency of making hydroxide aluminum particles with positive charges. However they have very different ability to affect the zeta potential value. Among the three cation ions, Ca^{2+} has the strongest ability, and K^+ follows, and Na^+ ranks the last in the list.

Influences of Caustic Ratio on Zeta Potential of Hydroxide Aluminum

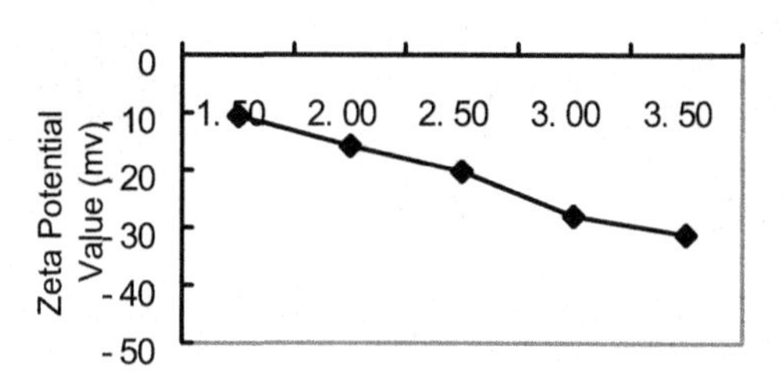

Figure 6 Relation Between Caustic Ratio and Zeta Potential Value

The above figure shows that Zeta potential of hydroxide aluminum Changes with caustic ratio. It can be seen that with the increasing of caustic ratio, the absolute Zeta potential value increases, which is harmful for particles to aggregate, and can just manifest the reason hydroxide aluminum particles aggregate more obviously in early stage of the precipitation process than in the late stage.

Influences of PH Value on Zeta Potential of Hydroxide Aluminum

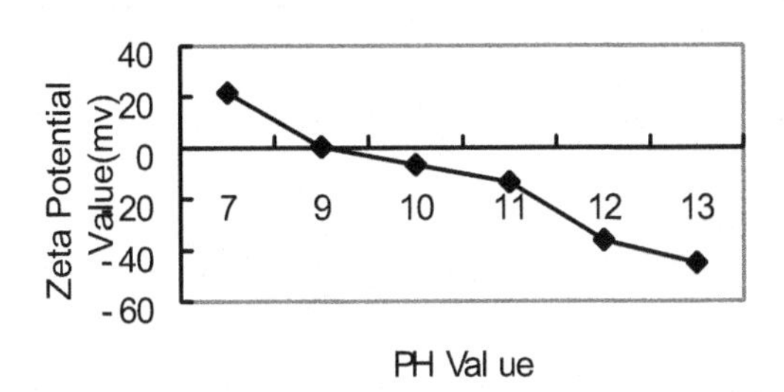

Figure 7 Relation Between Zeta Potential Value and PH Value

It can be concluded with the increases of OH^- concentration, negative charges particles bear also increase. For the industrial liquor in the alumina industry, the concentration of OH^- ion is higher than 1mol/l, so the absolute value of zeta potential should be very high, but due to the existences of many cations, its value can be reduced so that particles can collision and cohesion. From the above figure, it is deducted that the reason that particles are difficult to aggregate in higher caustic concentration lies not only in the lower supersaturation of sodium aluminate liquor, but also its leading to higher absolute value of Zeta potential.

Influences of Cations and Anions on Zeta Potential of Hydroxide Aluminum in Caustic Liquid

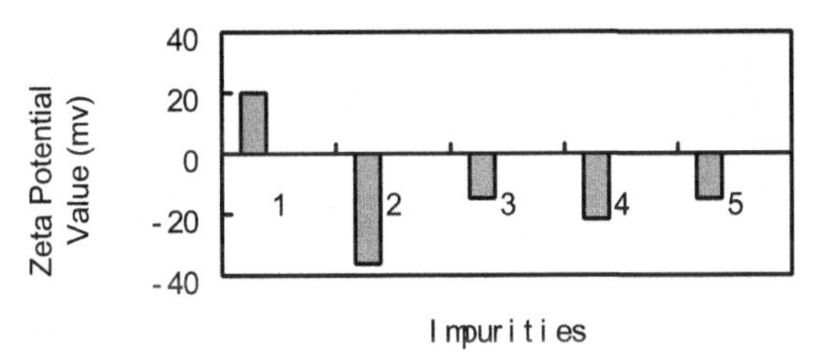

Figure 8 Relation Between Zeta potential Value and Impurities in Caustic Liquor

1-pure water 2- NaOH 3-NaOH+Na_2SO_4
4- NaOH+Na_2CO_3 5- NaOH+KCl

The above figure shows the influences of impurities on Zeta potential of particles in caustic liquor. In the test, the concentration of NaOH remains stable, and the concentration of anion s is the same. From figure 8, it can be proposed that in the caustic liquor, the existence of such impurities as Na_2SO_4 $Na_2CO_3$and KCL decreases the zeta potential absolute value of hydroxide aluminum, and they are helpful to increase the aggregation of particles.

Conclusions

1. OH^-、SiO_3^{2-}、CO_3^{2-}、$C_2O_4^{2-}$and SO_4^{2-} influence greatly on the Zeta Potential of hydroxide aluminum in pure water, and all of them make hydroxide aluminum particles have the tendency of bearing negative charges.

2. Among the three cation ions, Ca^{2+}, K^+, and Na^+,Ca^{2+} has the strongest ability to enable the particles to bear positive charges , and K^+ follows, and Na^+ ranks the last in the list. So in the caustic liquid, the increases of Ca^{2+} and K^+ are more useful for the aggregation of hydroxide aluminum, and are helpful to produce hydroxide aluminum with large size.

3. When retaining the same NaOH concentration, with the decrease of caustic ratio, the absolute potential value of hydroxide aluminum is decreasing, which just manifest the mechanism that hydroxide aluminum particles are more apt to aggregate in the early stage of precipitation process of seeded sodium aluminate liquor than in the last stage.

4. With the growth of concentration of NaOH in the liquor, the absolute Zeta potential value of hydroxide aluminum increases. In the scope of industrial alumina production, this rule is the same. Therefore, with the growth of caustic soda, aggregation of hydroxide aluminum particles become more difficultly, that is, higher concentration of NaOH is a barrier to produce coarse hydroxide aluminum.

5. In the caustic soda liquor, the existences of Na_2SO_4 ,Na_2CO_3 and KCl are useful for particles to aggregate.

References

[1]Marchal P , David R , Klein J P , et al . Crystallization and Precipitatlon Engineering(I) : An Efficient Method for Solving Population Balance in crystallization with Agglomeration[J] . Chemical Engineering Science , 1988 , 43(1) :

[2]Warren L.Fine Particles Process.Chapter 48.1980,947

EFFECTS OF POWER ULTRASOUND ON PRECIPITATION PROCESS OF SODIUM ALUMINATE SOLUTIONS

Jilai Xue, Shaohua Li, Jun Zhu, Baoping Song
School of Metallurgical and Ecological Engineering, University of Science and Technology Beijing
Xueyuan Road 30, 100083, Beijing, China

Keywords: Power ultrasound, Precipitation process, Sodium aluminate solution

Abstract

Sodium aluminate solutions for alumina production have been ultrasound treated to enhance the seeded precipitation in a laboratory Bayer process. Higher precipitation ratio was found with ultrasound treated solutions than those without ultrasound. More new nuclei are observed on the surface of gibbsite from precipitation process with ultrasound. Particle size distribution of gibbsite varies with ultrasound power level and processing time. More fine fractions are found with ultrasound than without ultrasound. Ultrasound can improve the precipitation efficiency and promote growth of gibbsite.

Introduction

Precipitation process is of great importance for both product quality and productivity in alumina production of Bayer process. For years, numerous efforts have been made to increase the precipitation ratio and improve the quality of alumina for smelter. Zhao et al [1, 2] found that the precipitation ratio increased in ultrasound treated sodium aluminate solution. Ruecroft et al [3] applied ultrasound for removal of the key impurity sodium oxalate to aid crystallization in the precipitation process.

In general, the effects of ultrasound are related to two aspects: one is ultrasound variables, such as power level and frequency [4], and the other is the chemical conditions that include Na_2O concentration, the molecular ratio of Na_2O/Al_2O_3, temperature of reaction and the seed content in the solution [5]. The nucleation, agglomeration and crystal growth of gibbsite in the solutions were also studied by SEM and particle size distribution analysis [6-11].

The primary goal of this work is to find out the required power level of ultrasound and processing time in making a positive effect on the precipitation process with quality products. In this paper, the results from a laboratory study of precipitation process in the ultrasound treated sodium aluminate solution will be presented, and the quality of the corresponding precipitation products will also be characterized.

Experimental

Figure 1 shows the experimental set-up used for a laboratory Bayer process. It includes a precipitation system and an ultrasonic system. The precipitation process occurred in a stainless steel crystallizer placed in a water tank with a temperature controller and a mechanical stirrer. In the ultrasonic system, the piezoelectric transducer was connected to a frequency generator (20 kHz); and the power capacity was from 50 W to 900 W.

The sodium aluminate solution was made up by dissolving industrial aluminum hydrate into industrial sodium hydroxide, and the Na_2O concentration was 150g/l. The solutions were then ultrasound treated for 5 minutes before the primary seeds (seed content, 250g/l) were added for the gibbsite precipitation processed at a constant temperature of 50 ^{0}C and at an agitation rate of 400 r/min.

The precipitation ratio and the molecular ratio of Na_2O/Al_2O_3 can be determined by a titration technique. The gibbsite filtered from the sodium aluminate solutions was examined using SEM microphotography.

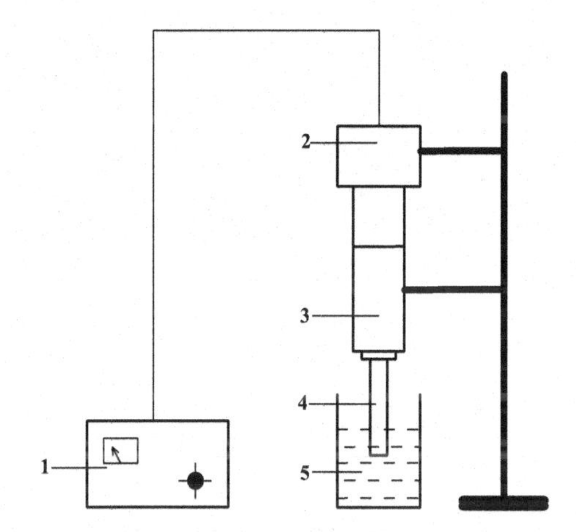

Figure 1. Schematic diagram of the experimental set-up
1—Signal generator, 2—Transducer, 3 and 4—Sonoprobe system
5—Sodium aluminate solution

Results and Discussion

Effects of ultrasound on precipitation ratio

The precipitation ratio, η, and the molecular ratio of Na_2O/Al_2O_3, α, can be calculated by

$$\alpha = \frac{n_{Na2O}}{n_{Al2O3}} \quad (1)$$

$$\eta = (1 - \frac{\alpha_a}{\alpha_m}) * 100\% \quad (2)$$

where n_{Na2O} and n_{Al2O3} can be obtained from the titration tests.

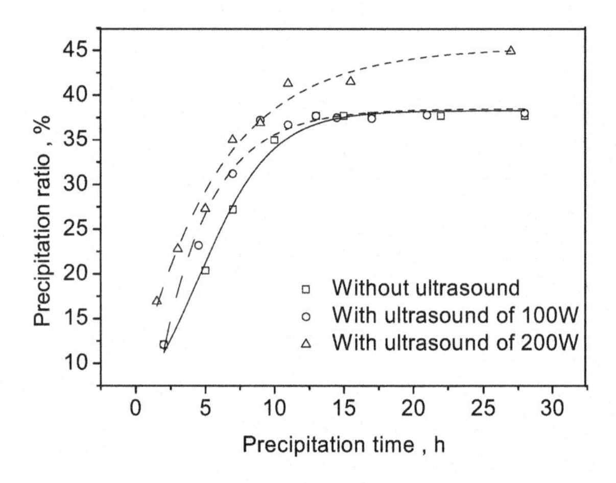

Figure 2. Effects of ultrasound power on precipitation ratio in sodium aluminate solutions with $\alpha_k = 1.52$

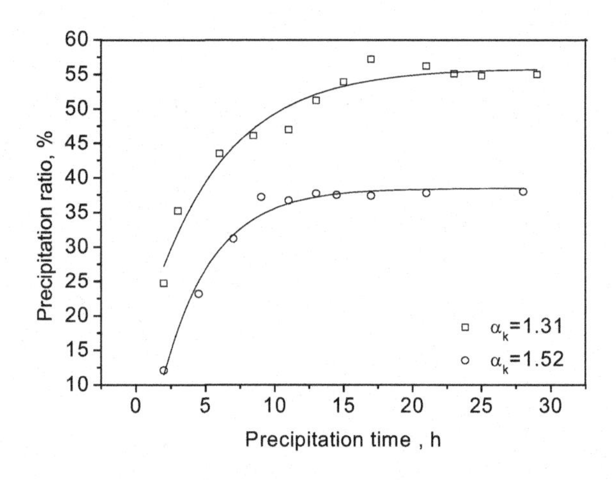

Figure 3. Effects of α_k on precipitation ratio with ultrasound power of 100W

a) Precipitation for 2 hours b) Precipitation for 7 hours c) Precipitation for 15 hours

Figure 4. SEM microphotographs of seeds from sodium aluminate solution without ultrasound

a) Precipitation for 2 hours b) Precipitation for 7 hours c) Precipitation for 15 hours

Figure 5. SEM microphotographs of seeds from sodium aluminate solution with ultrasound of 200 W

In Figure 2, three curves of precipitation ratio are given as a function of time, in which the sodium aluminate solutions have been treated with and without ultrasound. At 100 W of ultrasound power, the precipitation ratio shows only a very small increase, while at 200 W, such an increase becomes significant. This phenomenon is in a rough agreement with the results obtained by Zhao [1, 2]. However, the values of precipitation ratio may vary as the seeds and instrument system differ in quality and performance.

Figure 3 demonstrates that a higher precipitation ratio can be obtained with a higher molecular ratio of Na_2O/Al_2O_3. This indicates that the effects of ultrasound are also dependent on the chemical nature of the sodium aluminate solutions, which may play a role in selection of ultrasound variables and processing parameters [5, 6]. An appropriate combination of the ultrasound variables and the process parameters could yield better effect on the precipitation process.

The fact that the ultrasound can improve the precipitation ratio means a potential of increasing industrial productivity of alumina. With implementation of this technology, the process of seeded precipitation could be shortened with savings in terms of operation costs.

Effects of ultrasound on secondary nucleation

Figure 4 and Figure 5 are two sets of microphotographs of gibbsite from sodium aluminate solutions with and without ultrasound, respectively, where both had the identical conditions for sodium aluminate solutions and precipitation parameters. It is obvious that there are more tiny secondary nuclei on the surface of seeds with ultrasound (in Figure 5) than those without ultrasound (in Figure 4). The tiny particles on the surface of the seeds with ultrasound, as shown in Figure 6, was identified by using EDS probe as the $Al(OH)_3$, the same as the seeds.

Figure 6. SEM micrograph of tiny particles on the seed surface from precipitation process with ultrasound

Moreover, the number of the secondary nuclei increases with increasing the precipitation time (see Figure 5). This means that ultrasound can promote secondary nucleation during a seeded precipitation process.

With continuous agitation, the newly formed nuclei could get off the seed surface and enter into the sodium aluminate solution, so that they become secondary nuclei in the solution. The fresh nuclei may act as seeds for further decomposition of the sodium aluminate solution and thus increase the precipitation ratio, as early shown in Figure 2. However, the details of mechanism of the secondary nucleation promoted by ultrasound are still unclear at this stage, which needs further investigation.

Effects of ultrasound on particle size distribution

Figure 7 illustrates patterns of particle size distribution against precipitation time with and without ultrasound. In principle it is possible for particle size distribution to supply the information on secondary nucleation, agglomeration of fine particles and seed or nuclei growth during precipitation process in sodium aluminate solution.

The general pattern in Figure 7 shows that more fractions of larger particles are developed with increasing precipitation time, no matter with or without ultrasound. The largest fraction of the particle under this investigation was one with about 46 μm in size.

Figure 8 further demonstrate that particle size distribution of gibbsite can vary with ultrasound power level and processing time. For 7 hours, more fine particles can be seen on the curve with 100 W, which is in a rough agreement with the results from Chen [7], while for 15 hours, more fractions of larger particles are presented in the gibbsite with ultrasound power 200W, which has a tendency similar to the data from Zhao [2].

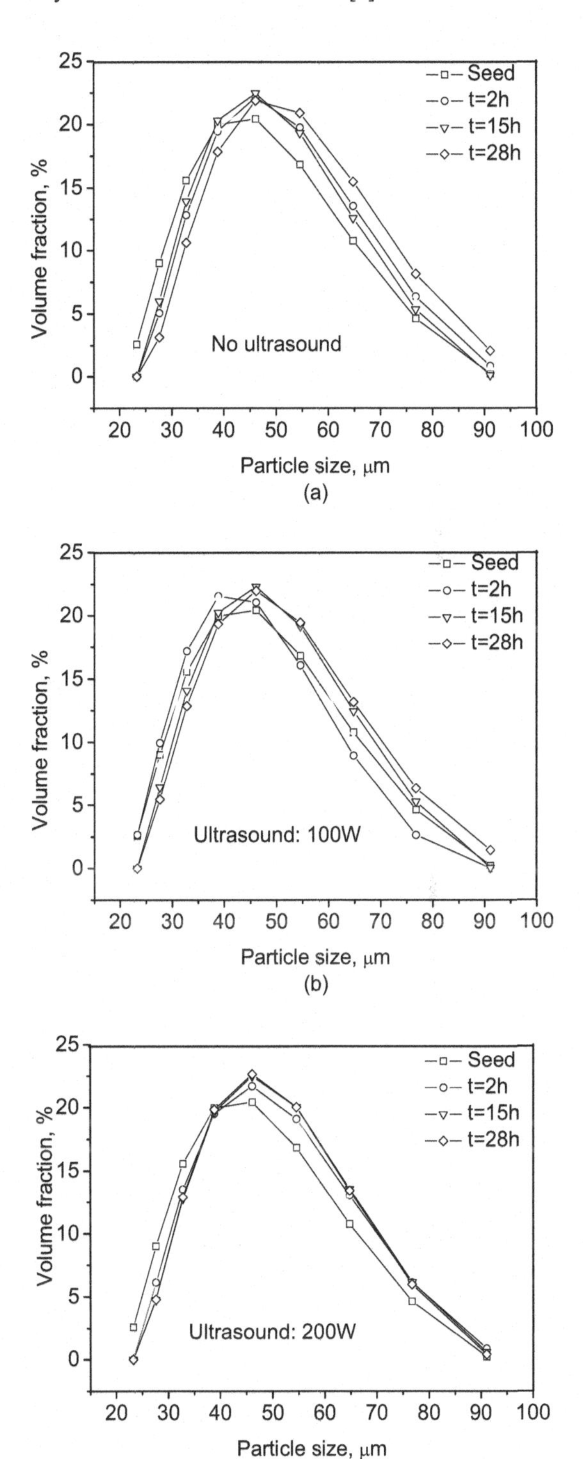

Figure 7. Particle size distribution against precipitation time for precipitation process with and without ultrasound

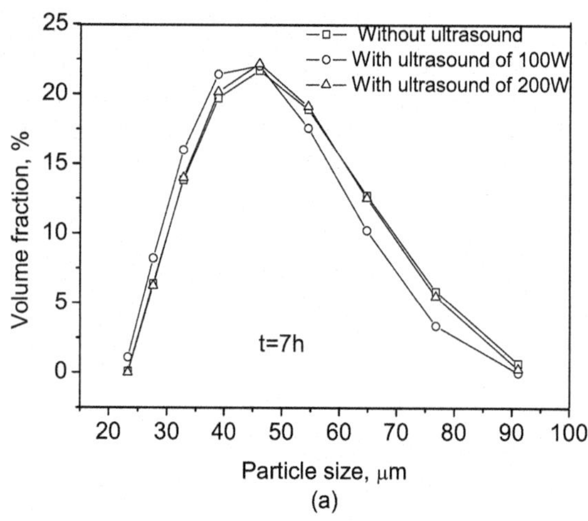

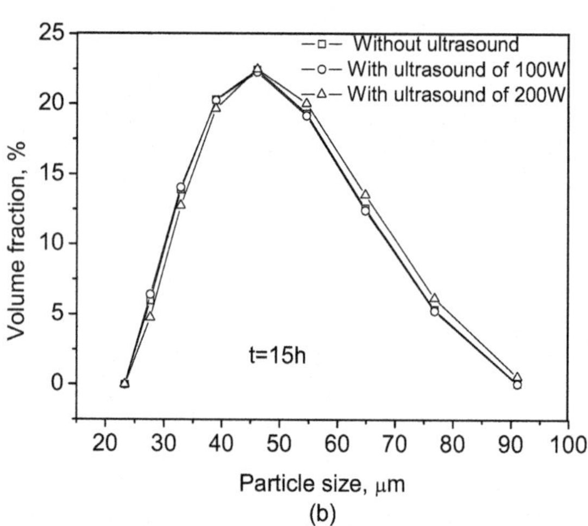

Figure 8. Particle size distribution against ultrasound power used in precipitation process with various periods of time

The authors believe that the pattern change in particle size distribution mentioned above could be associated with the precipitation time: at a relative early stage, more fine gibbsite are produced from the new nuclei formation accelerated by the ultrasound, while at a relative later stage, growth of gibbsite are the major tendency enhanced by the ultrasound power and longer processing time. This explanation can be partly supported by SEM observation in this work.

Conclusions

1. Precipitation ratio is improved in ultrasound treated sodium aluminate solutions.
2. More new nuclei are found on the surface of gibbsite from precipitation process with ultrasound.
3. Particle size distribution of gibbsite varies with ultrasound power level and processing time. More fine fractions are found with ultrasound than without ultrasound at a relative early stage, while more fraction of larger gibbsite at a relative later stage.

Acknowledgements

Financial support from National Natural Science Foundation of China (NSFC) and University of Science and Technology Beijing is gratefully acknowledged.

References

1. J. Zhao and Q. Chen, "Intensifying Precipitation Seeded in Bayer Process with Ultrasound," *Acta Metallurgica Sinica*, 2(2002), 171-173.

2. J. Zhao, "Study on Alumina Hydrate Precipitation under Ultrasound by Bayer Process," *Light Metals* 2002.

3. G. Ruecroft, D. Hipkiss and M. Fennell, "Improving the Bayer Process by Power Ultrasound Induced Crystallization of Key impurities," *Light Metals* 2005, 163-166.

4. J. Liu et al, "Effect of Ultrasound Frequency on the Precipitation Process of Supersaturated Sodium Aluminate Solution," *The Chinese Journal of Process Engineering*, 4(2)(2004), 130-135.

5. J. Zhao et al, "Effect of the Reaction Conditions on the Enhancement of $Al(OH)_3$ Precipitation Seeded from Sodium Aluminate Solution under Ultrasound," *Acta Metallurgica Sinica*, 38(2)(2002), 166-170.

6. G. Chen et al, "SEM Observation of Gibbsite Precipitation with Seeds from Sodium Aluminate Solutions Promoted by Ultrasound," *Tran. Nonferrous Met. Soc. China*, 13(3)(2003), 708-714.

7. J. Chen et al, "Kinetics of Crystal Growth on Seeded Precipitation of Sodium Aluminate Solutions with New Device," *Tran. Nonferrous Met. Soc. China*, 14(4)(2004), 824-828.

8. G. Chen et al, "Nucleation during Gibbsites Precipitation with Seeds from Sodium Aluminate Solution Processed under Ultrasound," *Tran. Nonferrous Met. Soc. China*, 14(2)(2004), 401-405.

9. P. Smith and G. Woods, "The Measurement of Very Slow Growth Rates during the Induction Period in Aluminum Trihydroxide Growth from Bayer Liquors," *Light Metals* 2002.

10. D. Rossiter et al, "Investigation of the Unseeded Nucleation of Gibbsite $Al(OH)_3$ from Synthetic Bayer Liquors," *J Crys Growth*, 191(1998), 525-536.

11. S. Kumru and H. Bale, "Aggregation in Aluminum Hydroxide Solution Investigated by Small X-ray Scattering," *J Applied Crystallography*, 27(5)(1994), 682-692.

STUDY ON THE OSCILLATION PHENOMENA OF PARTICLE SIZE DISTRIBUTION DURING THE SEEDED AGGLOMERATION OF SODIUM ALUMINATE LIQUORS

Jianguo Yin[1], Qiyuan Chen[1], Zhoulan Yin[1], Jiangfeng Zhang[2]
[1]School of Chemistry and Chemical Engineering, Central South University, Changsha 410083, China
[2]Institute of Technology and Economy in Nonferrous Metals, Beijing 100035, China

Keywords: particle size distribution, oscillation, sodium aluminate liquors, agglomeration

Abstract

Chaotic and Fractal theories were used to study the oscillation phenomena of particle size distribution (PSD) during the agglomeration process of sodium aluminate liquors. Conclusions were made as follows: PSD of particles is in macroscopic disorder, while it is regular in certain range; Curves of PSD appear to oscillate sometimes, but particles of certain range own general rules in the whole process. That is, the curves of PSD oscillating decrease with time for small and median particles, but increase for large particles; The minor change of initial conditions can be enlarged in the process and the particles size information will be affected. As a result, the butterfly effect of the PSD appears; curves of PSD have the space fractal characters for particles of certain ranges. Preciser structure of curves can be identified if smaller divided rule of particle size is adopted.

Introduction

Large-scale prebaked anode cell and purified fume technology of dry-process have been adopted all over the world in order to reduce the energy consumption and protect environment since the middle of 1970S. Sandy alumina, which has good mobility, adsorptive capacity and intensity, can meet all the above demands (1). Particle size distribution (PSD) of the products is strictly demanded for the physical and chemical performances of sandy alumina. Seeded precipitation of sodium aluminate liquors is one of the main steps for the production of alumina, and it is also the key step to obtain the sandy alumina during the Bayer process (2). Information of particle size in the precipitation process should be analysed in order to control the particle size of alumina and produce satisfactory products in the end. Our previous investigations showed that the changes of gibbsite particle size with time are very complicated and will appear to the oscillation phenomena (3).

Chaotic and Fractal theories have displayed powerful vitality in solving complicated scientific problems since their foundation, and they have been widely adopted in the scientific fields of mathematics, physics, chemistry, metallurgy, biology, philosophy, economics, weather forecast, geologic exploration, etc (4-14). Chaotic and Fractal theories were used to study the oscillation phenomena of PSD in the agglomeration process of sodium aluminate liquors in this paper and the basic rules about the changes of gibbsite particle size with time in the process were attained.

Experimental

Experimental methods

Experimental conditions were well controlled to ensure that the precipitation was in agglomeration process [15]. Samples were periodically taken out from the precipitation tank and centrifugated. Clear liquors were used to analyse the content of alumina and alkali, and the solid samples were washed for three times in hot distilled water for the PSD determination [16]. Metallurgical industry standard of YB-817-75 was adopted for the analysis of chemical components, and particle size analyzer of Mastersizer 2000 was used for the PSD determination.

Experimental apparatus

An ACD-□intelligent–controlled precipitation tank, made by Zhengzhou Research Institute of Chalco, was adopted in the precipitation experiments.

Liquors and seeds

The liquors were prepared from industrial alkali and aluminum hydroxide. Industrial sheet alkali was dissolved in half of the required volume of distilled water in a stainless steel vessel. Aluminum hydroxide from the Pilot Alumina Plant of Zhengzhou Research Institute of Chalco was slowly dissolved in hot alkali liquors with agitation. Then the liquors were diluted to certain volume with hot distilled water and filtered twice, which were ready for experiments.

Seeds from the Pilot Alumina Plant of Zhengzhou Research Institute of Chalco were passed through sieve of 45μm, and the minus mesh seeds were used for experiments.

Conditions of seeded precipitation

Experiments were performed under the conditions of alumina (Al_2O_3) concentration of $169.8 g\cdot l^{-1}$, molecular ratio (α_K, which was the ratio of Na_2O_K to Al_2O_3) of 1.36 or 1.45, agitation rate of 100r·min-1, seeds mass ratio (Ks) of 0.25 and temperature of 75.5□.

Results and discussion

Macroscopic disorder and microscopic regular for PSD in time and space dimensions

Changes of Volume Percentage (VP) with time for particles of small, median and large sizes were obtained in the experiments as shown in Fig.1. It can be seen that changes of VP with time are obviously different for particles less than 5 μm, median particles about 20 μm and large particles over 45 μm, and PSD of particles appears to be macroscopic disorder in time and space dimensions. Change of VP with time is non-monotonic for particles of certain size during the seeded precipitation of sodium aluminate liquors. First, particles of certain size are generated by agglomeration of smaller particles, which leads to the increase of their VP. On the other hand, they will agglomerate themselves and generate larger particles, which leads to the decrease of their VP. Generation and consumption are contrary parts of each other, and they both may be affected by some random and casual factors in the process. So oscillation of PSD appears easily.

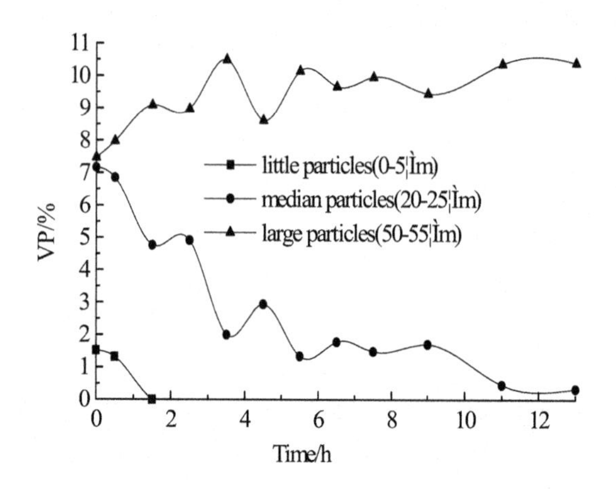

Fig.1 Change of VP with time for particles of different size

By using the related software of Mastersizer 2000, particle size data are processed according to a particle size interval of 0.5 μm. It is found that particles in certain size range obey some rules. Changes of VP with time for particles of 0-6 μm and 6-9.5 μm are shown in Fig.2 and Fig.3. Fig.2 and Fig.3 show that VP always goes down in the period of 0-1.5 hours for particles of 0-6μm, which illustrates that particles of this range are in their agglomeration process. Yet, VP changes non-monotonically with time for particles of 6-9.5μm. In the period of 0-0.5 hours, VP always goes down for particles of 6-7μm and remains constant for particles of 7-8μm and goes up for particles of 8-9.5μm. All these particles are agglomerated fully at the end of 1.5 hours. It may be concluded that most agglomeration takes place orderly, firstly for smaller particles and then for larger ones.

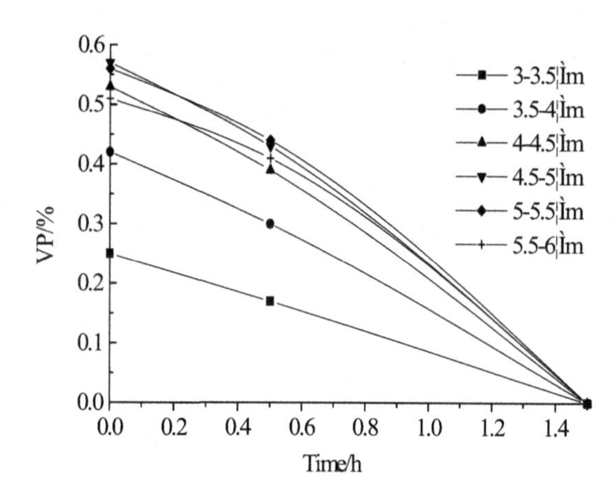

Fig.2 Change of VP with time for particles of 3-6 μm

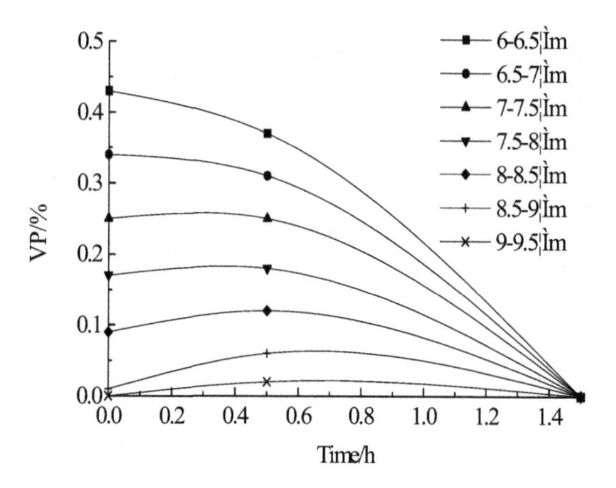

Fig.3 Change of VP with time for particles of 6-9.5 μm

Particles in other size ranges such as 15-17.5 μm, 18.5-24 μm, 29-30.5 μm, 45-50 μm and over 55 μm have different rules of particle size information. That is to say, particle size information in the whole range doesn't have the same rule and it appears to be macroscopic disorder in time and space dimensions. Yet it is discovered that particle size information of certain size range has its own rule by the application of Chaotic theory.

unstableness in some areas and stableness in whole areas for PSD

Change of VP with time for particles of 18.5-24 μm is shown in Fig.4. It demonstrates that PSD of this range shows obvious oscillation. VP of particles changes acutely with time and goes up and down alternately in the periods of 0-5.5 hours and 7.5-11 hours. VP remains almost stable in the periods of 5.5-7.5 hours and 11-16 hours, which demonstrates that the rates of consumption and generation are almost the same and reach a temporary balance. It is also discovered that for each curve there are peaks at 0, 2.5, 4.5, 9 hours and the peak values decrease gradually during the 16 hours of precipitation process, and remains almost unchangeable in the period of 11-16 hours in the end.

According to Chaotic theory, when a fluctuation makes a current system enter a state far from balance, the system reaches a

"crossing" point. Probability acts at or nearby the "crossing" points, yet determinism acts among the "crossing" points [17].
Supersaturated sodium aluminate liquors are in an unstable thermodynamics state which may cross balance state and reach metastable or unstable state. As a result, precipitation process is easily affected by some random and casual factors and their PSD information displays easily chaos and oscillation at the "crossing" points. The whole process, especially in most areas among the "crossing" points, is affected by decisive factors and owns universal rules. That is, the curves of PSD oscillating decrease with time for small and median particles, which means agglomeration process is dominating. The curves of PSD oscillating increase for large particles, which means generation rate dominate in the process.

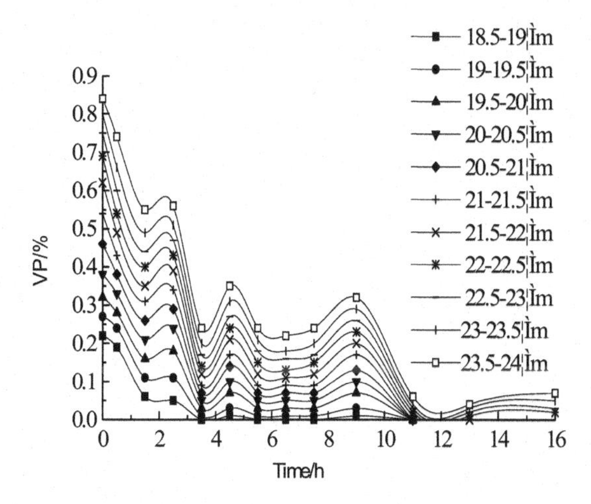

Fig.4 Change of VP with time for particles of 18.5-24 μm

Butterfly effect of particle size information

Butterfly effect was proposed by a famous meteorologist called Lorenz in 1979 as " a butterfly fans its wings in Brazil, which may cause a tornado in Texas of USA two months later". This effect shows that result has the extremely sensitive dependence to the initial condition, the minimum deviation of initial condition will be amplified in the process and be able to cause the result enormous difference [18].
Seeded precipitation process of sodium aluminate liquors is affected by many factors such as temperature, concentration of liquors, initial molecular ratio of liquors, seeds character, seeds mass ratio, agitation rate and the inwall states of tank, etc. It was found in our experiments that minor changes of these factors will affect the particle size information in the process greatly.
Influence of α_K on VP for particles over 45 μm in the products is shown in Fig.5.

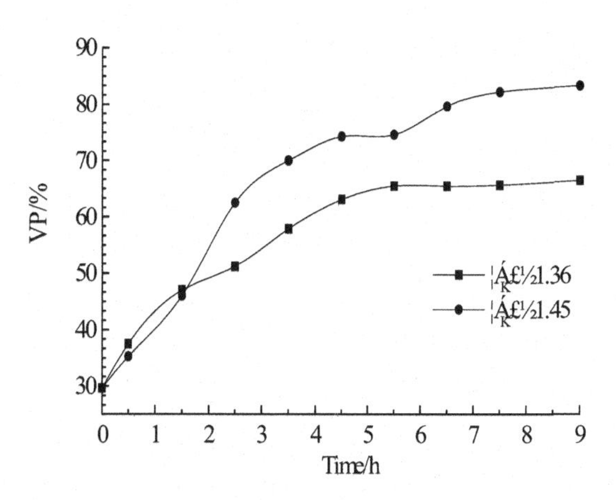

Fig.5 Influence of α_k on VP with time for particles over 45 μm in the products

It can be learned that the minor change of α_K (0.09, which is minor variation for numerical value) can cause large change of VP for particles over 45 μm in the products under the same experimental conditions. Amplitude of change will be enlarged with time, which is from zero at the beginning to 16.85% at the 9th hour. Minor changes of other factors in the precipitation process will also cause a large variation of PSD, which may be called the butterfly effect of PSD [19].

Fractal character of PSD graphs

Simplest description about Fractal is that certain structure or process of a subject is self-similar from the point of different space or time dimensions. So a lot of subjects, phenomena, characters or rules in the world which show self-similar in certain range can be investigated by Fractal theory [10]. The oscillation phenomena of PSD of alumina was studied by Fractal theory as follows.
20-25 μm $\subset$ 20-30 μm $\subset$ 0-30 μm is a mathematics relationship of containing among these particle size ranges, which belongs to fractal-whole of nestification. As shown in Fig.6, curves of VP with time are basically similar for particles among 20-25 μm, 20-30 μm and 0-30 μm , and there are obvious nested relationships among these curves. That is, each curve among these ranges has fractal character in space dimension.
PSD graphs of particles are slightly different when different divided rules are adopted. VP in the period of 1.5-2.5 hours decreases a little when the divided rules are 30 μm and 10 μm. Yet it increases a little when the divided rule is 5 μm. VP in the period of 5.5-7.5 hours remains almost constant when the divided rule is 0.5 μm, Yet a peak at 6.5 hour appears when the divided rules are 30 μm, 10 μm and 5μm. It can be concluded that preciser structure of the curves of PSD can be identified if smaller divided rule of particle size is adopted. Divided rule of 0.5 μm was used and good results were attained in our experiments when the precise structure of graphs were comprehensively considered. It is found that all curves of PSD own the similar space fractal character when the particle size ranges are 0-6 μm, 6-9.5 μm, 15-17.5 μm, 18.5-24 μm, 45-50.5 μm and 55.5-60 μm .

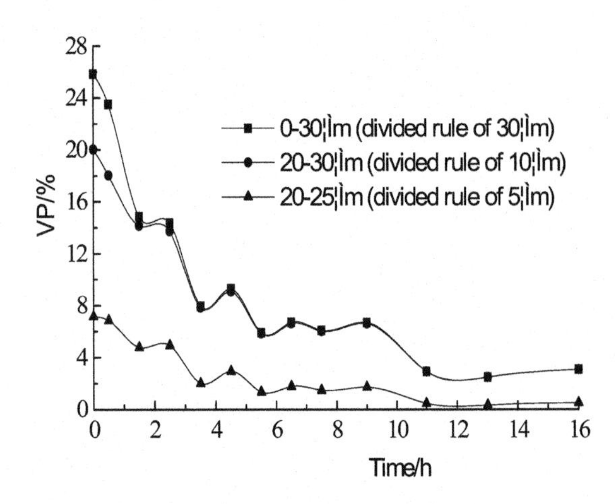

Fig.6 Influence of divided rules on VP with time

Conclusion

Chaotic and Fractal theories were used to study the oscillation phenomena of PSD in the agglomeration process of sodium aluminate liquors. Conclusions were made as follows:

1) PSD of particles is in macroscopic disorder, while it is regular in certain range.

2) Curves of PSD appear to oscillate sometimes, but particles in a certain range own general rules in the whole process. That is, the curves of PSD oscillating decrease with time for small and median particles, but oscillating increase for large particles.

3) The minor change of initial conditions such as temperature, concentration of liquors, initial molecular ratio of liquors, seeds character, seeds mass ratio, agitation rate, the inwall states of tank and so on can be enlarged in the precipitation process and the particles size information will be affected greatly. As a result, the butterfly effect of PSD appears.

4) Each curve of PSD has the space fractal character for particles of certain range. Preciser structure of curves can be identified if smaller divided rule of particle size is adopted. Divided rule of 0.5 μm was used in our experiments and good results were attained.

Acknowledgement

Our work is supported by National Priority Development Project Fundamental Research in China ('973' project) "fundamental research on extracting alumina from concentrated bauxite with high efficiency" (project number : 2005CB623702) and National Natural Science Fund of China project (project number: 20476107).

References

1. Liu Baowei, "Study on the problem of sandy alumina," World Nonferrous Metals, 10 (1999), 45-47.

2. Tong Libin et al, "Study of seed to decomposition process during production of sandy alumina," Journal of Shenyang Institute of Chemical Technology, 14(4)(2000), 269-272.

3. Li Wangxing et al, "Analysis on the particle size information in sodium aluminate solutions during seeded agglomeration," Light Metals, (2005), 235-237.

4. Pan Qing, "Knowledge about chaos," Journal of Hechi Teacher's College, 20(4)(2000), 95-98.

5. Tang Wei, Li Dianpu, and Chen Xueyun, "Chaotic theory and research on its application," Automatization of electric power, 4 (2000), 67-70.

6. Wang Xingyuan, Zhu Weiyong, "Chaos theory and its application in chemical reactions," Shenyang Chemical Engineering, 1 (1997), 48-51.

7. Shun Shijie, Teng Ronghou, and Ma Ruzhang, "Study of the state of Fe in Ni-Al crystallite alloy," Power Metallurgy Industry, 6(2)(1996), 7-9.

8. Chen Guohui et al, "Spectral dimension during the precipitation of gibbsite from the seeded caustic aluminate liquors," Journal of Central South University of Science and Technology, 33(2)(2002), 157-159.

9. Fang Yaomei, "Chaotic theory—a new theory which can change the way of one's minds," Tongji University Journal, 6(1)(1995), 88-94.

10. Kang Kuanying, "The foundation, development and its means of scientific methodology of fractal theory," Scientific Management Research, 16(6)(1998), 53-56.

11. Chen Jing, "Realizing chaos from the point of view of philosophy," Journal of Systemic Dialectics, 12(2)(2004), 22-24, 30.

12. Chang Qingying, "Chaos and forecast in market trading," Journal of China Agricultural University, 2 (2004), 39-43.

13. Hou Ronghua, "Necessity of research on economic chaos," Journal of Shanghai Maritime University, 23(1)(2002), 1-8.

14. Lu Zhiguang, Bai Liping, and Lu Li, "A study on long term disaster forecast model by using chaos theory," Journal of China Agricultural University, 7(3)(2002), 43-46.

15. Chen Xiaohu, "A study on precipitation conditions of high concentrated sodium aluminate liquors," Guizhou Science, 17(4)(1999), 266-269.

16. H. Watling, J. Loh, H. Gatter, "Gibbsite crystallization inhibition 1. Effects of sodium gluconate on nucleation, agglomeration and growth," Hydrometallurgy, 55(2000): 275-288.

17. I. Prigogine, I.Stengers, Order out of chaos: Man's new dialogue with nature (Toronto, Canada: Bantam Books, 1984).

18. Edward Lorenz, The Essence of Chaos (WA 98145-5096: University of Washington Press, 1993).

19. Min Qi, "Introduction to Chaos," Journal of Mengzi Teacher's College, 14(4)(1997): 48-53.

Light
Metals
2006
ADDENDUM

Light Metals 2006 *Edited by Travis J. Galloway* *TMS (The Minerals, Metals & Materials Society), 2006*

DIGESTER DESIGN USING CFD

Jennifer Woloshyn, Lanre Oshinowo, John Rosten
Hatch Associates Ltd.; 2800 Speakman Drive; Sheridan Science & Technology Park; Mississauga, Ontario, Canada L5K 2R7

Keywords: CFD, Simulation, Reactor, Modeling, Bauxite, Digester, Hydrodynamics, Kaolinite, Dissolution, RTD

Abstract

In the Bayer process, dissolution of gibbsite and kaolinite occur in the digester train. Understanding the hydrodynamics of the digester is key to improving the extent of dissolution, and thus the extraction of alumina and re-precipitation of silica. Deviation from ideal plug flow results in a miscalculation of the slurry retention time. The outcome may be a loss of undigested alumina to the red mud with a consequent reduction in extraction efficiency.

To address this issue, the hydrodynamics of a digester train were modeled using computational fluid dynamics (CFD). The impact of column aspect ratio and inlet configuration on the slurry residence time distribution (RTD) was investigated. The RTD was used to estimate reaction extents and evaluate the effect of design parameters on performance. The modeling approach allows the inclusion of the digestion chemistry to directly evaluate the yield. Results show that the choice of slurry inlet configuration significantly impacts performance.

Introduction

Globally, the Bayer process is the established method for the production of alumina from bauxite that has been in use for over a century. With large quantities of bauxite being processed and a highly competitive marketplace, incremental process improvements resulting in minor increases in efficiency have a significant benefit to the overall operating costs.

A 2001 review of the alumina industry, sponsored by industry leaders and the governments of Australian and the USA [1], advised that both continuous improvement through incremental changes and major advances through innovative step changes will be required for the alumina industry to remain viable over the next several decades. A number of technology initiatives were identified to bring these changes to fruition, including Computational Fluid Dynamics (CFD) modeling. To maximize the benefits of CFD, it should be employed to investigate "what-if" scenarios during the earliest possible stages of the design process [2].

In recent years, CFD has been widely used to understand and optimize process steps in the hydrometallurgical and, more specifically, the alumina industry [3,4,5]. The use of powerful and reliable modeling techniques as a supplement to traditional engineering practices allows process engineers to virtually prototype equipment. Process or geometry changes to unit operations can be applied to optimize the equipment design and process configuration. In this way, capital and operating costs may be reduced by reducing over-design. Although CFD simulation of process equipment or unit operations is becoming more widespread, no previous CFD study of bauxite digesters could be found in the literature.

For optimal reactor performance of the digester column, based on the characteristics of the digestion reactions, the flow field should approach plug flow. In ideal plug flow, the fluid moves in an orderly front through the vessel without any axial mixing along the flow path. This paper presents a case study for digester design using CFD to determine the flow characteristics of bauxite slurry within the digester vessel. CFD analysis can identify dead spots in which the digester volume is poorly utilized. By evaluating the residence time distribution (RTD) of the bauxite slurry the effect of the vessel aspect ratio and inlet configuration on the digester performance was quantified. The RTD results are then used to estimate the extent of the digestion in the train.

Bauxite Digestion

During the digestion process, alumina in the bauxite is dissolved in the Bayer liquor in the form of sodium aluminate. Gibbsite bauxites are digested at 140-150°C in a series of vertical, cylindrical pressure vessels. The iron mineral impurities are sparingly soluble, but another common impurity, kaolinite, dissolves. Gibbsite and kaolinite dissolution may be represented by the following equations [6]:

$$a\,NaOH + b\,Al_2O_3{\cdot}3H_2O \rightarrow c\,NaAlO_2 + d\,H_2O \qquad (1)$$

$$e\,NaOH + f\,Al_2O_3{\cdot}2SiO_2{\cdot}2H_2O \rightarrow g\,Na_2O{\cdot}SiO_2 + h\,NaAlO_2 + i\,H_2O \qquad (2)$$

Kaolinite dissolution is accompanied by reprecipitation of the silicate as a complex sodium aluminoslicate, known as desilication product (DSP). This reaction occurs at a much slower rate than the gibbsite dissolution and therefore controls the required digestion time [6]. Although most refineries attempt to deal with this issue by pre-desilication in holding tanks ahead of the digestion train, a digester train sized to accommodate the slow kinetics of kaolinite dissolution and reprecipitation as DSP ensures sufficient retention time for alumina digestion.

Digesters used in the Bayer process are typically sized based on throughput and digestion time under the assumption of plug flow. If the flow though the digester deviates from plug flow, then a quantity of slurry arrives at the digester outlet either earlier or later than the ideal retention time. If the slurry arrives early, then the bauxite digestion is premature, leading to losses of undigested alumina in the red mud and autoprecipitation in the washing train. If the slurry arrives late, then the column is over-sized. Each condition leads to increased operating or capital costs of the digester units. By applying CFD to the design of digesters, the hydrodynamics can be examined and design modifications can be made to approach plug flow conditions.

Deviation from the plug flow assumption is assessed by examining the residence time distribution (RTD) in response to a pulse input to the model. CFD has been previously shown to be an effective tool for predicting RTDs [7,8]. For the present case of the bauxite slurry flow through the digester, a homogeneous, non-settling slurry is assumed. Work is underway to incorporate a multiphase modeling approach, which is required to predict the flow of both the liquor and the bauxite, including the bauxite particle size distribution.

Digester Configuration

The digester configuration is of typical design — a vertical, cylindrical vessel with hemispherical heads. Alternate geometries (e.g. tube digesters) were not considered in this study. In the present configuration, the inlet slurry feed nozzle is located at the top and the outlet nozzle is located at the bottom of the vessel.

A single digester is studied to assess the impact of two geometric variables — the height-to-diameter ratio of the vessel and the slurry inlet configuration — on the performance indicator, the RTD. Two different height-to-diameter vessel aspect ratios are considered: 4:1 and 7.5:1. The volume of the digester is held constant (see Figure 1) for the applied throughput. Additionally, three slurry inlet configurations are considered: normal, central, and tangential, as shown in Figure 2. A deflector plate is added below the central inlet to distribute the slurry in the head of the vessel and reduce short-circuiting of the feed to the outlet at the bottom of the vessel. To predict the performance of the digestion train, a series of three digesters is also modeled.

The hydrodynamics of the digester are complex with turbulent, multiphase flow and liquid-solid reactions. These key complexities are not considered in the usual black-box approach of flow sheet simulation and can introduce uncertainty and risk into the process design. By employing CFD to investigate the hydrodynamics and to predict the performance of a digester during the initial design, many of these complexities are taken into account. The technical risks are mitigated and areas for design improvement can be readily identified and introduced to the design cycle project earlier rather than later.

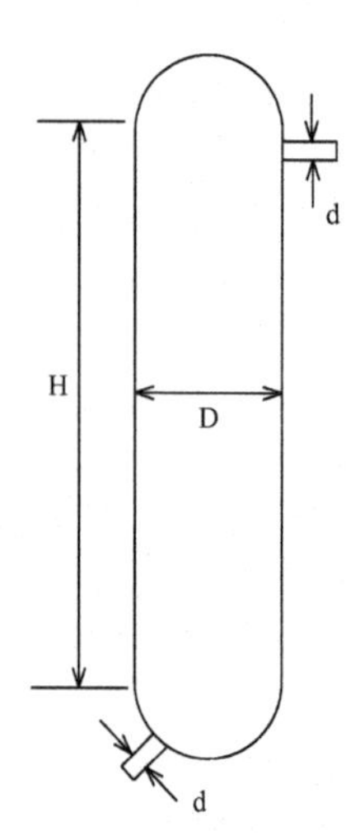

Aspect Ratio	4:1	7.5:1
H, m	23	33.3
D, m	5.2	4.44
d, m	0.6	0.6
Deflector*, m	1.52	1.30
V, m^3	562	562

*Applies to Central inlet configuration only. See Figure 2.

Figure 1:Digester column geometry and dimensions.

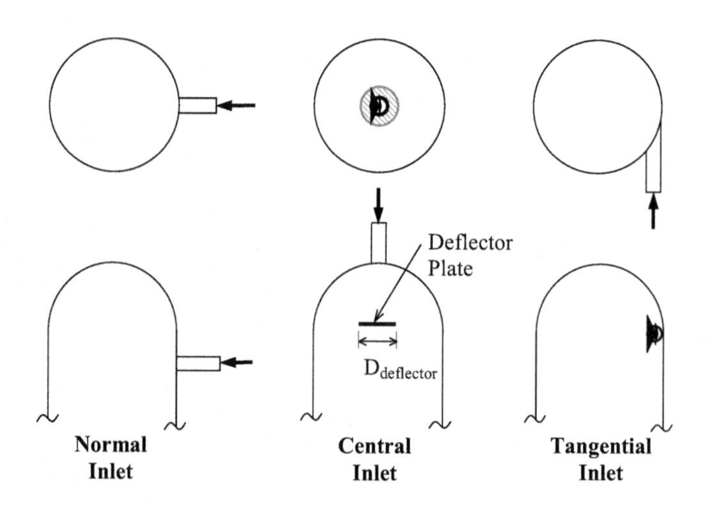

Figure 2: Schematic of the three slurry inlet nozzle configurations considered.

Modeling Methodology

In the first stage of modeling, the slurry hydrodynamics are studied to assess the influence of the vessel aspect ratio and inlet configuration leading to selection of preferred vessel geometry. In the second stage, the analysis was extended to estimate the digestion extent based on the RTD. A time-dependent, tracer technique is applied to find the RTD of the flow field. A pulse of a neutrally buoyant tracer is injected at the inlet to the digester and the concentration of the tracer is monitored at the outlet. The transient discharge profile is compared to the expected profile for ideal plug flow. The mean and variance of the concentration curve are used to quantify the deviation of the digester column from plug flow.

Numerical Methods and Boundary Conditions

The commercial code, FLUENT version 6.2, was used in this study. The 3D computational grid of the digester model consisted of approximately 200,000 hexahedral cells with symmetry applied for the normal and central inlet configurations. The grid size for the tangential inlet configuration was 400,000 hexahedral cells. Turbulence was modeled using the realizable k-ε turbulence model in the normal and central inlet digesters. For the tangential inlet configuration, the Reynolds stress turbulence model was applied due to the high degree of rotational, i.e., swirling, in the flow. The inlet slurry flow rate was 2840.5 m^3/h with an average slurry density of 1280 kg/m^3. The liquor has a specific gravity of 1.17 and the solids have a specific gravity of 2.98. The inlet slurry contains 6wt% solids. The inlet velocity boundary condition is 2.81 m/s. The outlet boundary was set to system pressure.

Slurry Hydrodynamics

The flow patterns with digester aspect ratios of 4:1 and 7.5:1 are shown in Figure 3 for the normal inlet configuration. The flow fields are similar, with large recirculation zones along the digester wall, below the inlet nozzle. These large recirculation zones indicate significant back-mixing within the digester and substantial deviation from plug flow. For the 4:1 aspect ratio, the recirculation zone extends along the full height of the column and into the bottom head. Increasing the aspect ratio increases the length that is available for the flow to recover downstream of the inlet. For the higher aspect ratio of 7.5:1, the flow becomes more aligned with the column geometry in a downward direction just upstream of the outlet. The digester was modeled without a vapor space in the vessel head contrary to typical digester operation. Neglecting the small headspace volume will not significantly alter the overall vessel circulation and the conclusions of the present study.

Figure 4 compares the flow fields in the 4:1 column for the three slurry feed nozzle configurations. Note that the flow patterns in the 7.5:1 columns with central and tangential inlets are similar to those shown for their 4:1 column counterparts. The inlet configuration greatly influences the flow patterns in the digester. As described above, the normal inlet nozzle leads to a large recirculation zone inside the digester. With a central inlet nozzle, a deflector plate was included to distribute the slurry jet and prevent short-circuiting through the vessel to the outlet. A recirculating wake develops immediately downstream of the deflector, but the flow develops into nearly uniform further downstream. The tangential inlet configuration gives rise to swirling flow throughout the height of the vessel with two counter-rotating vortices: an inner vortex traveling from the bottom upwards and a second vortex traveling back down the center of the digester.

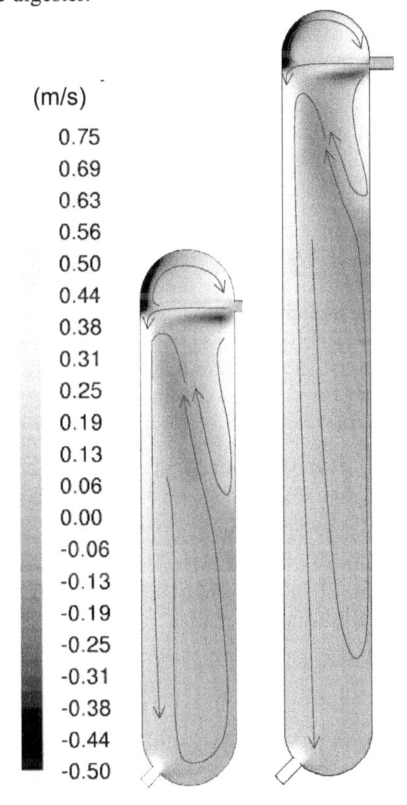

Figure 3: Comparison of axial velocity and the characteristic flow patterns in the digester for aspect ratios of 4:1 and 7.5:1.

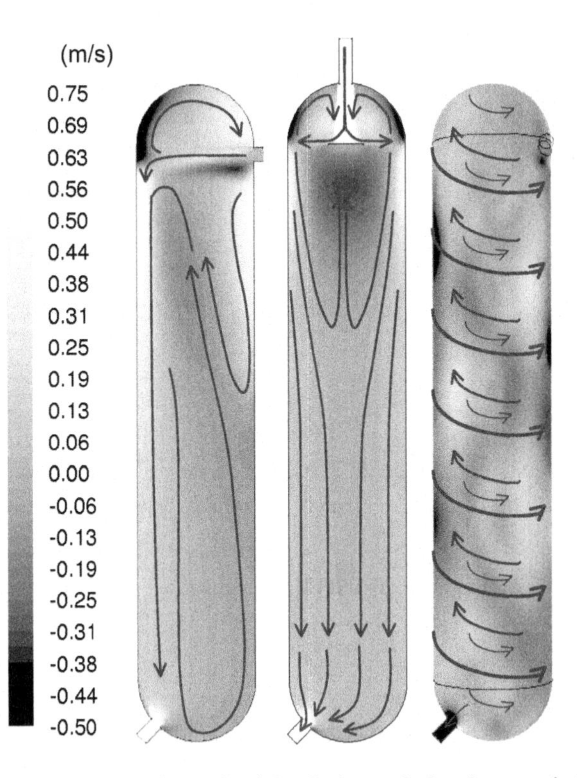

Figure 4: Comparison of axial velocity and the characteristic flow patterns in the digester with a normal, central, and tangential slurry inlets.

Figure 5 and Figure 6 show the tracer residence time distribution for digester column aspect ratios of 4:1 and 7.5:1, respectively. Based on the slurry throughput of 2840 m^3/h and a digester volume of 562 m^3, the mean retention time of the digester column is 712 s. The central inlet produces an RTD with a peak near the mean retention time. The normal inlet configuration produces an early, indistinct peak with a long tail that decays slowly indicating that some slurry short-circuits while the remainder is held up in the large region of recirculation. The tangential inlet RTD shows sharp peak followed by a very long tail indicating that most of the material short circuits to the outlet and the remaining material is held up in the swirling flow in the center of the digester. The column diameter decreases with an increase in aspect ratio, increasing the time for the slurry to swirl down the wall of the digester resulting in a long retention time.

Table I summarizes the mean time and variance for the RTD plots shown in Figure 5 and Figure 6. The lower variance for the central inlet configuration is an indication of less axial dispersion.

The bauxite particles exhibit single-particle terminal settling velocities exceeding the liquor superficial velocity in the digester. The residence times of the solid phase will likely differ from the liquid phase as the solid particles are expected to settle faster than the liquid flowing down the vessel. A multiphase simulation of the liquid and bauxite particle flow through the digester train (3 vessels) is required to predict hindered settling. Work is underway to include this effect and thereby obtain a more representative assessment of the bauxite (kaolinite) conversion

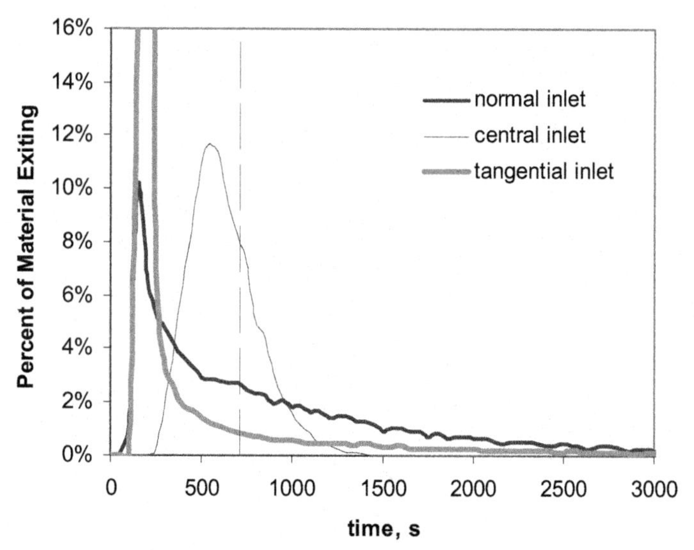

Figure 5: Single digester RTD. Aspect ratio = 4:1.

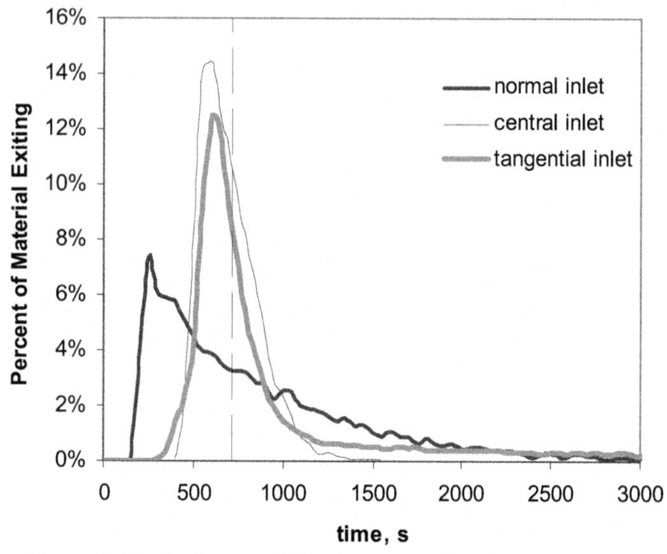

Figure 6: Single digester RTD. Aspect ratio = 7.5:1.

Table I: Influence of column aspect ratio and inlet nozzle configuration on mean residence time $\bar{t}$ and variance of the RTD (Results are for a single vessel).

	Aspect Ratio 4:1		Aspect Ratio 7.5:1	
Inlet.	$\bar{t}$, s	σ^2, s^2	$\bar{t}$, s	σ^2, s^2
Normal	940	856,000	830	420,000
Central	647	48,600	665	25,000
Tangential	505	691,000	1280	2,340,000

Extent of Digestion

The hydrodynamics and RTD of the digesters are related to the extent of dissolution of the silica-containing kaolinite to assess the extent of digestion and the performance of the vessel configurations. Compared to gibbsite dissolution, it is the slower desilication rate that controls the digestion time. A simple axial dispersion model [9] is used to estimate the extent of the kaolinite dissolution reaction based on the axial dispersion. The deviation from plug flow is characterized by the dispersion number D/uL where D is the axial dispersion coefficient, u is the superficial velocity and L is the characteristic length of the reactor/digester and is determined from:

$$\sigma_\theta^2 = \frac{\sigma^2}{\bar{t}^2} = 2\left(\frac{D}{uL}\right) - 2\left(\frac{D}{uL}\right)^2\left(1 - e^{-uL/D}\right) \tag{3}$$

The variance σ^2 and the mean residence time $\bar{t}^2$ are determined from the RTD (See Table I). The extent of the reaction, also known as the conversion, can then be determined from:

$$X = 1 - \frac{4a \cdot \exp\left(\frac{1}{2}\frac{uL}{D}\right)}{(1+a)^2 \cdot \exp\left(\frac{a}{2}\frac{uL}{D}\right) - (1-a)^2 \cdot \exp\left(-\frac{a}{2}\frac{uL}{D}\right)} \tag{4}$$

where $a = \sqrt{1 + 4k\tau(D/uL)}$

k = reaction rate constant

$\tau = V/Q$ = retention time of the column

The conversion of the kaolinite dissolution reaction can be estimated based on the following assumptions and simplifications:

1. The dispersion number is within the range $0.1 < D/uL < 1$. When the dispersion number is greater than 1, deviation from plug flow is large and the axial dispersion model is less accurate.
2. The reaction is first order with an estimated reaction rate constant, k, of 0.002 s^{-1}, based on laboratory studies of the bauxite digestion. The rate constant is a function of many variables including the bauxite ore, particle size distribution and morphology, and the digestion conditions: temperature and pressure. Though the relationship between the rate constant and the conversion in non-linear, the conversion is sensitive to the magnitude of the reaction rate.
3. Ideal plug flow is assumed in the pipes connecting the vessels. This assumption is reasonable because of the relatively small pipe diameter and the high-velocity flow.
4. The reaction is not mass-transfer limited, and relatively independent of the kaolinite concentration, because the reaction rate constant is small.
5. There is an availability of liquor (NaOH) for dissolution.

Table II and Table III show the reaction extent through a single vessel for different aspect ratios. Cases for which $D/uL > 1$ fall outside of the appropriate range for this conversion model. Conversions tabulated for $D/uL > 1$ are included for comparison purposes only. The superior hydrodynamic performance of the central slurry inlet nozzle configuration translates into improved digester performance with kaolinite conversion well above the other configurations.

Table II: Extent of the kaolinite dissolution reaction after one digester vessel. Aspect ratio = 4:1.

	Aspect Ratio 4:1		
	Normal Inlet	Central Inlet	Tangential Inlet
σ_θ^2	0.97	0.12	2.7
Dispersion, *D/uL*	10.4	0.062	9.5×10^8
Extent of Reaction, *X*	59%	73%	59%

Table III: Extent of the kaolinite dissolution reaction after one digester vessel. Aspect ratio = 7.5:1.

	Aspect Ratio 7.5:1		
	Normal Inlet	Central Inlet	Tangential Inlet
σ_θ^2	0.610	0.0565	1.43
Dispersion, *D/uL*	0.586	0.0291	6.5×10^7
Extent of Reaction, *X*	65%	75%	59%

RTD of Individual Solid Particles – Effect of Settling

The above analysis has assumed that the bauxite particles in the slurry follow the liquor flow. To assess the effect of solid settling on the extent of digestion, discrete particles were tracked through the digester in a Lagrangian frame of reference. The size distribution of the bauxite particles is shown in Table IV. Particles representing each diameter range were tracked through the 4:1 digester with central inlet. Figure 7 shows the RTD for the selected particles sizes. The RTD for the non-settling slurry is also plotted for comparison. The 100 µm particles, representing 50% of the solids by weight, follow the liquor closely. The larger particles begin to settle out of the liquor and exit the digester early.

Table IV: Slurry solid particle size distribution.

Diameter (µm)	wt%
100	50
450	30
600	10
800	5
1000	4

On the basis of the RTD of each particle size, the extent of the kaolinite dissolution was estimated once again. The overall, mass-weighted, kaolinite conversion is 89%. Based on this analysis, the settling solids decrease the conversion by nearly 10%. The influence of particle-particle collision has been neglected in the Lagrangian multiphase approach and could possibly overpredict the RTD's for the solids. A dense multiphase approach is being investigated to more accurately predict the transport of the solids phase through the digester. Additionally, the heterogeneous reactions between the caustic liquor and the bauxite ore will be incorporated to model the digestion reactions directly while coupled with a full multiphase, non-isothermal, hydrodynamic analysis to present a complete picture of the digester operation.

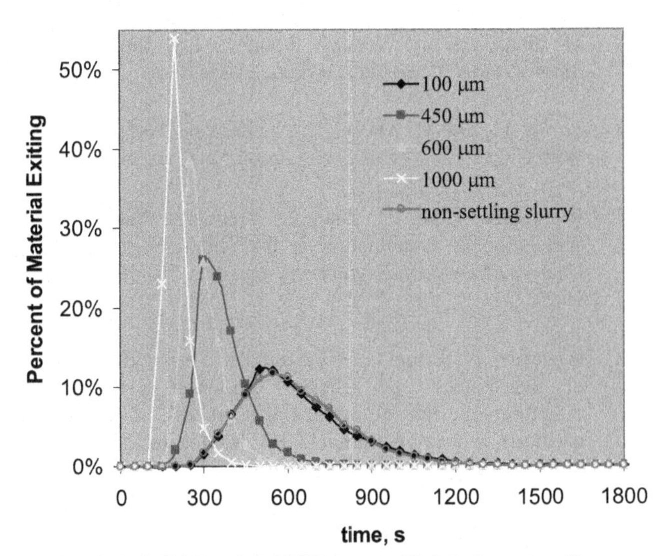

Figure 7: Solid Particle RTD (central inlet, Aspect ratio = 4:1).

Summary

CFD has been shown to be a valuable tool for the design of a bauxite digester by providing a cost-effective way to experiment with vessel design towards determining an optimal vessel configuration for digestion. The traditional black box approach does not account for effects of geometry such as the slurry inlet nozzle configuration or vessel aspect ratio. The inlet configuration has been shown to have a strong influence on performance in terms of the kaolinite conversion extent. The central inlet results a flow field most closely approaching plug flow and offers the best performance.

Further work is underway to extend CFD modeling methodology to include both multiphase representation of the slurry liquor and solid and the heterogeneous liquid-solid dissolution reactions. By modeling the multiphase hydrodynamics with reactions, the performance may be predicted with fewer assumptions and the design uncertainty may be further reduced.

Acknowledgements

The authors would like to thank Peter Bath and David Walker for reviewing this paper and contributing their expertise in the area of alumina processing.

References

1. AMIRA International Ltd., "Alumina Technology Roadmap," November 2001

(http://www.amira.com.au/documents/Alumina/AluminaTechnologyRoadmap.pdf).

2. S. Thornke, Experimentation Matters: *Unlocking the Potential of New Technologies for Innovation* (Cambridge, MA: Harvard Business School Publishing Corporation, 2003).

3. Kahane, R; T. Nguyen, and M. P., Schwarz, "CFD modeling of thickeners at Worsley Alumina Pty Ltd," *Applied Mathematical Modelling*, 26 (2002), 281-296.

4. Throp, E., and C. Marsh, C., "Separation Problem Solved With CFD," *The Chemical Engineer*, 2004, no. 756: 42-43.

5. Rousseaux, J.-M.; C. Vial, H. Muhr, E. Plasari, "CFD Simulation of Precipitation in the Sliding-Surface Mixing Device," *Chemical Engineering Science*, 56 (2001), 1677-1685.

6. Raghaven, N. S., and G. D. Fulford, "Mathematical modeling of the kinetics of gibbsite extraction and kaolinite dissolution/desilication in the Bayer process," *Light Metals*, ed. Barry Welch, (Warrendale, PA: The Minerals, Metals & Materials Society, 1998), 29-36.

7. Oshinowo, L, L. Gunnewiek, and K. Fraser, "CFD in Autoclave Circuit Design" *Hydrometallurgy 2003 – Fifth International Conference in Honor of Professor Ian Ritchie – Volume 1: Leaching and Solution Purification*, ed. C.A. Young, A.M. Alfantazi, C.G. Anderson, D.B. Dreisinger, B. Harris and A. James, (Warrendale, PA: The Minerals, Metals & Materials Society, 2003), 671-686.

8. Ekambara, K. and, J. B. Joshi, "CFD Simulation of Residence Time Distribution and Mixing in Bubble Column Reactors," *The Canadian Journal of Chemical Engineering*, 81 (2003), 669-676.

9. Levenspiel, O., *Chemical Reaction Engineering, 3rd Ed.*, (New York, NY: Wiley, 1999).

AUTHOR INDEX

Subject Index